Multi Frequency EPR Spectroscopy of Conjugated Polymers and Their Nanocomposites

Multi Frequency EPR Spectroscopy of Conjugated Polymers and Their Nanocomposites

Victor I. Krinichnyi

Russian Academy of Sciences
Moscow, Russia

CRC Press
Taylor & Francis Group
Boca Raton London New York

CRC Press is an imprint of the
Taylor & Francis Group, an **informa** business

CRC Press
Taylor & Francis Group
6000 Broken Sound Parkway NW, Suite 300
Boca Raton, FL 33487-2742

First issued in paperback 2019

ISBN-13: 978-1-4987-7964-7 (hbk)
ISBN-13: 978-0-367-88988-3 (pbk)

Library of Congress Cataloging-in-Publication Data

Names: Krinichnyi, Victor I.
Title: Multi frequency EPR spectroscopy of conjugated polymers and their nanocomposites / Victor I. Krinichnyi.
Description: Boca Raton : CRC Press, 2017. | Includes bibliographical references and index.
Identifiers: LCCN 2016038736 | ISBN 9781498779647
Subjects: LCSH: Conjugated polymers. | Polymers--Electric properties. | Electron paramagnetic resonance spectroscopy. | Nanocomposites (Materials)
Classification: LCC QD382.C66 K755 2017 | DDC 620.1/92042--dc23
LC record available at https://lccn.loc.gov/2016038736

**Visit the Taylor & Francis Web site at
http://www.taylorandfrancis.com**

**and the CRC Press Web site at
http://www.crcpress.com**

This book is dedicated to my wife, Tatiana Krinichnaia. Without her understanding and support, I would never have completed this project.

Contents

List of Symbols

A	Tensor of hyperfine interaction
A	Anisotropic hyperfine interaction constant (mT)
A	Orientation degree of chains
A_{fr}	Friction force
A_{ii}	Principal values of the hyperfine interaction tensor (mT)
A/B	Asymmetry factor of spectrum
$a = \frac{1}{3}\sum_3 A_{ii}$	Averaged (isotropic) hyperfine interaction constant (mT)
a	Lattice constant (nm)
A/B	Asymmetry factor of EPR spectrum
b	Lattice constant (nm)
$\mathbf{B}_0$	External magnetic field vector
B_0	External magnetic field modulus (1 T = 10^4 Oe, G)
B_1	Amplitude of magnetic term of microwave polarizing field
B_{hf}	Magnetic field due to spin hyperfine interaction
B_m	Amplitude of Zeeman magnetic modulation
B_{loc}	Local magnetic field strength
c	Light velocity ($c = 2.99792458 \times 10^8$ m/s)
c	Lattice constant (nm)
D	Diffusion coefficient (m^2/s)
$D_{1D,3D}$	Intra- and interchain diffusion coefficients, respectively (rad/s)
d	Dimensionality of a disordered system (1–3)
$d_{1D,3D}$	Intra- and interchain lattice constants (nm)
E	Energy (1 eV = $1.60217733 \times 10^{-19}$ J = 8.06557982×10^3 cm^{-1} = 2.41800×10^{14} Hz)
E_0	Energy distribution width (eV)
E_a	Activation energy (eV)
E_d	Potential of induced electric field
E_g	Energy gap (eV)
E_h	Hopping energy (eV)
E_p	Formation energy of polaron (eV)
E_{ph}	Energy of lattice phonons (eV)
E_r	Inner- and outer-sphere reorganization energy (eV)
E_t	Trap depth (eV)
e	Elementary charge ($e = 1.60217733 \times 10^{-19}$ Coul)
$G(\tau)$	Autocorrelation function
g	Splitting tensor
g	Landé splitting factor ($g_e = 2.00231930436153$ for free electron)

g_{ii}	Principal values of the splitting tensor
g_{iso}	Isotropic (averaged) g-factor equal to $1/3(g_{xx}+g_{yy}+g_{zz})$ or $1/3(2g_\perp+g_{\parallel})$
$g(\omega)$	Factor of line shape
$\mathbf{H}$	Hamiltonian of magnetic interaction
$h=2\pi\hbar$	Planck quantum constant ($h=6.6260755\times10^{-34}$ J·s)
I	Nuclear spin
I	Inertia moment
$\mathbf{I}_i$	ith term of nuclear spin operator
I_i	Intensity of ith spectral component
J	Total electron angular momentum
J_{af}	Antiferromagnetic exchange coupling constant
J_c	Exchange coupling coefficient
J_{ex}	Spin exchange interaction constant
$J(\omega)$	Spectral density function
k_B	Boltzmann constant ($k_B=1.380658\times10^{-23}$ J/K)
L	Total electron orbital momentum
L	Spin/charge wave localization function
l_i	Mean free path of spin/charge
M_0	Equilibrium magnetization of spin ensemble
$\mathbf{M}$	Total magnetization of spin ensemble
M_2	Second moment of a spectrum
M_d	Disturbing force
m	Orientation magnetic number
m_c	Mass of charge carrier
m_e	Mass of electron ($m_e=9.10938356\times10^{-31}$ kg)
m_s^*	Effective mass of soliton
m_u	Number of monomers in a polymer chain
N	Spin concentration (cm^{-3})
N_A	Avogadro's number (6.02214129×10^{23} mol^{-1})
N_s	Soliton width in polyacetylene
n	Spin/charge concentration per monomer unit
$n(\varepsilon_F)$	Density of states at the Fermi level ε_F
p_{cr}	Probability of spin cross-relaxation
p_{ff}	Spin flip-flop probability
$P(r, r_0, t)$	Translating spin motion propagator
Q	McConnell proportionality constant ($Q=2.3$ mT)
R	Spin/charge hopping distance
$\mathbf{r}$	Radius vector between dipoles
r_i	Separation between impurities
r_{NO}	Distance between N and O atoms in nitroxide radical
r_{ss}	Distance between spins (cm)
$\mathbf{S}$	Electron spin operator
S	Total electron spin of molecule

S	Single spin configuration
s	Saturation factor
T	Absolute temperature (K)
$T_{+,0,-}$	Triple spin configuration
T_0	Percolation constant of disordered system (K)
T_1	Electron longitudinal (spin-lattice) relaxation time
T_{1p}	Proton longitudinal (spin-lattice) relaxation time
T_2	Electron transverse (spin-spin) relaxation time
T_c	Characteristic temperature
T_g	Glass transition temperature (K)
t	Time (s)
t_C	Coulombic integral
t_{cc}	Resonant overlapping integral of C=C bond
t_h	Hopping integral (eV)
t_r	Intrachain transfer integral (eV)
$t_{\|}$	Intrachain–interchain coupling integral (eV)
$t_\perp$	Interchain coupling integral (eV)
$U(\omega t)$	Dispersion signal term
$g(\omega)\sin(\omega t - \varphi)$	In-phase ($\varphi = 0$), out-of-phase ($\varphi = -\pi$), and $\pi/2$-out-of-phase ($\varphi = \pm\pi/2$) (in phase quadrature) dispersion signal terms registered at φ shift in phase detector with respect to applied modulation
u_i	Amplitude of ith dispersion signal
V	Sample volume
v	Spin motion rate (cm/s)
v_F	Velocity of spin diffusion near the Fermi level (cm/s)
W	Spin/charge hopping energy (eV)
x, y, z	Axes of molecular coordinate system
y	Doping level
α	Constant of electron–phonon interaction
α, β	Spin levels
β	Medium *reactive* dispersion parameter
γ	Transition rate
γ_e	Gyromagnetic ratio for electron ($\gamma_e = 1.760859708_{39} \times 10^{11}$ T^{-1}s^{-1})
γ_p	Gyromagnetic ratio for proton ($\gamma_p = 2.675221900 \times 10^8$ T^{-1}s^{-1})
Δ_0	Half-bandgap
ΔB_{pp}	Linewidth from peak to peak
$\Delta B_{pp}^{\|L}$	Lorentzian peak-to-peak linewidth
$\Delta B_{pp}^{\|G}$	Gaussian peak-to-peak linewidth
$\Delta E_{\alpha\beta}$	Difference in electron α and β energy levels
ΔE_{ij}	Energy barrier between i and j sites
Δg	Shift of g-factor

Δm_e	Difference in orientation magnetic number for electron
Δm_I	Difference in orientation magnetic number for nuclear
$\Delta\omega$	Anisotropy of magnetic interaction
$\Delta\omega_{ij}$	Distance between spin packets (rad/s)
$\langle\Delta\omega^2\rangle$	Averaged constant of dipole spin interaction in a powder sample
δ	Gamma function
δ	Skin layer thickness
$\delta(\Delta B_{pp})$	Spectral line broadening
δB	Spectral diffusion distance
$\delta B(t)$	Time line shift
ε	Dielectric constant
ε_0	Dielectric constant for vacuum
ε	Parameter of medium nonlinearity
ε_F	Fermi energy level (eV)
ξ	Decay length for charge carries
η	Coefficient of dynamic viscosity
θ	Angle between spin vectors
θ	Torsion (dihedral) angle
ϑ	Angle between main radical axis and external magnetic field
κ	Dimensionality constant
κ	Medium dissipation parameter
λ	Spin–orbit coupling constant (eV)
μ_{eff}	Effective magneton
μ_B	Bohr magneton ($\mu_B = 9.27400968_{20} \times 10^{-24}$ J/T)
μ_0	Permeability for vacuum ($\mu_0 = 4\pi \times 10^{-7}$ V·s/A m)
μ	Dipole moment of spin-modified molecule
μ	Spin/charge mobility ($cm^2\,V^{-1}\,s^{-1}$)
ν	Charge diffusion rate
ν_e	Resonance frequency of free electron ($\nu_e = \omega_e/\pi$)
ν_{ph}	Photon frequency
ν_{ex}	Frequency of spin-packets exchange
$\rho(r)$	Density of unpaired electron at distance r from nucleus
Σ_{ij}	Lattice sum for powder
σ	Intrinsic conductivity
σ_{ac}	Alternating current electric conductivity (S/m)
σ_{dc}	Direct current electric conductivity (S/m)
τ	Effective relaxation time
τ_0	Spin separation in radical pair
τ_c	Correlation time of radical rotation
τ_m	Mechanical relaxation time
$\tau(R)$	Recombination time of spins separated by distance R
φ	Spin macroprobe orientation angle

χ	Static magnetic susceptibility of spin ensemble
χ'	Dispersion term of magnetic susceptibility
χ''	Absorption term of magnetic susceptibility
χ_C	Curie magnetic susceptibility
χ_P	Pauli magnetic susceptibility
$\Psi(r)$	Wave function for electron localization ($\Psi^2(r) = \rho(r)$)
ψ	Angle between external magnetic field and polymer chain
ω_e	Angular resonance frequency of electron transition between α and β levels
ω_{ex}	Exchange frequency between spin packets (rad/s)
ω_{hop}	Angular frequency of spin diffusion
ω_0	Hopping attempt frequency
ω_l	Angular libration frequency (rad/s)
ω_m	Angular frequency of Zeeman magnetic modulation (rad/s)
ω_p	Angular precession frequency of proton (rad/s)

List of Abbreviations

ac	Alternating current
AFNP	2-(Azahomo[60]fullereno)-5-nitropyrimidine
AMPSA	2-Acrylamido-2-methylpropane sulfonic acid
BDPA	1,3-*bis*-Diphenylene-2-phenyl allyl
BHJ	Bulk heterojunctions
BKdV	Burgers–Korteweg–de Vries (theory)
CB	Conjugated band
cis-PA	*cis*-Polyacetylene
CSA	Camphorsulfonic acid
CW	Continuous wave
DBSA	Dodecylbenzenesulphonic acid
DBTTF	Dibenzotetrathiafulvalene
dc	Direct current
DPPH	2,2-Diphenyl-1-picrylhydrazyl
ENDOR	Electron nuclear double resonance
EPR	Electron paramagnetic resonance
FIR	Far infrared band/irradiation with photon energy of 0.00124–0.0827 eV and wavelength of 1.5×10^4–1.0×10^5 nm
HFI	Hyperfine interaction
HFS	Hyperfine structure
HCA	Hydrochloric acid
HOMO	Highest occupied molecular orbital level
IR	Infrared band/irradiation with photon energy of 1.2–1.8 eV and wavelength of 700–1000 nm
LEPR	Light-induced electron paramagnetic resonance
LUMO	Lowest unoccupied molecular orbital level
MW	Microwave (frequency)
NIR	Near infrared band/irradiation with photon energy of 0.89–1.6 eV and wavelength of 770–1400 nm
NMR	Nuclear magnetic resonance
P3AT	Poly(3-alkylthiophene)
P3DDT	Poly(3-dodecylthiophene)
P3HT	Poly(3-hexylthiophene)
P3MT	Poly(3-methylthiophene)
P3OT	Poly(3-octylthiophene)
PA	Polyacetylene
PANI	Polyaniline
PANI-EB	Emeraldine base form of polyaniline
PANI-ES	Emeraldine salt form of polyaniline

PANI-LE   Leucoemeraldine form of polyaniline
PANI-PN   Pernigraniline form of polyaniline
PATAC   Poly(*bis*-alkylthioacetylene)
PC   Paramagnetic centers
$PC_{x1}BM$ [6,6]-Phenyl-C_{x1}-butanoic acid methyl ester
PCDTBT   poly[*N*-9''-hepta-decanyl-2,7-carbazole-alt-5,5-(4',7'-di-
                 2-thienyl-2',1',3'-benzothiadiazole)]
PdT   Polydithiophene
PMA   Poly(ortho-methoxyaniline)
PMNA   Poly(*m*-nitroaniline)
PP   Polyphenilene
ppm   Percents per a million cell units
PPP   Poly(*p*-phenylene)
PPS   Poly(*p*-phenylene sulphide)
PT   Polythiophene
PTTF   Polytetrathiafulvalene
QxD   Quasi-x-dimensionality
RT   Room temperature
SA   Sulfuric acid
SSH   Su-Schrieffer-Heeger (theory)
ST-EPR   Saturation transfer electron paramagnetic resonance
TCNQ   Tetracyano-*p*-quinodimethane
THA   Tetrahydroanthracene
TSA   *para*-Toluenesulfonic acid
TTF   Tetrathiafulvalene
trans-$(CD)_x$ *trans*-Polyacetylene deuterated
trans-PA   *trans*-Polyacetylene
UV   Ultravisible band/irradiation with photon energy of
                 3–120 eV and wavelength of 10–400 nm
VB   Valence band
Vis   Visible band/irradiation with photon energy of
                 1.8–3.2 eV and wavelength of 390–700 nm
VRH   Variable range hopping
X-, K-, W-, D-bands 30 mm, 8 mm, 3 mm, and 2 mm wavebands EPR
                 when a spin $S = \frac{1}{2}$ with $g_{iso} = 2$ and precession fre-
                 quency $\omega_e/2\pi = 9.7$ 24, 85 and 140 GHz is registered at
                 resonant magnetic field $B_0 = 0.34, 0.86, 3.3$, and 4.9 T,
                 respectively
X-ray   Röntgen radiation with photon energy of 0.1–100 keV
                 and wavelength of 0.01–10 nm

Preface

The past two decades have seen extraordinary progress in the synthesis and study of organic conjugated polymers and their nanocomposites. This was mainly due to the widespread utilization of such systems in molecular electronics and spintronics. One of the main scientific goals is to reinforce the human brain with computer ability. However, a convenient modern computer technology is based on three-dimensional silicon crystals, whereas the human organism consists of lower-dimensional biological systems. Therefore, the combination of a future computer based on organic conducting polymers of low dimensions with biopolymers is expected to considerably increase the power of human apprehension. This is why understanding the major factors determining specific spin charge-transfer processes in conjugated polymers is now a hot topic in organic molecular science.

The charge in such systems is transferred by topological excitations, solitons, and polarons characterized by spin and high mobility along polymer chains. This stipulated the use of electron paramagnetic resonance (EPR) spectroscopy as a unique direct tool for more efficient study and monitoring of reorganization, relaxation, and dynamic processes carried out in polymer systems. Twenty years have passed since the publication of the first book [122] in which the basic methodological approaches of millimeter waveband EPR spectroscopy in the study of various model, biological, and polymer systems were described. It was demonstrated that the study of such objects at higher registration frequencies allows increasing sufficiently the efficiency of the method to obtain qualitative new information on organic solids and to solve various scientific problems. During this time, the variety of EPR techniques was expanded. For example, the Bruker Corporation developed and started supplying scientific centers with EPR spectrometers operating at wide (1–263 GHz) wavebands. Besides, home-made millimeter waveband EPR spectrometers were constructed and widely used in different scientific centers.

This book's focus is on the use of the technique in conjunction with spin label and probe, steady-state saturation, saturation transfer, and conductometric methods in the study of initial and nanomodified conjugated polymers. Chapter 1 of this book discusses the fundamental properties of conjugated polymers in which a charge is transferred by topological distortions, solitons, polarons, and bipolarons. The theoretical background of magnetic resonance, relaxation, and dynamic parameters of such charge carriers in conjugated polymers is briefly explicated in Chapter 2. It could be a valuable introduction to students interested in EPR, particularly of conjugated polymers. The instrumentation and experimental details are described briefly in Chapter 3. Chapter 4 is devoted to the original data obtained by an X-band to D-band (30–2 mm, 9.7–140 GHz) EPR study of the nature, relaxation, and dynamics of paramagnetic centers delocalized on nonlinear charge carriers as well as the mechanisms of charge transfer in some conjugated polymers differently modified with nanoadducts. The use of some conjugated polymers as electron donors in organic composites is described in Chapter 5. Chapter 6 reveals the possibility to handle charge transport in some multispin polymer composites by using spin–spin exchange. Chapter 7 presents concluding remarks, including the prospects of the study of organic polymer systems for the further construction of novel elements of molecular electronics. Therefore, this book documents both background knowledge and the results of latest research in the field. Unique features include comparisons of data obtained at different microwave frequencies and magnetic fields. Coherent treatment of the subject by the leading Chernogolovka high-field EPR laboratory covers the theoretical background as well as state-of-the-art research both in terms of instrumentation and application to conjugated polymer systems.

The author hopes that the multifrequency EPR spectroscopy and related approaches will be of interest to students and scientists and will encourage them to apply EPR methods more widely to polymeric materials. This book covers a wide range of specific approaches suitable for analyzing processes carried out in polymer systems with paramagnetic adducts providing readers with knowledge of the underlying theory, fundamentals, and applications. These, no doubt, help bridge the gap between the chemistry and physics communities and stimulate research in this fascinating and important field. The goal of this book is not to make the reader an expert in the field, but rather to provide enough information about the EPR spectroscopic method for the reader to determine how the available approaches can be used to solve a particular polymer problem. This book reviews in detail the main experimental methodological approaches developed by our team for the study of various organic condensed systems. It provides an outlook for future developments and references for further reading. This information is essential for postdoctoral scientists, professionals, academics, and graduate students working in this field as well as analytical chemists and chemical engineers designing and studying novel molecular electronic objects. Besides, the author would feel well rewarded if this book helps resolve some of the problems of finding useful information on properties of conjugated polymers and their composites in the ever-growing scientific literature.

Acknowledgments

The author is very grateful to Professors Yakov S. Lebedev and Gertz I. Likhtenshtein, who introduced him to the field of EPR, and also to Professor Dr. Hans-Klaus Roth for introducing him to the field of conducting polymers and for the many discussions. Their early influence and mentoring are deeply appreciated. He sincerely thanks all his coworkers and graduate students for their excellent contributions to this book. He expresses sincere gratitude to the outstanding designer Grigory T. Rudenko, a veteran of World War II, for his inestimable contribution to the implementation of the present work. A constructive collaboration with Professor Oleg Y. Grinberg and Drs. Evgenija I. Yudanova and Nikolai N. Denisov is gratefully acknowledged. The author especially expresses sincere gratitude to his father and mother for their rare kindness and to his family for their love, as well as to relatives and friends for their help and support.

Author

Victor I. Krinichnyi received his higher education at Kazan State University, Kazan, Russia, the birthplace of electron paramagnetic resonance (EPR), after completion of a diploma in 1975 in "Microwave study of mechanism of molecular dynamics in liquids" at the Institute of Chemical Physics at the Russian Academy of Science (RAS) under the supervision of Professor L.A. Blumenfeld. Since then he was employed as an engineer, a principal engineer, and a young scientific researcher at the Institute of Chemical Physics in Chernogolovka, RAS. During this period, Dr. Krinichnyi, in Prof. Y.S. Lebedev's laboratory (the Institute of Chemical Physics RAS) participated in the pioneering joint collaboration and creation of the first multifunctional X-band (140 GHz, 5 T) EPR spectrometer with superconducting magnet. The first exemplificative D-band EPR studies of different organic and biological systems showed great efficiency of this method and allowed to obtain qualitative new information on these objects and solve various practical problems in physics, chemistry, molecular biology, and interdisciplinary sciences. The basic methodological approaches developed at this waveband EPR were then implemented in the first commercial 3 mm waveband Bruker ELEXSYS E680 spectrometer.

Dr. Krinichnyi's PhD (physics and mathematics) work, "2-mm Waveband EPR spectroscopy as a method of the study of paramagnetic centers in biological and organic polymers," was carried out under partial supervision of Professor Y.S. Lebedev and was successfully defended in 1986 at the Institute of Chemical Physics in Chernogolovka RAS. He was employed as a scientific researcher and a senior scientific researcher at the institute (formerly the Institute of Problems of Chemical Physics RAS). His SciD (physics and mathematics) thesis (habilitation) titled "High-resolution 2-mm wave band EPR spectroscopy in the study of biological and conducting polymers" was also successfully defended in 1992 at the same institute, where he works today as a leading scientific researcher and heads the High-Frequency EPR Group.

Research interests of Dr. Krinichnyi resulting from the practical application of multifrequency EPR spectroscopy are concerned with relaxation and dynamics of nonlinear charge carriers, solitons, and polarons in conjugated polymers and their nanocomposites, mechanism of charge transport in molecular crystals, spin phenomena in condensed systems, and organic molecular electronics, photonics, and spintronics. Since 1992, he is member of the International EPR (ESR) Society. Thirteen Russian and International scientific projects were carried out under his direct guidance as principal investigator. The original theoretical and experimental results obtained through these projects were reported at international summits, invited lectures at different European scientific centers, and described in more than a hundred scientific publications, including a monograph, 5 invited contributions in edited books, and 11 reviews.

1

Introduction

1.1 Properties of Conjugated Polymers

Various low-dimensional compounds can be attributed to organic conductors, such as molecular crystals based on charge-transfer complexes and ion-radial salts [1–5], modified fullerenes [6–8], phthalocyanine metal complexes [9], dyes [10], and metal-filled polymers [11,12]. These compounds are interesting from the scientific standpoint, concerning the fundamentals of charge-transfer processes. Various polymers and their composites with carbon nanomaterials can also be classified in to organic conductors [13–18].

During the past two or three decades, a new class of electronic materials, organic conjugated polymers, attracted great attention due to their unique capabilities, namely, flexible, solution processable, lightweight, and tunable electronic properties [3,19–31]. The particular interest to such systems was initiated in 1964 by the Little hypothesis [32] on principal possibility of synthesis of high-temperature superconductors based on conjugated polymers. Since that time, the investigation of conjugated polymers and their nanocomposites has generated entirely new scientific conceptions and a potential for their perspective application as an active material for the creation of components of organic molecular electronics [17,22–26,33–36], including sensors for solution and gas components [37–42], Schottky diodes [40,43–47], field-effect transistors [48–50], and photovoltaic devices [35,51–57]. One of the main scientific goals seems to reinforce the human brain with computer abilities. However, a modern convenient computer technology is based on 3D silicon crystals, whereas human organism consists of biological systems of lower dimensionality. So the combination of a future computer based on organic conjugated polymers of lower dimensionality with biopolymers is expected to increase considerably the power of human apprehension.

Figure 1.1 shows room temperature direct current (*dc*) conductivity σ_{dc} of some conjugated polymers in comparison with that of convenient conductors. The structures of some conjugated polymers used as an active matrix in molecular electronics are schematically presented in Figure 1.2. These materials have a highly anisotropic quasi-one-dimensional (Q1D) π-conjugated structure with delocalized charge carriers, which makes such systems

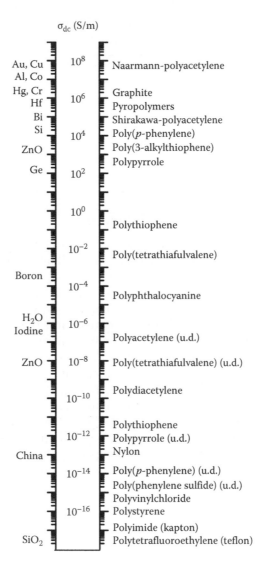

FIGURE 1.1

Room temperature direct current conductivity of undoped and highly doped conjugated polymers [3] in comparison with that of some convenient polymers and metals.

fundamentally different from traditional inorganic semiconductors (e.g., silicon and selenium) and from well-known convenient insulating polymers, for example, polyethylene, polyvinyl chloride, and polystyrene. Conjugated polymers are synthesized as film or powder and their conductivity can be varied by more than 12 orders of magnitude by chemical or electrochemical oxidation or reduction of their chains. Under the doping process, their *dc* conductivity increases from $\sim 10^{-15}$ to 10^{-10} S/m (insulator regime) up to $\sim 10^{-3}$–10^2 S/m

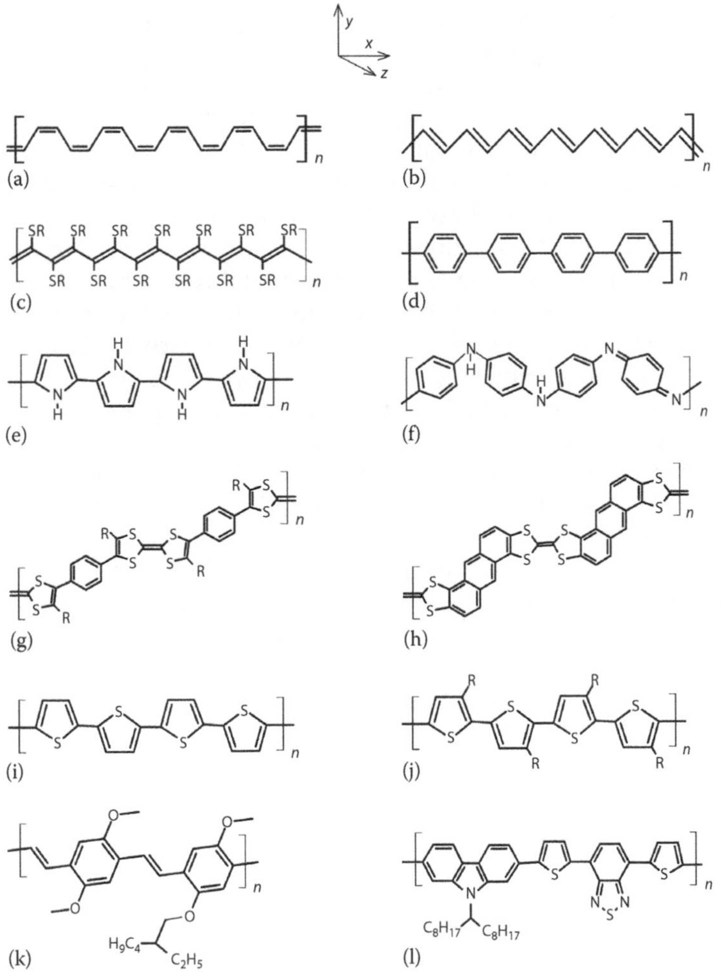

FIGURE 1.2
Schematic representation of undoped (a) *cis*-polyacetylene, (b) *trans*-polyacetylene, (c) poly(*bis*-alkylthioacetylene) (R = $-CH_3$, $-C_2H_5$, $-C_3H_5$), (d) poly(*p*-phenylene), (e) polypyrrole, (f) polyaniline, (g,h) poly(tetrathiafulvalene) with phenyl (PTTF-R-C_6H_4, R = $-CH_3$, $-C_2H_5$) and tetrahydroanthracene (PTTF-THA) bridges (i) polythiophene, (j) poly(3-alkylthiophene) (R = C_mH_{m+1}), (k) poly[2,5-dimethoxy-1,4-phenylene-1,2-ethenylene-2-methoxy-5-(2-ethylhexyloxy)-(1,4-phenylene-1,2-ethenylene)], and (l) poly(*N*-9'-heptadecanyl-2,7-carbazole-alt-5,5-(4',7'-di-2-thienyl-2',1',3'-benzothiadiazole)), which were extensively studied by electron paramagnetic resonance spectroscopy. The main molecular axes are shown schematically.

(semiconductor regime) and then up to ~10^4–10^8 S/m (metal regime) [3,58,59]. Electronic properties of such systems strongly depend on their structure, morphology, and quality, whereas the type of their conductivity is governed by the nature of the introduced counterion [60,61]. The introduction of anions, for example, HSO_4^- BF_4^-, ClO_4^-, J_3^-, and $FeCl_4^-$, into a polymer induces a positive charge on a polymer chain and thus leads to a *p*-type conductivity of the polymer. Conductivity of *n*-type is realized under polymer doping by Li^+, K^+, Na^+ and other ions of alkali metals.

In traditional 3D inorganic semiconductors, fourfold (or sixfold) coordination of each atom to its neighbor through covalent bonds leads to a crystalline structure. Electron excitations may be usually considered in the context of this rigid structure, leading to the conventional conception of generalized electrons or holes as dominant charge carriers. The situation with conjugated polymer semiconductors is quite different: once they form bulk nanocomposites with the dopants embedded, their Q1D structure becomes more susceptible to structural distortion. Therefore, electronic properties of conjugated polymers may be conventionally considered in the frames of bands theory [62–64] as well as of soliton and polaron one [65,66], based on Peierls instability [67], a characteristic of Q1D systems. Elementary charge and energy transfer phenomena occurring in such materials are crucial to the structural and conformational diversity of their polymer active matrix. So in order to construct organic electronic elements based on conjugated polymers, the correlations of their morphology, electronic properties, selectivity, sensitivity, etc., with magnetic, relaxation, and dynamic properties of spin charge carriers should be obtained and analyzed.

1.2 Solitons, Polarons, and Bipolarons in Conjugated Polymers

Polyacetylene (PA) is the simplest conjugated polymer. It can exist in *cis*- and *trans*-isomers (forms) (Figure 1.2), and the latter is thermodynamically more stable [68,69]. The morphology of this polymer mainly depends on the synthesis, the structure of the initial monomer, the nature of the introduced doping agents, and the film thickness. The chains in PA are situated parallel to one another, forming a fibril with the thickness of few dozens of nm and a length of few hundreds of nm. The longitudinal axes of such fibrils are arranged chaotically and can be partially oriented at the polymer stretching. Each fibril contains well-packed polymer chains, which exhibit crystalline structure. Thus, PA appears as a kind of polycrystal, consisting of *crystalline* fibrils with appropriate bandgap energy and conductivity. Three of four carbon valence electrons in PA occupy *sp²*-hybridized orbitals; two of σ-type bonds form 1D lattice, while the third one forms a bond with a hydrogen atom. In each isomer the last valence electron has the symmetry of $2p_z$ orbital, with its charge density oriented perpendicular to the plane defined by the other three electrons. Therefore, σ-bonds form a low-lying completely filled valence band (VB), while π-bonds form a partially filled conjugated band (CB) (Figure 1.3a). If the length of all bonds would be equal, pure *trans*-PA would be Q1D metal with a half-filled band. Such a system is unstable with respect to a dimerization distortion, in which adjacent CH groups move toward one another, forming alternating double and longer single bonds. The transition between C—C and C=C bonds does not require energy change, so the Peierls distortion opens up a substantial gap in the Fermi level. This twofold degeneration leads to the formation of nonlinear topological excitations (quasiparticles), namely, solitons on *trans*-PA chains

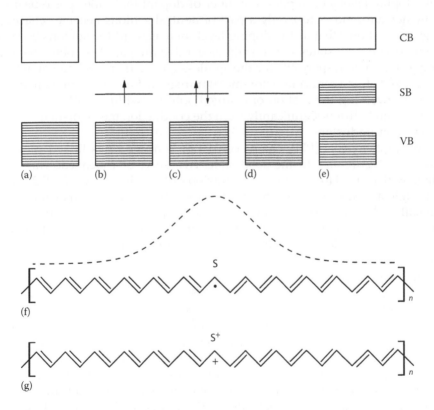

FIGURE 1.3
The formation of band structure, valence band and conjugated band divided by the midgap, of idealized *trans*-polyacetylene with no defects (a); the appearance of neutral soliton energy level in the polymer midgap (b); the evolution of band structure of *trans*-polyacetylene with the appearance of negatively (c) or positively (d) charged soliton at slow doping level, at higher doping level, where soliton states overlap and form soliton band (e) with quasi-metallic behavior; the formation of the spin neutral soliton S (f); and the spinless charged soliton S+ (g) on the *trans*-polyacetylene chain. The distribution of the spin on the soliton is shown by the dashed line.

corresponding to a break in the pattern of bond alternation [70,71] (Figure 1.3), whose energy level is localized at midgap, thus determining the fundamental properties of *trans*-PA. They are characterized by spin $S = \frac{1}{2}$ and high Q1D mobility along the polymer chains. Su-Schrieffer-Heeger (SSH) theory [70,71] predicts for a neutral soliton the width of $N_s = 14$ monomer units, an effective mass $m_s^* \cong 6m_e$, and activation energy of its Q1D motion $E_a = 2$ μeV.

Figure 1.3 exhibits schematically also the electronic configurations of a neutral and charged soliton formed in *trans*-PA. A single soliton in this polymer has a peculiar charge–spin relationship since a neutral soliton corresponds to a radical with spin $S = \frac{1}{2}$, while a negatively or a positively charged soliton loses the spin and becomes diamagnetic. Their energy level becomes completely filled (*n*-type doping) or empty (*p*-type doping). At a low doping level y (y implies a number of dopant molecules per repeating polymer unit, monomer), only a part of neutral solitons becomes charged. Figure 1.3 shows that as the doping level increases, all the solitons become spinless and their states start forming a band at midgap. The soliton band is suggested to be responsible for the spinless conduction mechanism, which is realized after a semiconductor–metal transition. Such a conduction mechanism, involving the motion of charged solitons within a filled or empty soliton band, differs significantly from the conduction mechanism in convenient semiconductors.

Poly(*p*-phenylene) (PPP) chains consist of benzene rings linked *via para*-position (Figure 1.2). In the solid state, two successive benzene rings are tilted with respect to one another by torsion (dihedral) angle $\theta \approx 23°$ [72]. Such an angle appears as a compromise between the effect of conjugation and crystal-packing energy, which would lead to a planar conformation, and the steric repulsing between ortho-hydrogen atoms, which would lead to a nonplanar conformation [73]. The globular structures of PPP consist of packed fibrils with a typical diameter of 100 nm.

The band structure of PPP is obtained as a result of the overlapping of π-orbits of the benzene rings (Figure 1.4). For this polymer, a resonance form can also be derived, which corresponds to a quinoid structure; however, in contrast with *trans*-PA, benzenoid and quinoid forms are not energetically equivalent, quinoid structure being substantially higher in energy. As a result, the solitons are trapped in a slightly doped PPP by the charges in polymer structure, and the other nonlinear excitations are formed, namely, polaron with spin $S = \frac{1}{2}$ (Figure 1.4) and energy levels lying in the gap above the VB and below the CB edges. At an intermediate doping level, the polaron pairs collapse and form spinless bipolarons (Figure 1.4). The width of the polaron and bipolaron is 4–5 and 5–5.5 monomer units, respectively [74–76]. As the doping level y increases, bipolaron states overlap forming bipolaron bands within the gap. At a high doping level, these bands tend to overlap and create new bandgap energy bands that may merge with the VB and CB allowing freedom for extensive charge transfer. The analogous band structure is formed also in the case of other PPP-like polymers. The energy

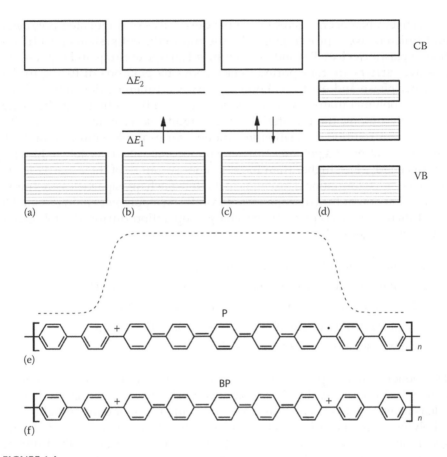

FIGURE 1.4
Evolution of an initial band structure of poly(p-phenylene) (a) during the polymer doping; (b) slight doping level with the appearance of polaron states in the midgap; (c) intermediate doping level, with the appearance of states of noninteracting bipolarons; and (d) high doping level, where bipolaron states overlap and form filled and semifilled bands with quasi-metallic behavior. The formation of a spin charge polaron P (e) and spinless bipolaron BP (f) on the chain of poly(p-phenylene) is shown. The distribution of the spin on the polaron is shown by the dashed line.

levels ΔE_1 and ΔE_2 of polarons depend on the polymer structure and the nature of a dopant molecule.

Among elements of organic electronics based on conjugated polymers, polymer:fullerene nanocomposites seem to be most suitable for polymer photovoltaics that stipulates their wide investigation [35,51–57]. Fullerene molecules embedded into a polymer matrix of such systems form the so-called bulk heterojunctions (BHJ) with polymer chains. In this case, a polymer matrix and adducts become electron donors (hole transporter, p-type material) and electron acceptors (electron transporter, n-type material), respectively. Nanocomposites of soluble derivatives of conjugated polymers and fullerene were proven [35] to be the most efficient systems for utilization in plastic devices. Beyond photoinduced charge exciting and separation, positive carriers are transported to electrodes by polarons diffusing in the polymer phase and electrons hopping between fullerene domains embedded into polymer matrix. A definitive advantage of BHJ is that it can be made by simply mixing these materials in an organic solvent and casting with well-known solution deposition techniques, for example, spin coating [77]. The illumination of a polymer:fullerene nanocomposite by visible light with photon energy $h\nu_{ph}$ higher than the π–π^* energy gap E_g of their matrix leads to the fast (~fs) formation of excitons in its BHJ. Such quasiparticles are transformed into ion radical quasipairs, polaron $P^{+\cdot}$ on a polymer chain (donor, D) and fullerene $F_{60}^{-\cdot}$ (acceptor, A) and charge separation during the following successive stages [78]:

1. Excitation of polaron on polymer chain: $D + A \xrightarrow{h\nu} D^* + A$.
2. Excitation delocalization on the complex: $D^* + A \xrightarrow{h\nu} (D + A)^*$.
3. Initiation of charge transfer: $(D - A)^* \rightarrow (D^{\delta+} - A^{\delta-})$.
4. Formation of ion-radical pair: $(D^{\delta+} - A^{\delta-})^* \rightarrow (D^{+\cdot} - A^{-\cdot})$.
5. Charge separation: $(D^{+\cdot} - A^{-\cdot}) \rightarrow D^{+\cdot} - A^{-\cdot}$.

The donor and acceptor units are spatially close but are not covalently bonded. At each step, the D–A system can relax back to the ground state, releasing energy to the *lattice* in the form of either heat or emitted light.

Polymer:fullerene BHJ are characterized by efficient light-excited charge generation at the interface between two organic materials with different electron affinity. Figure 1.5 illustrates the energy diagram of two exemplary intrinsic semiconductors, for example, regioregular poly(3-alkylthiophene) shown in Figure 1.2 and [6,6]-phenyl-C_{61}-butanoic acid methyl ester (PC$_{61}$BM), widely used in polymer:fullerene photovoltaic devices, before making a contact between them. A heterojunction formed by these materials inserted between a high work-function electrode (El_1) matching the highest occupied molecular orbital level of the donor (HOMO$_D$) and a low work-function electrode (El_2) matching the lowest unoccupied molecular orbital level of the electron acceptor (LUMO$_A$) should in principle act as a diode with rectifying current–voltage characteristics. Under forward bias (the low work-function electrode is biased negative in respect to the high work-function electrode), the electron injection into the LUMO$_A$ layer from the low work-function electrode as well as the electron extraction out of the HOMO$_D$ by the high work-function electrode is energetically possible and a high current may flow through the heterojunction. Under reverse bias (the low work-function electrode is biased positive in respect to the high work-function electrode), the electron removal from the electron donor and electron injection to the electron acceptor is energetically unfavorable. The formation of the polaron $P^{+\cdot}$

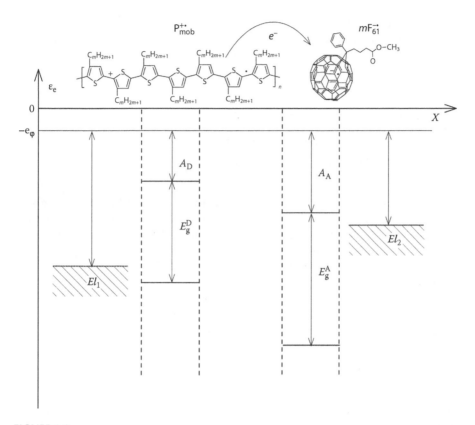

FIGURE 1.5
Schematic band diagram of two semiconductors with different electron affinities before making between them the bulk heterojunction composite. The electron donor (A_D) and electron acceptor (A_A) affinities are defined in relation to the electron energy in vacuum at the same electrical potential. E_g^D and E_g^A are the bandgap energies of the electron donor and electron acceptor, respectively. At the top, the poly(3-alkylthophene) (P3AT) and [6,6]-phenyl-C61-butanoic acid methyl ester ($PC_{61}BM$) are schematically shown as electron donor and electron acceptor, respectively. The appearance of the polaronic quasiparticle $P_{mob}^{+\bullet}$ with a spin $S = \frac{1}{2}$ and an elemental positive charge in a P3AT chain and ion radical $mF_{61}^{-\bullet}$ with an elemental negative charge and a spin $S = \frac{1}{2}$ on a $PC_{61}BM$ is shown as well. (From *Spectroscopy of Polymer Nanocomposites*, Ponnamma, D., Rouxel, D., Thomas, S., and Krinichnyi, V. I., EPR spectroscopy of polymer:fullerene nanocomposites, pp. 202–275, Chapter 9, Copyright 2016, with permission from Elsevier.)

and fullerene $F_{61}^{-\bullet}$ charge carriers is shown in Figure 1.5 as well. This process revealed by the time-resolved optical spectroscopy occurs in the femtosecond time domain [79,80], for example, near 100 fs in optimized polymer:fullerene BHJ [81], whereas the electron back transfer with charge annihilation is much slower possibly due to dynamics and relatively slow structural relaxation in such a system of lower dimensionality. Understanding of photoexcitation, recombination of charge carriers, and other electronic processes realized in conjugated polymers is of fundamental interest for both material characterization and molecular device fabrication.

1.3 Resonant Manipulation of Interacting Spins in Polymer Nanocomposites

The efficiency of energy conversion by organic polymer:fullerene quantum structures is governed partly by charge localization phenomena, which underlie their optical transitions. These molecular systems in which a charge is transferred by weakly spin–orbit coupled carriers are gaining particular interest because the use of spin carriers with additional spin degree of freedom opens an undoubted imperative to develop a remarkable new generation of electronic, *spintronic* devices with spin-assisted and, therefore, handling electronic properties. It is well known that the data processing in conventional computers is limited by the transport of elemental charge carriers, electrons, through silicon semiconductors. Exploiting the orientation of an electron spin rather than its charge takes a possibility to create spintronic devices that will be smaller, more versatile, and more robust than those currently making up silicon chips and circuit elements. This can be also used, for example, for encoding and transferring information more efficiently than using spinless ones [82]. Deposition of isolated tetracyano-*p*-quinodimethane (TCNQ) molecules each possessing a spin to a graphene monolayer leads to the formation in the latter of spatially extended spin-split electronic bands [83] that may be used in organic spintronic devices [84]. The careful controllable preparation, preservation, and manipulation of quantum states should form the backbone of the use of quantum information processing to create organic molecular spin-controlled electronic devices. Orientation of spin in charge carriers survives for a relatively long time (nanoseconds, compared to tens of femtoseconds during which electron momentum decays), which makes spintronic devices particularly attractive for memory storage and magnetic sensors applications and potentially for quantum computing where electron spin would represent a bit (called *qubit*) of information [85].

Spin feature of charge carriers stabilized and/or induced, for example, in the previously described polymer:fullerene BHJ can also be used for manipulation by electronic properties of such a system. Excitons formed in a polymer:fullerene BHJ under illumination at the intermediate step can be conversed into polaron pairs or donor–acceptor complexes, which then collapse into spin pairs of polarons and fullerene anion radicals. Besides, the triplet state of fullerene characterized by high electron spin polarization can also be easily photoexcited in such a nanocomposite. This can possibly be used as an active media for masers and other molecular devices [86]. Diffusing and meeting in a system bulk, the previous charge carriers recombine. Most organic hydrocarbon materials show only radiative recombination from the singlet state so that high triplet formation rates form a formidable loss channel in devices with light-emitting recombination [87]. However, this process itself is not a trivial phenomenon, of which microscopic details still remain unknown. For example, one would not necessarily expect the recombination of a charge carrier with just the first opposite carrier. In this case, both the charge carriers exchangeable spin-flip, so that their further recombination become dependent on their dynamics, number, polarization, and mutual distinct. For large separations, when the thermal energy exceeds the interaction potential, the charges can be considered as noninteracting. However, once the carriers become nearer than the inverted Coulombic interaction potential, their wave functions begin to overlap so that exchange interactions become nonnegligible. This can lead to the formation in organic semiconductors of exciton of either singlet or triplet configuration. It is important to note that an electron spin-flip arising from spin–orbit coupling may thus transform the singlet exciton into a triplet one with a lower energy level. This process prevails in the most planar structures with particular molecular symmetries [88]. So depending on various conditions, a singlet ($\uparrow\downarrow$–$\uparrow\downarrow$ or S) and three triplet ($\downarrow\downarrow$, $\uparrow\downarrow$+$\uparrow\downarrow$, $\uparrow\uparrow$ or T_+, T_0, T_-) spin configurations can be realized in these molecular devices [89].

Free charge carriers are generated through dissociation of either singlet or triplet polaron pairs in polymer:fullerene BHJ. The dissociation rates of these carrier pairs are spin dependent so that the sum number of charge carriers should depend on the initial spin state of the polaron pair. A spin configuration of such pairs transforms between singlet and triplet ($S \leftrightarrow T$) either randomly due to irreversible spin–lattice relaxation (with characteristic time T_1) or coherently by an external electromagnetic field, which induces electron spin precession and resonant reorientation. Such a coherent manipulation of the spin state requires phase coherence of the initiated spin with the initiating microwave (MW) field. Loss of phase coherence arising, for example, due to the intrinsic (homogeneous) properties of the spin system or extrinsic (inhomogeneous) characteristics provoking, for example, by the system disorder or anisotropic local field distribution is characterized by spin–spin relaxation time T_2 for homogeneous and T_2^* for inhomogeneous processes. If the charge carriers recombine with higher probability than own dissociation, they form excitons with single or triple multiplicity. This recombination process is strongly spin dependent due to the energetic difference between singlet and triplet states and, therefore, can be characterized by appropriate transition rates. So formed excitons may then decay, radiatively or nonradiatively, to the molecular ground state. This process involves charge transfer through a polymer:fullerene BHJ in which a two-step exciton dissociation is accompanied by the transfer of its energy from the donor to the acceptor followed by a polaron transfer to the donor [90]. In this photovoltaic system, even rarely distributed defects may initiate exciton dissociation and the formation of polaron pairs [91]. Undoubtedly, this process is also spin dependent because in such a system only singlet excitons can be photoinduced and the triplet excitons are bound stronger than the singlet ones, so they have different energetics. This means that in most photovoltaic systems, the triplet exciton dissociation rate can, therefore, be nulled and the contribution of singlet excitons should only be considered.

1.4 Multifrequency EPR of Spin Charge Carriers in Conjugated Polymer Systems

Different methods can be used for the study of charge carriers stabilized and excited in polymers and their nanocomposites [92]. Physics processes carried out in such objects may be identified and analyzed comparing results obtained by all possible methods at similar experimental conditions, for example, optical (fluorescence and phosphorescence) and/or electrically detected magnetic resonance [89]. The former can, in principle, sense singlet and triplet excitons in organic semiconductors [93]. However, there are generally other absorbing species present in such systems, namely, polaronic charge carriers, which themselves introduce efficient subgap optical transitions so that an unambiguous identification of excitations is not always possible. In some cases, incorporation of a heavy-metal atom into an organic complex can increase spin–orbit coupling, effectively mixing singlet and triplet levels. A strong modulation of the phosphorescence signal can occasionally be observed in the presence of an electric field, thus proving the presence of spin triplet polaron pair species, which are much more polarizable than the ultimate triplet exciton [94].

EPR spectroscopy was proved [89,95–103] to be one of the most widely used and productive physical direct methods in structural and dynamic studies of high-molecular systems, which contain free radicals, ion radicals, molecules in triplet states, transition metal complexes, and other paramagnetic centers (PC). The method is based on resonant absorption of microwave radiation by a paramagnetic sample due to the splitting of the energy levels in an external magnetic field. It allowed getting more detailed information on processes carrying out in various solids containing PC.

As the original properties of conjugated polymers are related to the existence of PC localized and/or delocalized along and between their chains, a great number of EPR experiments have been performed for the study of their magnetic, relaxation, and dynamic properties [104–106]. For more than 50 years, such EPR investigations are predominantly carried out at convenient X-band (3 cm, 9.7 GHz) EPR, that is, at registration frequency $\omega_e/2\pi = \nu_e \leq 10$ GHz and fields $B_0 \leq 330$ mT. However, at these wavebands, the signals of organic free radicals with g-factor lying near g-factor of free electron (g_e = 2.00232) are registered in a narrow magnetic field range and, therefore, can overlap the lines with close g-factors. Normally, the solitons and polarons in conjugated polymers demonstrate at low frequencies uninformative single spectra, so the line shape and concentration of these quasiparticles can only be directly measured. Besides, strong cross-relaxation of PC that is still perceptible at low magnetic fields [107] additionally complicates the registration and identification of such centers. Earlier, the new experimental techniques improving the efficiency of EPR spectroscopy in the study of solid-state systems were developed. They are electron spin echo spectroscopy [108–110], a method based on the effect of spin polarization, in which EPR signal is registered optically [111], methods of double electron-nuclear resonance [112,113], and MW saturation transfer EPR (ST-EPR) method [114]. However, most of these methods may be applied only to solve specific problems and investigate special objects.

Electron spin can be oriented parallel and antiparallel to a direction of the applied magnetic field forming a two-level quantum system. To probe resonance spin hopping between these pure quantum states, weak spin–orbit coupling is required, which limits the applicability of the convenient magnetic resonance methods for the study of spin-assisted charge transfer in inorganic spintronic materials [115]. In organic semiconductors, however, spin–orbit interaction is very weak due to low atomic order number, so that the information about spin orientation can be easily established by magnetic spectroscopic techniques [116]. It allows using cutting-edge spectroscopic methods for determination of spin polarization in spin-assisted processes and also for their controlled manipulation. EPR spectroscopy was shown [89] to be a powerful direct tool able to reveal about the underlying nature of spin carriers excited in quantum systems including multispin polymer systems with weak spin–orbit coupling. The differences in lifetime between the three excited-state triplet sublevels formed in such systems give rise to a spin-dependent buildup of macroscopic polarization [117], including spin charge carriers stabilized and initiated in conjugated polymers [118] and their fullerene-based nanocomposites for photovoltaic applications [35,55,57]. However, conventional X-band EPR technique operated at $B_0 \sim$ 330 mT is still less sensitive than, for example, the fluorescence-based one. Amplifying this method with, for example, optical and/or conductometric approaches, one can increase the sensitivity of the method to spin-assisted charge transfer in solids [87,89]. If the rate of formation of singlet excitons out of singlet polaron pairs is greater than the rate of formation of triplet excitons out of triplet polaron pairs, and effective mixing between singlet and triplet polaron pairs occurs, for example, due to spin–lattice relaxation, then the effective electroluminescence quantum yield of a singlet-emitting material should exceed 25% [87]. This becomes possible due to the adjustable preferential recombination via the singlet channel. Because a recombination of charge carriers was proved to be spin dependent, EPR method, for example, its electrically detected modification [89], can control the resulting current in photovoltaic devices. Indeed, the polaron and triplet populations are reduced as the MW energy matches the difference between the energies of parallel and antiparallel spins. This can only arise from increased formation of singlet excitons or, alternatively, increased suppression of triplet excitons by polarons through an Auger-like nonradiative recombination process. In this case, singlet and triplet polaron pairs are mixed randomly and such a mixing is a function of applied magnetic field. One can conclude that polaron pairs are preferentially lost to singlet excitons under paramagnetic resonance rather than forming singlet and triplet excitons with comparable probability. The fractional change in polaron-to-triplet exciton populations can be related directly to the ratio between singlet and triplet exciton formation rates. It makes possible to investigate the EPR effect on the induced MW absorption features. According to an alternative theory [119], the interaction between polarons and excitons increases under paramagnetic resonance that leads to a quenching of singlet excitons by promoting nonradiative decay to the ground state accompanied by the expense of a polaron or a triplet exciton. There are, therefore, no simple picture cleaning spin resonance-assisted processes in organic photoexcited systems; however, they undoubtedly can be governed by spin-dependent exciton–charge interactions and consistent with the spin-dependent polaron pair recombination model [120]. Besides, such processes are governed also by the energy of initiating photons [57]. Therefore, manipulation by electronic spin-assisted processes in organic devices should require synchronous imbalance in the exciton formation rate and effective mixing between singlet and triplet polaron pairs. Besides, EPR spectroscopy becomes a powerful method for the study of an exciton collapse into a pair of spin charge carriers, charge separation, transfer, and recombination in organic donor:acceptor systems. Indeed, the evidence for a successful charge transfer is based on the fact that excitons generated by light absorption have zero spin and therefore cannot be detected by EPR spectroscopy. If the exciton is split at the interface between donor and acceptor, polaron and anion radical are created which have half-integer spin, leading to the appropriate EPR signal. The amount of light-induced charge carriers can be determined by comparing the EPR spectrum in the dark and under illumination.

Except X-band (of frequencies extending from 8 up to 12 GHz), such investigations are also carried out at K-band (18–26.5 GHz), Q-band (30–50 GHz), W-band (75–110 GHz), and D-band (110–170 GHz) EPR. It was shown [121–124] that the registration of organic radicals in different solids, especially in conjugated polymers at high spin precession frequencies $\nu_e \geq 100$ GHz corresponding to fields $B_0 \geq 3$ T, enables to increase considerably the precision and descriptiveness of the EPR method. At D-band EPR ($\nu_e = 140$ GHz, $B_0 \approx 5$ T), it was investigated in detail the structure, relaxation, dynamics, and other specific characteristics of radical centers and their local environment and elementary charge-transfer processes in different solids, organic polymers, and biopolymers among them. It should, however, be noted that the advantages of the high-frequency/field EPR spectroscopy are limited in practice by a concentration sensitivity that decreases with increasing MW frequency. Nevertheless, multifrequency EPR spectroscopy seems to be a powerful method for the detailed study of conjugated polymers and their nanocomposites with interacting spin packets.

1.5 Summary

This book summarizes the main results obtained in the investigation of various conjugated polymers and their nanocomposites at 9.7–140 GHz waveband EPR in the Institute of Problems of Chemical Physics, Chernogolovka, Russia, in cooperation with some European groups. Some PA samples were also investigated at higher spin precession frequencies (up to 430 GHz) at the Services National des Champs Intense, CNRS, Grenoble, France. Multifrequency EPR study of other systems is described in detail, for example, in [98,102,122,125–133] and is excluded from consideration.

Chapter 2 of this book presents succinctly theoretical fundamentals of EPR spectroscopy necessary for the interpretation of experimental results obtained at the study of low-dimensional systems. In Chapter 3 are shortly described EPR techniques, approaches, and standards used at the study of conjugated polymers and their nanocomposites with domestic and initiated spin charge carriers. The advantages of multifrequency EPR spectroscopy combined with the steady-state saturation of spin packets, spin label and probe, and saturation transfer methods are demonstrated in Chapter 4 where the results of the study of PC stabilized in some exemplary conjugated polymers and their nanocomposites are discussed. Analogous study of spin charge carriers photoinitiated in some polymer:fullerene nanocomposites is described in Chapter 5. Chapters 6 and 7 of this book demonstrates a possibility to use multifrequency EPR spectroscopy for a detailed study of spin-assisted charge transfer in multispin polymer nanocomposites.

2

Theoretical Backgrounds of Paramagnetic Resonance and Spin Transfer in Conjugated Polymers

2.1 Magnetic Parameters of Organic Radicals Directly Obtained by EPR Spectroscopy

All paramagnetic compounds possess an unpaired electron, whose interactions with external magnetic field and with electron and nuclear spins are generally described by the Hamiltonian:

$$H = \gamma_e \hbar m \mathbf{B}_0 - \gamma_e \hbar \mathbf{S} \sum_{i=1} a \mathbf{I}_i - \gamma_e \gamma_p \hbar^2 \mathbf{S} \mathbf{I}_i \int \psi^*(x) \frac{1 - 3\cos^2 \theta}{r^3} \psi(x) d\tau, \qquad (2.1)$$

where

γ_e and γ_p are the gyromagnetic ratio for electron and proton, respectively
$\hbar = h/2\pi$ is the Planck constant
$\mathbf{S}$ is the electron spin operator
$\mathbf{I}_i$ is the ith term of nuclear spin operator
$a = 1/3(A_{xx} + A_{yy} + A_{zz})$ is an isotropic hyperfine interaction (HFI) constant with the main terms A_{ij}
θ is the angle between the external magnetic field $\mathbf{B}_0$ vector and $\mathbf{r}$ vector, which is the vector between dipole moments of electron spin and the ith nucleus
x is the electron spin coordinate
$\psi(0)$ is a wave function proportional to a probability of the localization (or the density) of an unpaired electron near the interacting nucleus

The a value is not equal to zero only for the electrons, possessing a nonzero spin density $\rho(0) = [\psi(0)]^2$, that is, for s-electrons.

The first term of the Hamiltonian characterizes the Zeeman interaction of an unpaired electron characterized by the Landé splitting factor g. If the resonance condition

$$\hbar\omega_e = \gamma_e \hbar B_0 = g\mu_B B_0 \qquad (2.2)$$

is fulfilled (here $\omega_e = 2\pi\nu_e$ is the electron precession frequency and μ_B is the Bohr magneton), an unpaired electron absorbs an energy quantum and is transferred to a higher excited state. The second term of Equation 2.1 defines the nuclear interaction and the third one is the contribution of the nuclear Zeeman interaction.

One of the most significant characteristics of a spin reservoir is its Landé g-factor, which is the ratio of electron mechanic momentum to a magnetic moment. It is stipulated by the distribution of spin density in a radical fragment, the energy of excited configurations, and spin–orbit interaction. Free s-electron with $S = J = 1/2$ and $L = 0$ (here J and L are the electron total angular and orbital moments, respectively) is characterized by g_e-factor equal to 2; however, the relativity correction yields $g_e = 2.00232$. $L \neq 0$ for p- and d-electrons; therefore, their g-factor mainly differs from g_e and varies in a wide range.

If a nucleus is located, the intensity of magnetic field, induced by electron orbital motion, is equal to zero for s-electrons ($L = 0$) or differs from zero in the case of p- and d-electrons ($L \neq 0$). Thus, the nuclei of atoms, having p- or d-electrons, are affected by a strong magnetic field, induced by an electron orbital magnetic moment.

As the g-factor depends on a paramagnetic molecular orientation in an external magnetic field. Therefore, according to the perturbation theory [98], it should be defined by a **g** tensor of the second order [134–136]:

$$\begin{vmatrix} g_{xx} & & \\ & g_{yy} & \\ & & g_{zz} \end{vmatrix} = \begin{vmatrix} 2\left(1 + \dfrac{\lambda\rho(0)}{\Delta E_{n\pi^*}}\right) & & \\ & 2\left(1 + \dfrac{\lambda\rho(0)}{\Delta E_{\sigma\pi^*}}\right) & \\ & & 2 \end{vmatrix}, \qquad (2.3)$$

where
   λ is the spin–orbit coupling constant
   $\rho(0)$ is the spin density
   $\Delta E_{n\pi^*}$ and $\Delta E_{\sigma\pi^*}$ are the energies of the unpaired electron $n \rightarrow \pi^*$ and $\sigma \rightarrow \pi^*$
   transitions, respectively

Relation (2.3) is fulfilled for a spin localized on a single nucleus when $\rho(0) = 1$. In most cases, a spin is delocalized when $\rho(0) < 1$; this is the reason why the value $\Delta g_{ij} = 2\lambda/\Delta E_{ij}$ should additionally be multiplied by a coefficient $\rho(0)$. It is reasonable that the sign of Δg_{ij} is generally defined by that of ΔE_{ij}. If the depinning of the orbital moment is stipulated by the excitation of the filled electron shells ($n \to \pi^*$ and $\sigma \to \pi^*$ transitions), then $\Delta E_{ij} < 0$ and $\Delta g_{ij} > 0$. If the orbital motion takes place due to the interenergetic transitions of unfilled electron shells ($\pi^* \to \sigma$ transitions), then $\Delta E_{ij} > 0$ and $\Delta g_{ij} < 0$. In the latter case, $\Delta E_{\pi\sigma^*}$ is large; this is the reason why the g_{zz} value of most radicals lies close to g_e. The radicals, including heteroatoms (O, N, F, S, Cl, etc.), have a small energy of $n \to \pi^*$ transition. The constant of electron orbital interaction with these nuclei is greater than that with carbon nucleus, so then the g_{xx} and g_{yy} values deviate from g_e and the inequality $g_{xx} > g_{yy} > g_{zz} \approx g_e$ always holds for radicals of this type. Since the backbone of the polymer can be expected to lie preferably parallel to the film substrate [137], the lowest principal g-value is associated with the polymer backbone. The macromolecule can take any orientation relative to the z-axis, that is, the polymer backbone direction that is derived from the presence of both the g_{xx} and g_{yy} components in the spectra for all orientations of the film. Thus, the g-factor anisotropy is a result of inhomogeneous distribution of additional fields along the x and y directions within the plane of the polymer σ-skeleton rather than along its perpendicular z direction (see Figure 1.2). The increase in spectral resolution and determination of all spin Hamilton parameters of various solids can be reached at millimeter waveband EPR [123,128,130,138,139].

Harmonic librations of polymer chains with localized polarons can modulate the charge-transfer integrals in polymer composites as it is typical for organic molecular ordered systems [140]. This should change the effective g-factor as

$$g = g_0 + \frac{A}{\hbar\omega_l}\coth\left(\frac{\hbar\omega_l}{2k_BT}\right), \tag{2.4}$$

where
  g_0 and A are constants
  $\omega_l = \omega_0\exp(-E_l/k_BT)$ is the librational frequency
  E_l is the energy required for activation of such a motion
  k_B is the Boltzmann constant
  T is the temperature

In some radicals, a cloud of an unpaired electron is partially or completely localized near a nucleus with the nonzero spin value. The interaction of magnetic moments of electron and nucleus spins results in the appearance of a multiple hyperfine structure rather than a single signal in a spectrum. The splitting and multiplexity of the spectrum and the relative intensity of its components supply direct information about configuration of an electron cloud in a radical. Therefore, the constant of such HFI is another important parameter of PC.

Magnetic moments μ_I of nuclear spins induce an additional magnetic field in the place of an unpaired electron location that can enhance or attenuate an external magnetic field B_0 depending on the distribution of nuclear spins. Such additional intramolecular field causes a splitting of both spin energy levels into sublevels, so then the resonance condition (2.2) should be realized at different B_0 values. According to the selection principle for difference in orientation magnetic number for electron $\Delta m_e = 1$ and nuclear $\Delta m_I = 0$ spin transitions, this additional field results in the correlation of possible resonance transitions and interacting nuclear spins.

It should be emphasized that these speculations are correct in the case of a full localization of unpaired electron on a nucleus when $\rho(0) = 1$. If the electron is delocalized over several nuclei of polymer units, its constant of isotropic HFI with each of them is defined by spin density on the ith nucleus, which is described by the McConnell relationship $a_i = Q\rho(0)$, where Q is the proportionality factor equal to 2.2–3.5 mT for different radicals, and $\Sigma_i \rho(0) = 1$.

Polaron pairs photoinduced in organic semiconductors demonstrate, as was analyzed [120], a superposition of two Gaussians contributing by distinct positive and negative charge carriers. Obviously, there cannot be a single model explaining spectral line shape, because exchange, hyperfine, spin–orbit, or spin-dipolar interactions could jointly contribute to the observed EPR spectra at comparatively low spin precession frequency [120]. In organic semiconductors, only hydrogen nuclear can hyperfine interact with electron spin, so that initiating some of the magnetic field effects. Hyperfine fields B_{hf} should affect the effective paramagnetic resonance of a solitary spin charge carrier, because an additional magnetic field induced by the nuclei in the vicinity of this carrier should serve to broaden the selection criteria for allowed MW photon energies $\hbar\omega_e$ in Equation 2.2 [141]. The mutual position of polarons in a domestic pair controls both the exchange and dipole–dipole intrapair coupling. The effective average B_{hf} value experienced by the electron and polaron (hole) differs slightly, so that one may expect to distinguish these carriers in their magnetic resonance response to a magnetic term B_1 of MW field. If the B_{hf} exceeds B_1, the resonance condition fulfilled only for one of the charge carriers, whereas at the opposite case, when $B_{hf} < B_1$, both spin carriers can paramagneticly resonate [89]. In this case, the precession of both polaron spins results in a halving of the period in which the polaron pair mutates from the triplet to the singlet and back to the triplet manifold. At pulse spin handling, this doubles the Rabi oscillation frequency, so it appears to be possible to detect a joint electron–hole precession at both the harmonics of the Rabi frequencies [142]. This means that charge carriers with different B_{hf} should precess and, therefore, be registered at B_1 characteristic for this species. It is important to note that

EPR techniques can, in principle, accurately detect and identify the nature of HFIs, even at high B_0 [120].

The next important characteristic of a paramagnetic system is static paramagnetic susceptibility χ proportional to an effective number of PC. The minimal number of registered spins N_{min}, that is, the sensitivity, depends on the spin precession frequency ω_e

$$N_{min} = \frac{k_1 V}{Q_0 k_f P^{1/2} \omega_e^2},$$

(2.5)

where

k_1 is constant
V is a sample volume
Q_0 is an unloaded quality factor of a cavity
k_f is the filling coefficient
P is an MW power applied to a cavity input

With k_f and P being constants, $N_{min} \propto (Q_0 \omega_e^2)^{-1}$ and $Q_0 \propto \omega_e^{-1/2}$, that is, $N_{min} \propto \omega_e^{-\alpha}$, where $\alpha = 1.5$ [143]. The latter parameter α can in practice be varied from 0.5 to 4.5 [143] depending on the spectrometer modification, registration conditions, and sample size. Generally, this parameter of N_s PC with $S = 1/2$ consists of the Pauli susceptibility term of the Fermi gas χ_P as well as temperature-dependent contributions of localized Curie PC χ_C and the term χ_{ST} coming due to a possible singlet–triplet spin equilibrium in the system [144–146]:

$$\chi = \chi_P + \chi_C + \chi_{ST} = N_A \mu_{eff}^2 n(\varepsilon_F) + \frac{N_s \mu_{eff}^2}{3k_B T} + \frac{c_{af}}{T} \left[\frac{\exp(-J_{af}/k_B T)}{1 + 3\exp(-J_{af}/k_B T)} \right]^2,$$

(2.6)

where

N_A is Avogadro's number
$\mu_{eff} = \mu_B g \sqrt{S(S+1)}$ is the effective magneton
$n(\varepsilon_F)$ is the density of states per an energy unit (eV) for both spin orientations per a monomer unit at the Fermi level ε_F
$N_s \mu_{eff}^2 / 3k_B = C$ is the Curie constant per mol-C/mol-monomer
c_{af} is a constant
J_{af} is the antiferromagnetic exchange coupling constant

The contributions of the χ_C and χ_P terms to the total paramagnetic susceptibility depend on various factors, for example, on the nature and mobility of charge carriers, which can vary at the system modification. For interpretation of magnetic susceptibility of some polymer with couple-exchanged spin pairs, the Kahol–Clark model [147,148] can be more suitable. According to this approach, $N_s/2$ spin pairs randomly distributed in a polymer matrix can interact with the exchange coupling coefficient J_c. A small value of J_c corresponds to spin localization in a strongly disordered matrix. Higher value of J_c arises at an overlapping of wave functions of spin pairs in more ordered regions. In this case, effective spin susceptibility of such interacting spins should depend on temperature as [148]

$$\chi(T) = \frac{Ca_d}{3k_BT}\left[3+\exp\left(-\frac{2J_c}{k_BT}\right)\right]^{-1} + \frac{C(1-a_d)}{3}\left\{\frac{J_c}{3k_BT}+\ln\left[3+\exp\left(-\frac{2J_c}{k_BT}\right)\right]\right\},$$

(2.7)

where a_d is a fraction of spin pairs interacting in disordered polymer regions.

If two opposite charge carriers can be initiated in polymer matrix, the number of such PC depends on their further recombination. Let a polaron possessing a positive charge multihops along a polymer chain from one initial site i to another available site j close to a position occupied by a negatively charged guest spin. An effective recombination of these charges mainly stipulates by polaron hopping along a polymer chain, and its transfer from the chain to a site where a guest charge is localized. Polaronic dynamics in undoped and slightly doped conjugated polymers is highly anisotropic [57,118,122]. Therefore, the probability of a charge transfer along a polymer chain exceeds considerably that of its transfer between polymer macromolecules.

According to the tunneling model [149], positive charge on a polaron can tunnel and recombine with its negative charge during the time

$$\tau(R_{ij}^|) = \frac{\ln X}{v_{pn}}\exp\left(\frac{2R_{ij}}{a_0}\right),$$

(2.8)

where

R_{ij} is the spatial separation of sites i and j

a_0 is the effective localization (Bohr) radius

X is a random number between 0 and 1

v_{pn} is the attempt to jump frequency for positive charge tunneling from polymer chain to fullerene

The charge can also be transferred by the polaron tunneling through the energy barrier $\Delta E_{ij} = E_j - E_i$, so then

$$\chi(R_{ij}, E_{ij}) = \chi_0 \frac{\ln X}{v_{pp}}\exp\left(\frac{2R_{ij}}{a_0}\right)\exp\left(\frac{\Delta E_{ij}}{k_BT}\right),$$

(2.9)

where v_{pp} is the attempt frequency for a hole tunneling between the polymer chains. The values in the couples v_{pn}, v_{pp}, and R_{ij} may be different due, for instance, to the different electronic orbits.

Undoubtedly, both spin systems are characterized by different localization radii, which are a function of the structure and dynamics of spin–spin interacting pair. The nearest-neighbor distance of such pair with the typical radiative lifetime τ_0 changes with time t as

$$R_0(t) = \frac{a}{2} \ln\left(\frac{t}{\tau_0}\right). \tag{2.10}$$

Assuming that photoexcitation is turned off at some initial time $t_0 = 0$ at a charge carrier concentration n_0 and taking into account a time period of geminate recombination $t_1 - t_0$, one can write for concentration of charge carriers

$$n(R) = \frac{n}{1 + \frac{4\pi}{3} n_1 \left(R_0^3 - R_1^3\right)}, \tag{2.11}$$

where
R_0 is specified by Equation 2.10
$R_1 = R(t_1)$ describes the distance between the nearest-neighbor charge carriers at time t_1 after, which a solely nongeminate recombination is assumed
n_1 is the charge carrier concentration at time t_1

It follows from Equation 2.11 that the time dependence of residual carrier concentration does not follow a simple exponential decay but shows a more logarithmic time behavior. After a very long time, that is, at large R_0, one obtains $n(R_0) = (3/4\pi)R_0^{-3}$, which is independent of the initial carrier density n_1 and also n_0. It is seen from Equation 2.8 that charge carriers have comparable long lifetimes, which are solely ascribed to the large distances between the remaining trapped charge carriers. The excited carrier concentration n_1 can be determined directly from EPR measurements, whereas both the a and τ_0 values can be guessed in a physically reasonable range. Finally, the concentration of spin pairs should follow the relation [150]

$$\frac{n(t)}{n_0} = \frac{\frac{n_1}{n_0}}{1 + \left(\frac{n_1}{n_0}\right)\frac{\pi}{6} n_0 a^3 \left[\ln^3\left(\frac{t}{\tau_0}\right) - \ln^3\left(\frac{t_1}{\tau_0}\right)\right]}. \tag{2.12}$$

Spin initiation can be accompanied by a reversible formation of spin traps in a polymer matrix. In this case, the number of spins initially excited at $t = 0$ has to be governed by some factors. One of them is the number and distribution of such spin traps. At the latter step, polaronic charge carrier can either be retrapped by a vacant trap site or recombined with an opposite guest charge.

Sequential trapping and retrapping of a polaron reduces its effective energy. This results in its further localization into a deeper trap and in the increase in number of localized polarons with the time. This process can be described in terms of the Tachiya's approach [151] of charges' recombination during their repeated trapping into and detrapping from trap sites with different depths in an energetically disordered semiconductor. Analyzing an EPR spectra, it becomes possible to separate the decay of mobile and pinned spin charge carriers excited in the polymer matrix. The traps in such a system should be characterized by different energy depths and energy distribution E_0. This approach predicts the following law for decay of charge carriers [151]

$$\frac{n(t)}{n_0} = \frac{\pi\alpha\delta(1+\alpha)v_d}{\sin(\pi\alpha)}t^{-\alpha},\tag{2.13}$$

where
  n_0 is the initial number of polarons at $t=0$
  δ is the gamma function
  $\alpha = k_B T/E_0$
  v_d is the attempt jump frequency for polaron detrapping

A positive charge on a polaron is not required to be recombined with the first negative charge on a subsequent acceptor. Thus, the probability of annihilation of charges can differ from the unit. Positively charged polaron Q1D hopping from site i to site j with frequency ω_{hop} may collide with the acceptor located near the polymer chain. While polaron is mobile, the molecule of acceptor can be considered as translatively fixed, but librating near its own main molecular axis. In this case, the spin flip-flop probability p_{ff} during a collision should depend on the amplitude of exchange and ω_{hop} value as [152,153]

$$p_{ff} = k_{ff} \cdot \frac{\alpha^2}{1+\alpha^2},\tag{2.14}$$

where
  k_{ff} is constantly equal to 1/2 and 16/27 for PC with spin $S=1/2$ and $S=1$, respectively
  $\alpha = (3/2)2\pi J_{ex}/\hbar\omega_{hop}$, and J_{ex} is the constant of exchange interaction of spins in a radical pair

In real polymer systems weak and strong exchange limits can be realized when the increase of ω_{hop} may result in the decrease or increase in exchange frequency, respectively. If the ratio $J_{ex}/\hbar$ exceeds the frequency of collision of both spins, the condition of strong interaction is realized in the system. This leads to the direct relation of spin–spin exchange interaction and polaron diffusion frequencies, so then $\lim(p) = 1/2$. In the opposite case, $\lim(p) = 9/2\left(\pi/\hbar\right)^2 \left(J_{ex}/\omega_{hop}\right)^2$. It is evident that the longer the tunneling times and/or the lesser the probability p_{ff}, the less ion-radical pairs can recombine and, therefore, higher spin susceptibility should be reached. Assuming multistep activation hopping of polarons along polymer chains with the frequency $\omega_{hop} = \omega_{hop}^0 \exp(-\Delta E_{ij}/k_B T)$ and the absence of dipole–dipole interaction between acceptors, one can get the following equation for paramagnetic susceptibility of such system:

$$\chi_p(T) = \chi_{pn} + \chi_P^0 \frac{\hbar}{J_{ex}}\left(\alpha + \frac{1}{\alpha}\right), \tag{2.15}$$

where ΔE_{ij} is the activation energy required for polaron diffusion.

In contrast with a solitary and isolated spin with δ-function absorption spectrum, the spin interaction with their own environment in a real system leads to the change in line shape and width. So by analyzing the shape and intensity of an experimental spectral line, it is possible to get direct information about electronic processes taking part in the system under study. It is known that an electron spin is affected by local magnetic fields, induced by another nuclear and electron n r_{ij}-distanced spins [154]

$$B_{loc}^2 = \frac{1}{4n}\gamma_e^2\hbar^2 S(S+1)\sum_{i,j}\frac{\left(1-3\cos^2\theta_{ij}\right)}{r_{ij}^6} = \frac{M_2}{3\gamma_e^2}, \tag{2.16}$$

where M_2 is the second moment of a spectral line. If a line broadening is stipulated by local magnetic field fluctuating faster than the rate of interaction of a spin with nearest environment, a first derivative of the Lorentzian line with a width between positive and negative peaks ΔB_{pp}^L and maximum intensity between these peaks $I_L^{(0)}$ is registered [155]:

$$I_L^| = \frac{16}{9}I_L^{|(0)}\frac{(B-B_0)}{\Delta B_{pp}^L}\left[1+\frac{4}{3}\frac{(B-B_0)^2}{\left(\Delta B_{pp}^L\right)^2}\right]^{-2}. \tag{2.17}$$

At slower fluctuation of an additional local magnetic field, the line is defined by Gaussian function of distribution of spin packets [155]:

$$I_G^| = \sqrt{e}I_G^{|(0)}\frac{(B-B_0)}{\Delta B_{pp}^G}\exp\left[-\frac{2(B-B_0)^2}{\left(\Delta B_{pp}^G\right)^2}\right]. \tag{2.18}$$

Exchange interactions between the spins, realized in some real paramagnetic system, may result in the appearance of more complicated line shape, described by a convolution of the Lorentzian and Gaussian distribution function. This takes a possibility from the analysis of such a line shape to define the distribution, composition, and local concentrations of spins in the system. For example, when the equivalent paramagnetic centers with concentration n are arranged in the system chaotically, their line shape is described by the Lorentzian distribution function with the width $\Delta B_{pp}^{L} = 4\gamma_e \hbar n$, whereas Gaussian distribution with the width $\Delta B_{pp}^{G} = 4\gamma_e \hbar n$ is characteristic for PC arranged regularly [156]. In the mixed cases, the line shape transforms to Lorentzian at a distance from the center $\delta B \leq 4\gamma_e \hbar / r^3$ (here, r is the distance between magnetic dipoles) and with the width $\Delta B_{pp}^{L} = 4\gamma_e \hbar n$ in the center, and it becomes of Gaussian type on the tails at $\delta B \geq \gamma_e \hbar / r^3$ and with the width of $\Delta B_{pp}^{G} = \gamma_e \hbar \sqrt{n/r^3}$. An exchange between spin packets separated by $\Delta \omega_{ij}$ distance with frequency ω_{ex} should lead to an additional line broadening as [134]

$$\Delta B_{pp} = \Delta B_{pp}^{0} + \frac{\Delta \omega_{ij}^{2}}{8\gamma_e \omega_{ex}}, \tag{2.19}$$

where ΔB_{pp}^{0} is the linewidth at the absence of an interaction between PC.

Analyzing the linewidth of an EPR spectra of PC stabilized in a polymer backbone, one should take into account the lower dimensionality of such systems. The rate of charge hopping between two adjacent Q1D units can follow a semiclassical Marcus theory adopted for conjugated polymers [157,158]:

$$\omega_{hop} = \frac{4\pi^2}{\hbar} \frac{t_r^2}{\sqrt{4\pi E_r k_B T}} \exp\left(-\frac{E_r}{4k_B T} \right), \tag{2.20}$$

where
   t_r is the electronic coupling between initial and final states (intrachain transfer integral)
   E_r is both the inner- and outer-sphere reorganization energies of charge carriers due to their interaction with the lattice phonons

The t_r value decreases slightly with temperature, whereas its distribution broadens a line due to thermal motion of polymer units [159] similar to that happening in organic crystals [160,161].

Except fast electron spin diffusion, EPR line can also be broadened by the defrosting of molecular dynamic processes, for example, oscillations or slow torsion librations of the polymer macromolecules and/or their side groups. The approach of random walk treatment [162] provides that the Q1D, Q2D, and Q3D spin diffusion along and between the main axis with the frequencies of ν_{1D} and ν_{3D} respectively, changes the linewidth of a spin packet in the motionally narrowed regime as [163]

$$\Delta B_{pp} \approx \frac{\gamma_e^{1/3}(\Delta B_{pp}^{(0)})^{4/3}}{v_{1D}^{1/3}}, \tag{2.21}$$

$$\Delta B_{pp} \approx \frac{\gamma_e(\Delta B_{pp}^{(0)})^2}{\sqrt{v_{1D}v_{3D}}}, \tag{2.22}$$

$$\Delta B_{pp} \approx \frac{\gamma_e(\Delta B_{pp}^{(0)})^2}{v_{3D}}, \tag{2.23}$$

respectively, where $\Delta B_{pp}^{(0)}$ is an initial linewidth of immobilized spins far away from the saturation regime. In accordance with this theory, at the transition from Q1D to Q2D and then to Q3D spin motion, the shape of the EPR line should transform from Gaussian to Lorentzian. Indeed, the EPR line shape due to dipole or hyperfine broadening is normally Gaussian. The spin motion or exchange alters additionally the initial EPR line shape. For spin Q3D motion or exchange, the line shape becomes close to the Lorentzian shape, corresponding to an exponential decay of transverse magnetization with time t, proportional to $\exp(-\eta t)$; for a Q1D spin motion, this value is proportional to $\exp[-(\rho t)^{3/2}]$ [164]. Here, η and ρ are constants. So in order to determine the type of spin dynamics in the Q1D system, the anamorphosis $I_0^{\downarrow}/I(B)$ versus $[(B - B_0)/\Delta B_{1/2}]^2$ [164] (here, $\Delta B_{1/2}$ is the half-width of an integral line) is usually analyzed [165–167]. It, however, requires integrating of the first derivative of signal absorption that leads to an additional error. Besides, at millimeter waveband, EPR charge carriers frequently show EPR spectrum with integral-like and derivative lines due to the anisotropy of the g-factor (see the following example). So in order to analyze what line of such spectra reflects the spin motion, both the anamorphoses $I_0^{\downarrow}/I(B)$ versus $[(B - B_0)/\Delta B_{1/2}]^2$ and $I_0^{\downarrow}/I(B)$ versus $[(B - B_0)/\Delta B_{pp}]^2$ appeared to be more informative. Figure 2.1 demonstrates how the shape of a single line changes at the defrosting of spin Q1D and Q3D mobility. This approach takes a possibility to evaluate a dimension of the system under study, say, from an analysis of temperature dependence of its EPR spectrum linewidth.

Exchange interaction of a mobile PC with other spins in the Q1D polymer system additionally broadens its effective EPR spectrum as [152,153]

$$\delta(\Delta\omega) = p_{ff}\omega_{hop}n_g, \tag{2.24}$$

where
p_{ff} is the same probability as in Equation 2.14
n_g is the number of guest PC per each polymer unit

Spin packets in some polymer systems can be characterized by extremal dependency $\delta(\Delta\omega)(T)$ with characteristic temperature T_c (see the following example). According to the spin exchange fundamental concepts [152], this fact should evidence the realization of strong and weak spin–spin exchange

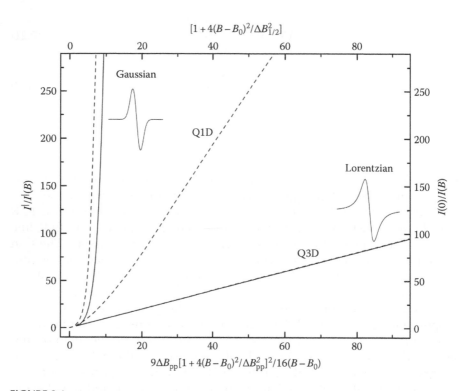

FIGURE 2.1

$I_0^!/I(B)$ versus $[(B-B_0)/\Delta B_{1/2}]^2$ and $I_0^!/I(B)$ versus $[(B-B_0)/\Delta B_{pp}]^2$ anamorphoses for electron paramagnetic resonance liner in the case of spin localization and Q1D and Q3D diffusion. The Q3D spin diffusion gives Lorentzian line shape and localized spin gives Gaussian line shape. (From Krinichnyi, V.I., *Appl. Phys. Rev.*, 1(2), 021305/01, 2014. With permission.)

interaction at $T \le T_c$ and $T \ge T_c$, respectively. An additional reason of the line broadening can be spin trapping at the temperature lower T_c. The combination of Equations 2.14, 2.20, and 2.24 yields for effective line broadening

$$\delta(\Delta B_{pp}) = \frac{\pi t_r^2 n_g(T)}{\hbar \sqrt{\dfrac{E_r k_B T}{\pi}}} \cdot \frac{\exp\left(-\dfrac{E_r}{4k_B T}\right)}{1 + \left[\dfrac{3J_{ex}}{2t_r^2} \sqrt{\dfrac{E_r k_B T}{\pi}} \exp\left(\dfrac{E_r}{4k_B T}\right)\right]^{-2}}. \qquad (2.25)$$

Since the charge carriers possessing spin and/or conduction electron move rapidly in conjugated polymers, the EPR linewidth generally becomes narrower than in the case of localized spins whom linewidth is governed by the unresolved hyperfine splitting due to nearby nuclei carrying a spin, proton, and fluorine. For example, the X-band EPR linewidth of neutral soliton confined within short *trans*-segment in real *cis*-PA is near to 1 mT [168,169]; however, this value decreases down to ca. 0.1 mT in entire *trans*-PA [170] where soliton can diffuse freely (see the following example). In the case of the fast motion of spins, their linewidth is governed by electron relaxation and motional modulation of local magnetic fields [171,172]. Besides, the linewidth of a polymer depends also on the number of spins on the polymer chain n_s and surrounding a chain n_c [173]:

$$\Delta B_{pp} = \frac{4M_2}{5\omega_0 n_s}\left[21\ln\frac{\omega_0}{\omega_e} + 18\ln n_c\right], \qquad (2.26)$$

where
  ω_0 is a hopping attempt frequency
  ω_e is a frequency of electron spin precession

The ordering of conjugated polymers also affects the linewidth and distribution of spin charge carriers stabilized in their matrix. Such systems are mainly characterized by inhomogeneous network with differently ordered phases. So then the spin charge carriers with the small anisotropy of *g*-factor stabilized in conjugated polymers and their nanocomposites normally demonstrate isotropic Lorentzian broadening due to unresolved hyperfine couplings of a spin with own nearest environment and anisotropic Gaussian broadening as a result of structural variation in magnetic parameters of these carriers. An appropriate convolutional spectral broadening, therefore, should contain the respective terms sensitive to the value and direction of an external magnetic field. This is a reason why the shape of EPR spectra of conjugated polymer systems is transformed from Lorentzian to Gaussian with the increase of registration frequency.

The observation of spin excitation in conjugated solids is complicated by the fact that the magnetic term B_1 of the MW field used to initiate resonance sets up eddy currents in the material bulk. These currents effectively confine the magnetic flux to a surface layer of thickness of order of the *skin depth*. This phenomenon affects the absorption of the MW energy incident upon a sample and results in less intensity of electron absorption per unit volume of material for large particles than for small ones. This leads also to the appearance of asymmetry in EPR spectra due to the so-called Dysonian term [174]. When the skin layer thickness δ becomes comparable or thinner than a characteristic size of a sample, for example, due to the increase of its alternating current (*ac*) conductivity σ_{ac}, the time of charge carrier diffusion through the skin layer becomes essentially less than a spin relaxation time, and the Dysonian line with characteristic asymmetry factor *A*/*B* (the ratio of intensities of the spectral positive peak to negative one) is registered as it is shown in the inset of Figure 2.2. Such distortion of line shape is accompanied by the line shift into the higher magnetic fields and the drop of sensitivity of EPR technique. PC with Dysonian line shape are commonly registered in some highly conjugated inorganic substances [175], organic conjugated ion-radical salts [1], *trans*-polyacetylene [176], polythiophene [177], poly(*p*-phenylene) [178], polyaniline [118,179–187], and other conjugated polymers [104,105,188].

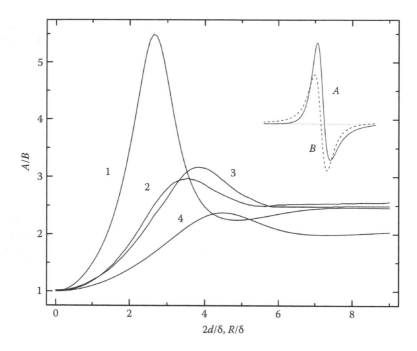

FIGURE 2.2
The dependence of the spectrum asymmetry factor A/B on the thickness of skin-layer formed on the conjugated plate (1), shank with square (2) and circular (3) section, and sphere (4). The inset shows electron paramagnetic resonance spectra of paramagnetic centers in insulating (dashed line, $D/A = 0$, $A/B = 1$) and conjugated (solid line, $D/A = 0.8$, $A/B = 2.2$) materials. (From Krinichnyi, V.I., *Appl. Phys. Rev.*, 1(2), 021305/01, 2014. With permission.)

Generally, the Dysonian line consists of dispersion $\chi^{|}$ and absorption $\chi^{||}$ terms; therefore, one can write for its first derivative the equation

$$\frac{d\chi}{dB} = A\frac{2x}{(1+x^2)^2} + D\frac{1-x^2}{(1+x^2)^2}, \qquad (2.27)$$

where $x = 2(B - B_0)/\sqrt{3}\Delta B_{pp}^L$. Organic polymers are usually synthesized as powders and films. Appropriate coefficients of absorption A and dispersion D in Equation 2.27 for skin layer on the surface of a spherical powder particle with radius R and intrinsic conductivity σ_{ac} can be calculated from the following equations [189]

$$\frac{4A}{9} = \frac{8}{p^4} - \frac{8(\sinh p + \sin p)}{p^3(\cosh p - \cos p)} + \frac{8\sinh p \sin p}{p^2(\cosh p - \cos p)^2}$$

$$+ \frac{(\sinh p - \sin p)}{p(\cosh p - \cos p)} - \frac{(\sinh^2 p - \sin^2 p)}{(\cosh p - \cos p)^2} + 1, \qquad (2.28)$$

$$\frac{4D}{9} = \frac{8(\sinh p - \sin p)}{p^3(\cosh p - \cos p)} - \frac{4(\sinh^2 p - \sin^2 p)}{p^2(\cosh p - \cos p)^2}$$

$$+ \frac{(\sinh p + \sin p)}{p(\cosh p - \cos p)} - \frac{2\sinh p \sin p}{(\cosh p - \cos p)^2}, \tag{2.29}$$

where

$p = 2R/\delta$

$\delta = \sqrt{2/\mu_0 \omega_e \sigma_{ac}}$

μ_0 is the magnetic permeability for vacuum

In case of the formation of skin layer on the flat plate with a thickness of $2d$, the earlier coefficients can be determined from the following relations [189]:

$$A = \frac{\sinh p + \sin p}{2p \ (\cosh p + \cos p)} + \frac{1 + \cosh p \cos p}{(\cosh p + \cos p)^2}, \tag{2.30}$$

$$D = \frac{\sinh p - \sin p}{2p \ (\cosh p + \cos p)} + \frac{\sinh p \sin p}{(\cosh p + \cos p)^2}, \tag{2.31}$$

where $p = 2d/\delta$.

From the analysis, it was revealed that the line asymmetry parameter A/B is correlated with the coefficients A and D of Equation 2.27 simply as $A/B = 1 + 1.5$ D/A independently on the linewidth. Figure 2.2 shows how changes the A/B ratio with the thickness of skin layer formed on the surface of different shape. Thus, it appears to be possible to determine directly an intrinsic conductivity of conjugated polymer systems from their Dysonian spectrum.

2.2 Electron Relaxation of Spin Packets in Conjugated Polymers

Electron spins being in the thermal equilibrium in an external magnetic field are distributed to energy levels according to the Boltzmann law. If this equilibrium is somehow disturbed for a certain time, for example, by the increase of MW field, the magnetic moments of the spins relax with the longitudinal (spin–lattice) T_1 and the transverse (spin–spin) T_2 relaxation times. These values are described by the well-known Bloch equations for a slow resonance passage and absence of MW saturation [134,156,190].

A spin-packet shape is assigned by the following set of time characteristics: T_1, T_2, $(\gamma_e \Delta B_{1/2})^{-1}$, ω_m^{-1}, $(\gamma_e B_m)^{-1}$, $(\gamma_e B_1)^{-1}$, and $B_1/(dB/dt)$, where ω_m and B_m are the angular frequency and intensity of *ac* modulation field, respectively.

The first three of them are stipulated by the origin of the substance, and the remaining ones are the instrumental parameters. If the parameters of a spectrum registration satisfy certain inequalities, it becomes possible to analyze qualitatively the behavior of a magnetization vector **M**. If the saturation factor $s = \gamma_e B_1 \sqrt{T_1 T_2} = \gamma_e B_1 \tau \ll 1$ (here, τ is the effective relaxation time), τ does not exceed precession time $(\gamma_e B_1)^{-1}$ of **M** vector near **B**$_1$ one, so then the line of spin packet can be described by the classic analytical expressions, Equations 2.17 and 2.18. For this case, the analysis of the possible line shape distortion is sufficiently elucidated in the literature (see, e.g., [191,192]).

The saturation of spin packets is realized as the opposite condition $s \geq 1$ holds. In this case, the intensity $I^{(0)}$ and linewidth $\Delta B_{pp}^{(0)}$ of their nonsaturated spectrum increase nonlinearly with the increase of B_1 value as [143]

$$I = I^{(0)} B_1 \left(1 + \gamma_e^2 B_1^2 T_1 T_2\right)^{-3/2}, \qquad (2.32)$$

$$\Delta B_{pp} = \Delta B_{pp}^{(0)} \sqrt{1 + s^2} = \Delta B_{pp}^{(0)} \sqrt{1 + \gamma_e^2 B_1^2 T_1 T_2}, \qquad (2.33)$$

and the qualitative characteristics of passage effects are defined by the time of resonance passage $B_1/(dB/dt)$ to the effective relaxation time τ ratio. If $B_1/(dB/dt) > \tau$, the spin system comes to equilibrium when the sinusoidal modulation field is at one end of its excursion with a sweep rate going to zero. **M** and **B**$_0$ vectors are parallel and remain unchanged during a sweep, assuming that adiabatic condition $\gamma_e \omega_m B_m \ll \gamma_e^2 B_1^2$ holds. In this case, the spectrum exhibiting the integral function of spin-packet distribution is observed at the modulation method of a spectrum registration instead of the traditional first derivative of absorption signals or dispersion. This phenomenon was called as passage effect and is widely considered in the literature [114,193–197] in terms of fast passage of saturated spin packets. In real solids, the interaction between spin packets can be realized. The probability p_{cr} of cross-relaxation decreases strongly with the increase of a polarizing MW frequency $\omega_e \propto B_0$ according to [107]

$$p_{cr} = 2\pi \sqrt{M_2} \exp\left(-\frac{\omega_e^2}{8\pi^2 M_2}\right). \qquad (2.34)$$

This value decreases strongly with the ω_e increase, so then the spin packets become noninteracting and, therefore, can be saturated at lower B_1 values. Besides, the electron relaxation time of PC in some solids can increase with the increase in ω_e. This is another reason for the appearance of fast passage effects at D-band [125,126,198], than at lower operation frequency EPR, on registering PC in different solids of lower dimensionality.

The dispersion or absorption signal is detected as the projection of **M** on +x-axis. If the system comes to equilibrium again at the other end of the sweep, then **M** is initially oriented along +z-axis, **M** and **B**$_{eff}$ become antiparallel during the passage, and the signal appears as the projection of **M** on

$-x$-axis. Thus, the signal from a single spin packet is registered with $\pi/2$-out-of-phase shift with respect to the field modulation. The first derivative of, for example, dispersion signal U is generally written as [199]

$$U(\omega t) = u_1 g'(\omega)\sin(\omega t) + u_2 g(\omega)\sin(\omega_e t - \pi) + u_3 g(\omega)\sin(\omega t \pm \pi/2), \quad (2.35)$$

where

u_1, u_2, and u_3 are the in-phase and quadrature dispersion terms, respectively
$g(\omega)$ is a factor of line shape

It is obvious that $u_2 = u_3 = 0$ without MW saturation of a spin packet. With the saturation being realized, the relative intensity of u_1 and u_3 components is defined by the relationship between the relaxation time of a spin packet and the rate of its resonance field passage. If the resonance passage is fast enough and the modulation frequency is comparable or higher than the effective relaxation rate τ^{-1}, the magnetic field change becomes too fast and the magnetization vector of the spin system does not reorientation in time for the $\mathbf{B}_1$ vector. At adiabatic condition, $\omega_m B_m \ll \gamma_e B_1^2$ such a delay allows spin to *see* only an average applied magnetic field, and the first derivative of a dispersion signal is mainly defined by the integral $u_2 g(\omega_e)$ and $u_3 g(\omega_e)$ terms of Equation 2.35 with $u_2 = M_0 \pi \gamma_e^2 B_1 B_m T_2/2$ and $u_2 = M_0 \pi \gamma_e^2 B_1 B_m T_2/(4\omega_m T_1)$.

When the effective relaxation time is less than the modulation period, $\tau < \omega_m$, but exceeds the passage rate, $\tau > B_1/(dB/dt)$, the magnetization vector has time to relax to the equilibrium state during one modulation period; therefore, the dispersion signal of a spin packet is independent on the relationship of its resonant field and an external field $\mathbf{B}_0$. The sign of a signal is defined by that from what side the resonance is achieved. In this case, the $\pi/2$-out-of-phase term of a dispersion signal is also registered as an integral function of spin-packet distribution, and the first derivative of a dispersion signal is mainly defined by $u_1 g'(\omega_e)$ and $u_3 g(\omega_e)$ terms of Equation 2.35, with $u_1 = M_0 \pi \gamma_e^2 B_1 B_m$ and $u_3 = M_0 \pi \gamma_e^2 B_1 B_m T_1 T_2/2$. Thus, both the times of spin relaxation can be estimated from the terms of a dispersion spectrum of saturated spin packets at an appropriate phase shift in phase detector.

EPR spectra of saturated PC appear as an integral (almost always Gaussian) distribution function of spin packets. Figure 2.3 shows the change of the shape of the in-phase and quadrature components of the D-band dispersion signal obtained for anisotropic PC at the B_1 term increase. The relaxation times of adiabatically saturated PC in π-conjugated polymers can be determined separately from the analysis of the u_1-u_3 components of its dispersion spectrum as [200,201]

$$T_1 = \frac{3\omega_m(1+6\Omega)}{\gamma_e^2 B_{1_0}^2 \Omega(1+\Omega)}, \quad (2.36)$$

$$T_2 = \frac{\Omega}{\omega_m} \quad (2.37)$$

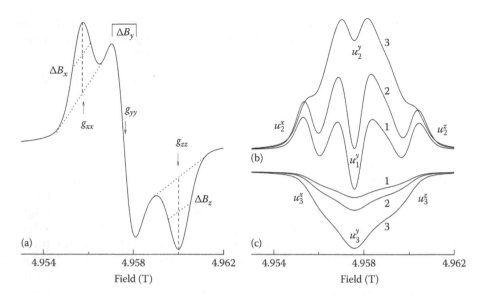

FIGURE 2.3

In-phase absorption (a) and in-phase (b) and $\pi/2$-out-of-phase (c) dispersion spectra of paramagnetic centers with $g_{xx} = 2.00401$, $g_{yy} = 2.003126$, $g_{zz} = 2.002322$, $\Delta B_x = \Delta B_y = \Delta B_z = 0.62$ mT obtained for $B_1 = 0.5$ μT (1), $B_1 = B_{1_0}$ (2), and $B_1 = 10$ μT (3) at D-band electron paramagnetic resonance. (From Krinichnyi, V.I., *Appl. Phys. Rev.*, 1(2), 021305/01, 2014. With permission.)

(here, $\Omega = u_3/u_2$, B_{1_0} is the polarizing field at which the condition $u_1 = -u_2$ is valid, in-phase dispersion spectrum 2 at Figure 2.3) at $\omega_m T_1 > 1$ and

$$T_1 = \frac{\pi u_3}{2\omega_m u_1}, \tag{2.38}$$

$$T_2 = \frac{\pi u_3}{2\omega_m (u_1 + 11u_2)} \tag{2.39}$$

at $\omega_m T_1 < 1$. The amplitudes of u_i components are measured in the central point of the spectra when $\omega = \omega_e$. It derives from the formulas that the determination of B_1 value in the cavity center is not required for the evaluation of relatively short relaxation times.

2.3 Spin Diffusion in Conjugated Polymers

Spin relaxation can be accelerated by molecular and spin dynamics existing in real organic solids. Indeed, the diffusion of electron spin S_1 shown

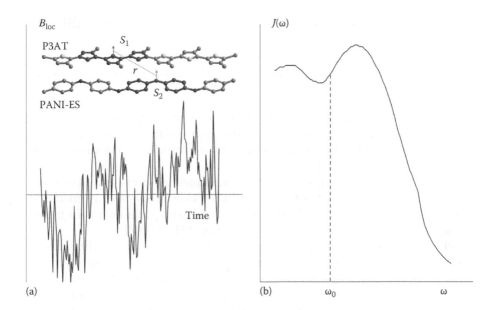

FIGURE 2.4
Schematic explanation of the fluctuating local field B_{loc} (a) and the spectral density function $J(\omega)$ (b) producing by spin S_2 stabilizing, for example, on the polyaniline chain and diffusing near spin S_1 localized on neighboring chain of the same polymer, for example, poly(3-alkyl-thiophene), and separated by the distance r (inset in (a)). (From Krinichnyi, V.I., *Appl. Phys. Rev.*, 1(2), 021305/01, 2014. With permission.)

in Figure 2.4 along and between polymer chains induces a local magnetic field $B_{loc}(t)$, fluctuating rapidly with time near the location of another electron (and/or nuclear) spin shown as S_2 due to their dipole exchange (and/or hyperfine) interaction. According to the theory of magnetic resonance, such spin diffusion is characterized by a translating motion propagator $P(r, r_0, t)$. Its $P_{tr}(r, r_0, t)dr$ value characterizes the probability of that, if the jth particle is located in r_0 point with respect to the ith particle at the initial moment $t = 0$, then it is located in $r + dr$ range with respect to a new location of the ith particle at the $t = \tau$ moment.

An analytical form of the motion propagator depends on the diffusion dynamics model applied to condensed systems. It can be determined solving a well-known Brownian diffusion equation:

$$\frac{dP(r, r_0, \tau)}{dt} = \mathbf{D}\Delta P(r, r_0, \tau), \tag{2.40}$$

with the initial condition $P(r, r_0, t) = \delta(r - r_0)$, where $\mathbf{D} = [D_i]$, $D_i = v_i c_i$ is the diffusion coefficient, v_i is a diffusion rate, and c_i is a proportionality constant, considering the discreteness of a real system, i is a unit vector of a molecular coordinate system. The solution of Equation 2.40 for the case of Q1D spin diffusion is as follows:

$$P(r, r_0, \tau) = \frac{1}{\sqrt{4D_{1D}\tau}} \exp\left[-\frac{(r-r_0)^2}{4D_{1D}c_{1D}^2\tau}\right] \exp(-D_{3D}\tau), \qquad (2.41)$$

where D_{1D} and D_{3D} are, respectively, the coefficients of spin diffusion along and between polymer chains and c_{1D} is the appropriate lattice constant. The latter exponent multiplier of Equation 2.41 is introduced because of spin interaction hopping probability.

The diffusion motion has an ordered character in solids and is realized over their lattice centers. For this case, the motion propagator is estimated with a lattice sum and depends on the symmetry and parameters of a lattice. Some lattice sums have been calculated earlier [171,202,203].

Let us consider how the experimental electron relaxation time values enable the establishment of molecular dynamics in condensed media in terms of the models described earlier. The relationship between electron relaxation times and parameters of molecular mobility is defined by the Hamiltonian of interaction and the accepted model of molecular mobility. This relationship can be generally written as $T_{1,2} = f[J(\omega)]$, where $J(\omega)$ is a function of spectral density, which is of significance in EPR relaxation theory. The Fourier transformation of autocorrelation function $G(\tau) = B_{loc}(t)B_{loc}(t+\tau)$ of a fluctuating field performs as $J(\omega)$ spectral components (harmonics) summation within ω variation range of $0 < \omega < \infty$:

$$J(\omega) = \int_{-\infty}^{+\infty} \overline{G(\tau)}\exp(-i\omega\tau)d\tau. \qquad (2.42)$$

A vinculum in Equation 2.42 indicates $G(\tau)$ value averaging over spin ensemble. Therefore, if $G(\tau)$ function varies for each molecule and every t moment, the average function value is, however, equal for all molecules and does not depend on t. Oscillations, coming into a resonance at a frequency multiple of ω_0, induce the transitions between ground and excited states of the spin and result in the acceleration of spin–lattice relaxation. The spin system having given an energy quantum to a lattice reservoir comes back to equilibrium with a lattice after characteristic time period T_1. Therefore, the measured T_1 value is to be proportional to a power $J(\omega)$ or spectral density of time correlation function $G(t)$ of a fluctuating local field. Thus, a frequency dependence of $J(\omega)$ spectral density can be obtained on measuring T_1 value as a function of frequency.

The time correlation function may be conventionally expressed with a motion propagator [171]

$$G(\tau) = \iint A(r_0, t)P(r, r_0, \tau)F(r_0)F^*(r)drdr_0 \qquad (2.43)$$

for a homogeneous system and with

$$G(\tau) = c_i \sum_{r}^{n} \sum_{r_0}^{n} A(r_0, t) P(r, r_0, \tau) F(r_0) F^*(r) dr dr_0 \tag{2.44}$$

for a discrete system. Here, $F(t)$ is a random time function; $F(r)$ is a function of probability of two spin locations at r distance at t moment; and $A(r, t)$ is a probability of spin location at r distance at t moment, which is equal to spin concentration n per a polymer unit. $F(r_0)F^*(r)$ product depends on a dipole and a scalar spin interaction. The motion propagator $P(r, r_0, t)$ is mainly defined by a dimension of spin dynamics. Therefore, the spin motion dimension is also reflected in the spectral density function $J(\omega)$, which appears as the Fourier image of $G(r, r_0, t)$ (see Equation 2.42).

The function of spectral density for Q1D translation diffusion is more complex. The Fourier image of $G(r, r_0, t)$ is [204]

$$J_{1D}(r, r_0, \omega) = J_{1D}(\omega) f_{1D}(r - r_0, \omega), \tag{2.45}$$

where

$$J_{1D}(\omega) = \frac{1}{\sqrt{4D_{1D}^{\shortmid} D_{3D}}} \sqrt{\frac{1 + \sqrt{1 + \left(\omega / 2D_{3D}\right)^2}}{1 + \left(\omega / 2D_{3D}\right)^2}}$$

$$= \begin{cases} \left(2D_{1D}^{\shortmid}\omega\right)^{-1/2}, & \text{at } D_{1D}^{\shortmid} \gg \omega \gg D_{3D} \\ \left(2D_{1D}^{\shortmid}D_{3D}\right)^{-1/2}, & \text{at } \omega \ll D_{3D} \end{cases}, \tag{2.46}$$

$$f_{1D}(r - r_0, \omega) = \exp(-xu) \left(\cos xu - \frac{v}{u} \sin xu \right), \qquad x = c_{1D} |r - r_0| \sqrt{2D_{3D}/D_{1D}^{\shortmid}},$$

$$u = \sqrt{\frac{1 + \sqrt{1 + \left(\omega / 2D_{1D}^{\shortmid}\right)^2}}{2}}, \ v = \sqrt{\frac{1 - \sqrt{1 + \left(\omega / 2D_{1D}^{\shortmid}\right)^2}}{2}}, \ D_{1D}^{\shortmid} = 4D_{1D}/L^2, \text{and } L \text{ is the}$$

number of monomer units along that a spin is delocalized.

Note, that for Q2D spin motion $J_{2D}(\omega) = \ln(4\pi^2 D_2/\omega)/2\pi\sqrt{D_1 D_2}$ [205].

Equation 2.46 is used for analyzing the short-range scalar coupling at a low frequency range when $\omega \ll D_{1D}c^2/(r_0 - r)^2$. At this frequency range, the spectral density of autocorrelation function appears as

$$J(\omega) = n J_{1D}(\omega) \sum_{r}^{n} \sum_{r_0}^{n} F(r_0) F^*(r) f_{1D}(|r - r_0|) = n J_{1D}(\omega) \sum_{r}^{n} \sum_{r_0}^{n} \frac{(1 - 3\cos^2 \theta)^2}{r_1^3 r_2^3},$$

$$\tag{2.47}$$

where

  n is in the present case the probability of spin location at r_1 at initial time moment

  θ is the angle between $\mathbf{r}_1$ and $\mathbf{r}_2$ vectors

As $f_{1D}(r - r_0, \omega) \approx 1$ at $\omega \ll D_{1D}c^2/(r_0 - r)^2$, therefore Equation 2.45 can be used for the analysis of dipole coupling. If the interaction of neighboring spins is characterized by intrachain and interchain coupling integrals $t_{\|}$ and $t_{\perp}$, respectively, the rate of the interchain diffusion is [164]

$$v_{3D} \approx (t_{\perp}/t_{\|})^{1/3} t_{\perp}/\hbar. \tag{2.48}$$

A spectral density function for rotational diffusion with correlation time τ_c is $J_r(\omega) = 2\tau_c/(1 + \tau_c^2 \omega_e^2)$.

Let the general relationships for relaxation times be written assuming a realization in spin system of both dipole and scalar HFI between electron and nuclear spins. A dipole interaction is always more strong and anisotropic, whereas a scalar HFI is isotropic and significantly weaker. Other interactions do not depend on spin dynamics and yield a small contribution to T_1 and T_2; therefore, they may be neglected.

For the case of dipole–dipole interaction between equal unpaired electrons, the expressions for electron relaxation rates are [134,171]

$$T_1^{-1} = \frac{2}{3}\gamma_e^4 \hbar^2 S(S+1)n\left[J(\omega_e) + J(2\omega_e)\right], \tag{2.49}$$

$$T_2^{-1} = \frac{2}{3}\gamma_e^4 \hbar^2 S(S+1)n\left[J(0) + 10J(\omega_e) + J(2\omega_e)\right], \tag{2.50}$$

where $n = n_1 + n_2/\sqrt{2}$, with n_1 and n_2 being the concentrations of mobile and localized spins, respectively, per a chain unit. The coefficient $1/\sqrt{2}$ in n is used because two spins diffuse independently with respect to one another. In polycrystalline samples, these values are averaging over spin ensemble as

$$T_1^{-1} = \langle\Delta\omega^2\rangle\left[2J(\omega_e) + 8J(2\omega_e)\right], \tag{2.51}$$

$$T_2^{-1} = \langle\Delta\omega^2\rangle\left[3J(0) + 5J(\omega_e) + 2J(2\omega_e)\right], \tag{2.52}$$

where $\langle\Delta\omega^2\rangle = (1/10)\gamma_e^4 \hbar^2 S(S+1)n \sum\sum (1 - 3\cos^2\theta)^2 r_1^{-3} r_2^{-3}$ is the average constant of the spin dipole interaction in a powder-like sample. If one assumes for the simplicity that the spins are situated near the units of the cubic lattice with concentration of the monomer units N_c and constant $r_0 = (8/3N_c)^{-1/3}$, then the average lattice sum for powder can be simplified as $\sum\sum (1 - 3\cos^2\theta)^2 r_1^{-3} r_2^{-3} = 6.8 r_0^{-6}$ [156].

In the case of partially oriented system with orientation degree A, these equations consist of two terms:

$$T_1^{-1} = A\langle\Delta\omega^2\rangle\left[2P_1 J(\omega_e) + 8P_2 J(2\omega_e)\right]$$

$$+ (1 - A)\langle\Delta\omega^2\rangle\left[2P_1^| J(\omega_e) + 8P_2^| J(2\omega_e)\right], \tag{2.53}$$

$$T_2^{-1} = A\langle\Delta\omega^2\rangle\left[3P_0 J(0) + 5P_1 J(\omega_e) + 2P_2 J(2\omega_e)\right]$$

$$+ (1 - A)\langle\Delta\omega^2\rangle\left[3P_0^| J(0) + 5P_1^| J(\omega_e) + 2P_2^| J(2\omega_e)\right], \tag{2.54}$$

where P_i and $P_i^|$ are attributed to oriented and chaotically situated chains, respectively.

The earlier equations reveal that spin transitions are induced by dipole–dipole interaction on the first and second harmonics of the polarizing field. The equations for contributions into relaxation rates under a normally realized $\omega_e \gg \omega_I$ condition are

$$T_1^{-1} = \frac{1}{3}\gamma_e^2\gamma_p^2\hbar^2 I(I+1)n \sum\sum J(\omega_e)\left[2 + \frac{S(S+1)(\langle I_z\rangle - I_0)}{I(I+1)(\langle S_z\rangle - S_0)}\right], \tag{2.55}$$

$$T_2^{-1} = \frac{1}{30}\gamma_e^2\gamma_p^2\hbar^2 I(I+1)n \sum\sum\left[4J(0) + 3J(\omega_I) + 13J(\omega_e)\right]. \tag{2.56}$$

The average value Σ_{ij} should be estimated for each type of interaction.

The isotropic HFI contribution to electron relaxation rates yields

$$T_1^{-1} = \frac{1}{3}\gamma_e^2\gamma_p^2\hbar^2 I(I+1)n_p J\left(\omega_e - \omega_I\right)_e\left[1 - \frac{S(S+1)(\langle I_z\rangle - I_0)}{I(I+1)(\langle S_z\rangle - S_0)}\right], \tag{2.57}$$

$$T_2^{-1} = \frac{1}{6}\gamma_e^2\gamma_p^2\hbar^2 I(I+1)n_p\left[J(0) + J(\omega_e - \omega_I)\right], \tag{2.58}$$

where n_p is the number of nuclear spins per polymer unit.

The contribution of spin–lattice relaxation, including $(\langle I_z\rangle - I_0)/(\langle S_z\rangle - S_0)$ multiplier, corrects for cross-relaxation between electron and nuclear spins, stipulated by the Overhauser effect [171]. This contribution may be ignored in strong magnetic fields, because the relation $T_1 \approx T_2$ realized at $\omega_e \to 0$ is converted to $T_1 > T_2$ at $\omega_e \to \infty$. T_1 is always a function of the first and second harmonics of the resonance frequency, whereas the expression for T_2 includes the frequency-independent term $J(0)$, called as a secular broadening $(T_2^{-1})|$. Frequency-dependent terms of the equations for T_1^{-1} define the so-called spin lifetime or nonsecular broadening $1/T_1'$. T_1/T_1' ratio is equal to 10/7, 10/8, and 10/5 for dipole, anisotropic, and isotropic HFI, respectively, and enables the identification of the relaxation mechanism in each specific case by measuring relaxation times at several resonance frequencies [204]. It should be noted that the polaron Q1D diffusion coefficient D_{1D} can also be determined by using indirect nuclear magnetic resonance method from the proton ^{1}H spin–lattice relaxation time T_{1p} from [205]

$$T_{1p}^{-1}(\omega_p, \omega_e) = \frac{1}{3}\gamma_e^2 \hbar^2 S(S+1)n\left[\frac{3}{5}d^2 J(\omega_p) + \left(a^2 + \frac{7}{5}d^2\right)J(\omega_e)\right], \qquad (2.59)$$

where

  d and a are the dipolar and scalar electron–nuclear coupling constants, respectively

  ω_p is the proton angular precession frequency

  $J(\omega_p)$ is the appropriate spectral density function for such a motion

2.4 Molecular Dynamics in Conjugated Polymers

Various methods are used for the study of molecular dynamics in conjugated polymers [58]. The method of spin label and probe based on the introduction into the polymer of a reporter radical group [206–208] is one of the powerful tools for the study of structure, conformation, dynamics, and other properties of the PC and their nearest microenvironment. With the introduction of a spin reporter group, one respects not to disturb a radical microenvironment in the object under study. In practice, however, nitroxide radicals, nitrogen oxide, paramagnetic ions of transition metals, lanthanide ions, peroxide radicals, etc., usually used as labels and probes can misrepresent the data obtained.

The method of paramagnetic probe can be applied to the investigation of various processes, for example, structurization and destruction, orientation, interdiffusion, and crystallization of polymers, which are accompanied by the change of segmental mobility in their bulk. The method can also be used to investigate the topochemical characteristics of charge carriers in metal-containing polymer complexes [209]. Winter et al. [210] showed that the spin probe technique is also applicable for investigating macromolecular dynamics and magnetic spin–spin interactions in organic polymer semiconductors. If PC motion is stipulated by a thermal activation with the energy E_a, the temperature dependence of correlation time of a paramagnetic probe rotation is well described by the Arrhenius equation

$$\tau_c = \tau_0 \exp(E_a/k_B T). \qquad (2.60)$$

Macromolecules and their units or side groups can diffuse in condensed systems with $E_a \sim 0.08$–0.16 eV [135,136,154]. Preexponential factor τ_0 lies near to 10^{-12}–10^{-13} s, that is, close to respective times of orientational swings of molecules in a condensed phase [135,136,154]. For polymer systems, E_a and τ_0 vary in 0.19–0.76 eV and 10^{-13}–10^{-20} s ranges, respectively [135,136,154]. A linear dependency of $\ln\tau_0$ on E_a can hold for polymers, indicating the possible compensation effect and a cooperative character of molecular motion in these systems. In contrast with nitroxide radicals, polarons with slighter anisotropic magnetic parameters become themselves as stable spin labels because they are native and localized on the chains of conjugated polymers. Therefore, the closest environment of such PC remains undisturbed, and the results obtained on structure and spin and molecular motion should be more accurate and informative. The method allows to determine anisotropic molecular motions of the polymer macromolecules near their main axes and then find possible correlations (modulation) involving the anisotropic spin charge carrier's transfer and macromolecular dynamics in the system under study.

The method of spin label and probe enabled the study of molecular dynamics in condensed media, occurring with $10^{-7} > \tau_c > 10^{-11}$ s. However, as in highly viscous compounds, in most low-dimensional polymer systems, the molecular processes often take place with characteristic correlation time $\tau_c \geq 10^{-7}$ s. Such a dynamics is studied by the saturation transfer EPR (ST-EPR) method [114], which is based on the fast passage of MW saturated spin packets and broadens the correlation time range up to 10^{-3} s. If three critical frequencies are introduced, $\Delta\omega$, τ_c^{-1}, and T_1^{-1} (here, $\Delta\omega$ is the magnitude of the anisotropy of the magnetic interaction), then the region of the superslow tumbling domain is characterized by the inequalities $\Delta\omega\tau_c \gg 1$ and $100T_1 > \tau_c > 0.01T_1$. The first inequality shows that the EPR absorption spectrum is the same as that obtained by using a rigid powder. The second inequality determines a condition for a diffusion of MW saturation across the spectrum, since rotational diffusion is comparable to spin–lattice relaxation rate. For nitroxide radicals, this inequality is valid in the range of $10^{-3} > \tau_c > 10^{-7}$ s. The most sensitive to such molecular motions are the $\pi/2$-out-of-phase first harmonic dispersion and second harmonic absorption spectra. According to the ST-EPR method [114], if the inequality

$$\tau_c \leq \frac{2}{3\pi^2 T_1 \gamma_e^2 B_1^4} \frac{\sin^2 \vartheta \cos^2 \vartheta \left(B_\perp^2 - B_\parallel^2\right)^2}{B_\perp^2 \sin^2 \vartheta + B_\parallel^2 \cos^2 \vartheta} \qquad (2.61)$$

holds for a correlation time of radical rotation near, for example, x-axis, the adiabatic condition $dB/dt \ll \gamma_e B_1^2$ can be realized for the radicals oriented by x-axis along $\mathbf{B}_0$ and cannot be realized for the radicals with other orientations. This eliminates MW saturation of spectra of radicals, whose y- and z-axes are oriented parallel to the field $\mathbf{B}_0$ and consequently to the decrease of their contribution to the total ST-EPR spectrum. Therefore, slow spin motion should lead to an exchange of y and z spectral components and to the diffusion of saturation across the spectrum with the average transfer rate [114]

$$\left\langle \frac{d(\delta B)}{dt} \right\rangle = \sqrt{\frac{2}{3\pi^2 T_1 \tau_c}} \frac{\sin \vartheta \cos \vartheta \left(B_\perp^2 - B_\parallel^2\right)}{\sqrt{\left(B_\perp^2 \sin^2 \vartheta + B_\parallel^2 \cos^2 \vartheta\right)}}, \qquad (2.62)$$

where
   δB is the average spectral diffusion distance
   $B_\perp$ and $B_\parallel$ are the anisotropic EPR spectrum component arrangements along the external magnetic field $\mathbf{B}_0$
   ϑ is the angle between this field and a radical axis

More detailed information about the motion of radical and its microenvironment is expected to be obtained when $T_1/\tau_c \leq 1$. Indeed, fast motion causes rapid averaging over all angles which excludes the details of individual random walk process. On the other hand, the motion near the rigid lattice limit prevents the particle walk far enough to give much insight into the random walk process. In principle, it has already been argued [114] that anisotropic motions can be studied more efficiently from the analysis of second-harmonic absorption $\pi/2$-out-of-phase spectra at $T_1/\tau_c \leq 1$.

As a rule, such approaches are used for the study of various condensed systems at $\omega_e/2\pi \leq 10$ GHz [206,207,211,212]. However, at these frequencies, the main terms of their **g** tensors cannot be determined separately, so the change in the HFI components' intensity and splitting can only be analyzed. Low-frequency EPR spectroscopy is limited for the study of PC in conjugated polymers due to insufficient resolution between the spectra of mobile polarons and spins stabilized in polymer matrix. Robinson and Dalton showed theoretically [213] that the sensitivity of the method to the anisotropic molecular rotations can be increased at K-band EPR. However, the anisotropy of a resonance field stipulated by a **g** tensor is comparable to that of HFI at this waveband; therefore, the overlapping of lines attributed to different main radical orientations remains almost unchanged, and consequently, the difficulties of the registration of anisotropic superslow molecular processes and the separation of dynamic and relaxation processes are not eliminated in condensed media. The literature concerning the effect of the motion of π-conjugated chains onto interchain charge transfer is sparse.

Let us consider the advantage of a millimeter waveband EPR spectroscopy for an investigation of molecular mobility of PC with anisotropic magnetic parameters in different conjugated polymers. As the spectral resolution of a D-band EPR is enhanced, the possibility of this method in a more detailed study of superslow molecular motion was proved to be widened. It was demonstrated [214] that, for example, at D-band EPR, all terms of anisotropic *g* and *A* tensors of nitroxide radicals are registered separately. It allows to determine subtle features of structure, conformation, dynamics, and other parameters of different condensed systems [125–127,198]. At this waveband, the correlation times of radical rotation near its, for example, a longer *x*-axis and a shorter *y*- or *z*-axis can be determined from the broadening of corresponding spectral components $\delta\Delta B_i$, by using the following simple expressions:

$$\gamma_e\delta(\Delta B_x)=2/\tau_{c2}, \tag{2.63}$$

$$\gamma_e\delta(\Delta B_y) = \gamma_e(\Delta B_z) = \tau_{c1}^{-1} + \tau_{c2}^{-1}. \tag{2.64}$$

At millimeter waveband EPR, the separate registration of all components of ST-EPR spectra becomes possible. Besides, the first harmonic π/2-out-of-phase dispersion spectrum term of Equation 2.35 appeared to be also informative for the registration of anisotropic superslow radical and spin-modified rotations of macromolecule fragments [132]. Figure 2.5 exhibits how it changes the ratio of appropriate amplitudes of this term with the correlation time $\tau_c^{x,y}$ of superslow anisotropic motion near *x*- and *y*-axes at different spin–lattice relaxation time [122,215]. This means that registering both terms at D-band EPR of adiabatically saturated dispersion signal of PC with anisotropic magnetic parameters, one can determine separately both the relaxation times and then all the parameters of superslow macromolecular dynamics. The correlation time of such anisotropic motion can be calculated from a simple equation [118,122]:

$$\tau_c^{x,y} = \tau_{c0}^{x,y}\left(u_3^{x,y}/u_3^{y,x}\right)^{-\alpha}, \tag{2.65}$$

where α is a constant determined by an anisotropy of *g*-factor.

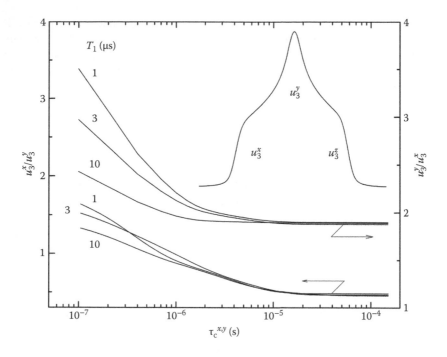

FIGURE 2.5

Logarithmic plots of u_3^i/u_3^j ratio calculated from the $\pi/2$-out-of-phase first harmonic dispersion spectrum of paramagnetic centers with $g_{xx} = 2.00401$, $g_{yy} = 2.003126$, and $g_{zz} = 2.002322$ and $A_{xx} = A_{yy} = A_{zz} = 0.62$ mT rotating near x- and y-axes versus an appropriate correlation time $\tau_c^{x,y}$ at different spin–lattice relaxation times. (From Krinichnyi, V.I., *Appl. Phys. Rev.*, 1(2), 021305/01, 2014. With permission.)

The methods described earlier can also be used for the study of solved conjugated polymers. These substances, however, are characterized by strongly anisotropic, structural, and conformational parameters. If one modifies them with a nonspherical probe, then its effective rotation frequency seems to be defined not only by a rotation of the whole but mainly by the intramolecular motion of an active fragment in the frames of either the Brownian anisotropic continuous diffusion or the hopping-like diffusion of Maxwell resilient substances. It implies that at fast interactions of spin probe with their own microenvironment, the latter can be considered as an elastic body. The correlation time τ_c of the probe rotation should therefore be significantly smaller than that of a mechanical relaxation of such an interaction and does not appear as a sufficiently informative parameter and characteristic of a dynamic process. The correct interpretation of the results, obtained by using the spin probe method, therefore implies the preliminary knowledge of structural and other properties of the system under study.

Krinichnyi et al. [216] showed that such an investigation can additionally be used for the so-called method of spin macroprobe, in which a comparatively massive paramagnetic with the characteristic sizes of about 10–100 µm is used as a probe. The method is based on the analysis of simultaneous rotation in the system under study of both nitroxide radical and a single microcrystal of a suitable ion-radical salt, for example, (dibenzotetrathiafulvalene)$_3$ PtBr$_6$, (DBTTF)$_3$PtBr$_6$. It was alighted that (DBTTF)$_3$PtBr$_6$ being placed into a strong magnetic field is oriented by own main crystal axis when the total dipole moment of its PC tends to be oriented along the direction $\mathbf{B}_0$ to attain the minimum energy of spin interaction with the magnetic field. Such a reorientation process is easily registered by its EPR line shift into lower magnetic fields. The time of crystal orientation is a function of many parameters, such as its weight, size, effective magnetic momentum, as well as medium viscosity and external magnetic field strength.

Figure 2.6 shows D-band EPR spectrum of frozen nujol/*tert*-butylbenzene mixture (1:10) in which a nitroxide radical is solved and (DBTTF)$_3$PtBr$_6$ single microcrystal is placed. First, the spin macroprobe is oriented by its main crystallographic axis along the magnetic field at the temperature

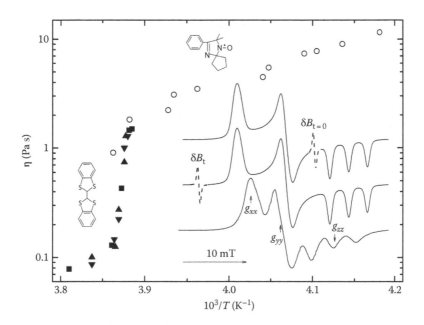

FIGURE 2.6
D-band electron paramagnetic resonance spectra of nitroxide radical with g_{xx}, g_{yy}, and g_{zz} (solid line) and single crystal (dibenzotetrathiafulvalene)$_3$PtBr$_6$ (dashed line) introduced into frozen and heat softened nujol/*tert*-butylbenzene mixture (1:10) at $t = 0$ (upper spectrum), t at $T \leq T_g$ (middle spectrum), and $t \to \infty$ at $T \leq T_g$ (lower spectrum) time period (inset) and the Arrhenius dependency of dynamic viscosity η of the mixture obtained by both spin probe (open symbols) and spin macroprobe (filled symbols) methods. In the study, three single crystals with different sizes were used as spin macroprobe. (From Schlick, S.: *Advanced ESR Methods in Polymer Research*. 307–338. 2006. Wiley-VCH Verlag GmbH & Co. KGaA. Reproduced with permission.)

significantly exceeding the temperature T_g of the matrix glass transition. Then the spin-modified sample is frozen down to $T \leq T_g$, turned by 90° (upper spectrum in Figure 2.6), and then it is slowly warmed up to $T \cong T_g$. Once the matrix becomes softer, the EPR line of the spin macroprobe begins to move along the spectrum as it is shown on the middle spectrum of Figure 2.6. This line shift follows an exponential law and attains the maximal value $\delta B \cong 20$ mT for this sample at $t \rightarrow \infty$. This process is accompanied by a narrowing of spectral components of nitroxide radical and their shift to its spectral center (lower spectrum in Figure 2.6). The change in the nitroxide radical spectrum shape is attributed to its Brownian diffusion rotation at least at low-temperature region whose correlation time was determined from Equation 2.60 to be $\tau_c = 4.7 \times 10^{-20} \exp(0.57 \text{ eV}/k_B T)$. The preexponential factor obtained exceeds considerably the inverted frequencies of molecular orientation oscillations in the condensed phase. Besides, the activation energy of radical rotation is higher than that typical for a pure solvent. This can be a result of correlation of the radical mobility and a segmental mobility of nujol extended molecules $(-CH_2-CH_2-)_n$, where $n > 10$.

The value characteristic of the orientation process of spin macroprobe is the time-dependent angle that changes as

$$\varphi(t) = \arccos[\delta B(t)/\delta B(t \rightarrow \infty)], \tag{2.66}$$

where $\delta B(t)$ are the EPR line shifts at a current time t reaching maximum value $\delta B(t \rightarrow \infty)$ at $t \rightarrow \infty$ limit. The dynamics of such a system can be described by the differential equation for a one-point-fixed oscillator:

$$I\ddot{\varphi} + A_{fr}(t) + M(t) = 0, \tag{2.67}$$

where
$I = mb^2/12$
$A_{fr}(t) = 6\pi\eta b^3 \omega$
$M_d(t) = NV\mu_0\mu_B B_0 \sin \varphi$ are the moments of inertia, friction, and disturbing force, respectively
m, b, V, and N are the mass, characteristic size, volume, and bulk spin concentration of the microcrystal, respectively
η is the coefficient of dynamic viscosity of the matrix

Characteristic time of rotating of such dissipative system in a magnetic field exceeds significantly $I/(6\pi\eta b^3)$ ratio, therefore in the case of the motion on small angles Equation 2.67 appears as

$$\dot{\varphi} + \tau_m^{-1}\varphi = 0, \tag{2.68}$$

whose solution is

$$\varphi(t) = \varphi_0 \exp(-t/\tau_m), \tag{2.69}$$

where $\tau_m = 6\pi\eta b^3/(NV\mu_O\mu_B B_0)$ is the mechanical relaxation time for such reorientational process.

The dependency presented allows the determination of dynamic viscosity of the system under consideration by using the Stokes equation

$$\tau_c = 4\pi\eta r_{NO}^3/(3k_B T). \tag{2.70}$$

Figure 2.6 depicts the temperature dependency of the viscosity of the model system, determined by the method of convenient spin probe from Equations 2.63 and 2.64 and by the method of spin macroprobe from Equation 2.70. It was noted in the foregoing that the hydrodynamic radius of nitroxide radical depends on the structure of the environment molecules and therefore varies in a wide range. Combining $\eta(T)$ dependencies, obtained by both methods, one obtains $r_{NO} = 0.14$ nm, close to the theoretical value $r_{NO} = 0.15$ nm for such PC of close structure.

As it would be expected, $\eta(T)$ dependency is of activation character and is proportional to the product $T\exp(E_a/k_B T)$ at $T \leq 256$ K. At these temperatures, a spin probe reflects the defrosting of motion of the radical microenvironment in a glass-like matrix. The matrix appeared to be liquid in a structural aspect in this temperature range, and it is solid in a mechanical aspect. $T_g \cong 256$ K is a point of graded change of the matrix viscosity (see Figure 2.6). The matrix begins melting mechanically at this temperature and gradually transforms into viscous-flow state, absorbing a certain quantity of heat, called *melting heat*. In an ideal case, the phase transition II is realized as hopping at T_g glass transition temperature. In this case, however, such a process proceeds in a limited temperature range. This can be attributed to finite molecule sizes of a matrix and probe and also to the dispersion in the extended organic molecule lengths in the matrix. The change in viscosity seems to be also of an activation character at $T \geq 260$ K.

Extrapolating $\eta(T)$ dependency obtained by method of spin microprobe to higher temperature, one obtains $\eta = 0.11$ Pa s at RT, that is, it falls within a range of viscosity for different oils. However, such an extrapolation yields $\eta = 5.6 \times 10^{-3}$ Pa s at RT, by considering η jump at T_g. This confirms once more the correctness of the joint application of the complementary methods described at investigating relaxation and dynamic processes in various condensed systems. The method of spin macroprobe successfully enriches the classic method of spin probe and enables the possibility of a more accurate determination of dynamic viscosity in a wider temperature range. It becomes possible to characterize the mechanical losses in solutions of conjugated polymers and other condensed media in a stationary regime and to establish their glass transition point in the case of extremal mechanical losses in these systems.

The dependency $\tau_m \propto b^3 B_0^{-1}$ is derived from Equation 2.69. Considering the higher concentration and absolute sensitivities of the D-band EPR spectroscopy, one can draw a conclusion that the application of the method at this waveband widens the range of dynamic viscosity (and likely the other parameters as well) measurement in condensed systems by at least two orders of magnitudes as compared with other bands and a higher magnetic field strength.

2.5 Charge Carrier Transfer Mechanisms in Low-Dimensional Systems

Electrodynamic properties of conjugated polymer systems whose conductivity can vary by 12–15 orders of magnitude are defined by various parameters of a system. In a macroscopic conductivity measurement, a superposition of different processes should be taken into account [217]:

1. Q1D diffusion of charge carriers along the polymer chains
2. Q3D hopping of charge carriers between the polymer chains through amorphous phase
3. Tunneling between conjugated regions separated by less conjugated parts of fibrils
4. Fluctuation-induced tunneling between fibrils

It is quite obvious that the contributions of these processes to the total conductivity depend on the structure, morphology, method of synthesis, and doping level of a polymer [218]. As the topological excitations, namely, solitons, polarons, and bipolarons, can have a major effect on the electronic states accessible for electrons transferred through a polymer, they can dramatically affect transport properties of polymer nanocomposite.

Despite the numerous reports on transport properties of organic conjugated polymers, no general agreement on the charge transport mechanism prevails. This is largely due to the complex structural and morphological forms that conjugated polymers are characterized. Generally, there are disordered systems whose charge transport can usually be considered by a percolation phenomenon. The specific nature of the mentioned charge carriers stabilized in these systems reflects the importance of electron–phonon interaction, but in addition, their Coulomb interaction characterized by appropriate integral may play a significant role. A conjugated carbon chain is a Q1D conductor; therefore, dimensionality is an important factor affecting electronic properties of conjugated polymers. Dimensionality starts to play a role when the film thickness becomes comparable to the charge carrier hopping length. Another factor strongly influencing the structure and electronic properties of the polymer characteristics is the doping process.

The temperature dependence of conductivity gives important information about the electrical conduction mechanism. In both insulators and semiconductors, there is an energy gap between the VB and the CB (see Figures 1.3 and 1.4). In the absence of thermal energy, no electrons are excited to the CB, and these systems become insulators. Lightly doped or undoped polymers show strong temperature dependences of the conductivity. RT conductivity is low and the electronic properties resemble those of insulators or poor semiconductors. Their charge transport is described as occurring via hopping between the states of the localized charge carriers. The electronic properties of moderately doped polymers are typically more difficult to interpret, because there may be several transport processes carrying out in parallel. In disordered systems, there are tunneling and hopping charge transfer characterized by an appropriate temperature dependence of the conductivity. In a metal, the VB and CB overlap and so negligible thermal energy is needed for an electron to move into a vacant state. Highly doped polymer systems also demonstrate complex mixture of the charge-transfer mechanisms, in which contribution depends on the system doping level and homogeneity. Besides, metallic behavior of the conductivity can be realized for the highly doped polymers as well.

The conductivity of a conjugated polymer due to dynamics of N spin charge carriers can be calculated from the modified Einstein relation

$$\sigma_{1,3D}(T) = Ne^2\mu = \frac{Ne^2 D_{1,3D} d^2_{1D,3D}}{k_B T}, \qquad (2.71)$$

where
  e is the elementary charge
  μ is the charge carrier mobility
  $d_{1D,3D}$ are the intra- and interchain constants, respectively

Several models have been proposed to describe a temperature dependence of conductivity experimentally obtained for conjugated polymers.

Kivelson proposed [219] a phenomenological model for phonon-assisted hopping of electrons between soliton sites in undoped and slightly doped *trans*-PA. In the frames of this model, charged solitons are coulombically bound to charged impurity sites. The excess charge on the soliton site makes a phonon-assisted transition to a neutral soliton moving along another chain. If this neutral soliton is located near charged impurity at the moment of charge carrier transfer, the energy of the charge carrier remains unchanged before and after the hop. In this case, a charge transfer should be governed by the probability that the neutral soliton is located near the charged impurity, and its initial and final energies lie within k_BT; hence, both the *dc* and *ac* conductivities can be determined as [219]

$$\sigma_{dc}(T) = \frac{k_1 e^2 \gamma(T) \xi \langle y \rangle}{k_B T N_i r_i^2} \exp\left(\frac{2k_2 r_i}{\xi}\right) = \sigma_0 T^n, \tag{2.72}$$

$$\sigma_{ac}(T) = \frac{N_i^2 e^2 \langle y \rangle \xi_{\parallel}^3 \xi_{\perp}^2 \omega_e}{384 k_B T} \left[\ln \frac{2\omega_e m_u}{\langle y \rangle \gamma(T)}\right]^4 = \frac{\sigma_0 \omega_e}{T} \left[\ln \frac{k_3 \omega_e}{T^{n+1}}\right]^4, \tag{2.73}$$

where
  $k_1 = 0.45$, $k_2 = 1.39$, and k_3 are constants
  $\gamma(T) = \gamma_0 (T/300\ \text{K})^{n+1}$ is the transition rate of a charge between neutral and charged soliton states
  $\langle y \rangle = y_n y_{ch}(y_n + y_{ch})^{-2}$, y_n, and y_{ch} are, respectively, the concentrations of neutral and charged carriers per a monomer unit
  $r_i = (4\pi N_i/3)^{-1/3}$ is the typical separation between impurities, in which concentration is N_i
  $\xi = (\xi_{\parallel} \xi_{\perp}^2)^{1/3}$, $\xi_{\parallel}$, and $\xi_{\perp}$ are dimensionally averaged, parallel and perpendicular decay lengths for a charge carrier, respectively
  m_u is the number of monomer units per a polymer chain

The decay length of a carrier wave function perpendicular to the chain can be determined from the relation [220,221]

$$\zeta_{\perp} = \frac{b}{\ln(\Delta_0/t_{\perp})}, \tag{2.74}$$

where $2\Delta_0$ is the bandgap and $t_{\perp}$ is the hopping matrix element estimated as [222]

$$t_{\perp}^2 = \frac{h^4 v_{ph}^3 D_{3D}}{4\pi^2 E_p^2} \exp\left(\frac{2E_p}{h v_{ph}}\right), \tag{2.75}$$

where E_p is the polaron formation energy lying near 0.1 eV for disordered conjugated polymers. In this case, a weak coupling of the charge with the polymer lattice is realized when hops between the states of a large radius take place. Note that this mechanism of charge transfer can be realized not only in conjugated polymers but also in modified convenient polymers, for example, in $MnCl_2$-filled polyvinyl alcohol films [223].

If the coupling of the charge with the lattice is stronger, multiphonon processes dominate. The mobility ultimately becomes simply activated when the temperature exceeds the phonon temperature characteristic of the highest energy phonons with which electron states interact appreciably. In this case, the strong temperature dependency of the hopping conductivity is more evidently displayed in *ac* conductivity.

A comparatively strong temperature dependency for polymer conductivity can be described in the frames of the Elliot model of a thermal activation of charge carriers over energetic barrier E_a from widely separated localized states in the gap to close localized states in the valence and conjugated band tails [224]. Both σ_{dc} and σ_{ac} terms of total conductivity are defined mainly by a number of charge carriers excited to the band tails; therefore,

$$\sigma_{dc}(T) = \sigma_0 \exp\left(-\frac{E_a}{k_B T}\right), \tag{2.76}$$

$$\sigma_{ac}(T) = \sigma_0 T \omega_e^\kappa \exp\left(-\frac{E_a}{k_B T}\right), \tag{2.77}$$

where
$0 \leq \kappa \leq 1$ is a constant reflecting the dimensionality of a system under study
E_a is the energy for activation of a charge carrier to extended states

As the doping level increases, the dimensionality of the polymer system rises and activation energy of charge transfer decreases. An approximately linear dependency of κ versus E_a was registered [225] for some conjugated polymers. At the same time, Parneix et al. [226] obtained $\kappa = 1 - \alpha k_B T/E_a$ dependency, for example, for slightly doped poly(3-methylthiophene) with $\alpha = 6$, $E_a = 1.1$ eV. Therefore, κ value can be varied in a 0.3–0.8 range and it reflects the dimensionality of a system under study.

E_a value can depend on the conjugation length and conformation of the polymer chains. Indeed, $E_a \propto n_d$ and $E_a \propto n_d^{1/2}$ dependencies were obtained [227,228] for *trans*-PA with the concentration n_d of sp^3 defects at different doping levels. $E_a \propto (n_\pi + 1)/n_\pi^2$ dependency was estimated [229] for relatively short π-conjugated systems having n_π delocalized electrons. In order to reduce the bandgap, the planarity of the polymer chains may be increased [230,231], for example, by introducing an additional hexamerous ring to a monomer unit [232].

When electrons are strongly coupled to bosonic quasiparticles, phonons, excitons, etc., self-trapped states can be formed in a conjugated polymer. If the size of the polaron is comparable with the lattice spacing, the charge transfer is then dominated by hopping processes: the quasiparticle has to overcome a potential barrier and loses its quantum coherence at each hop, giving rise to an activation law Equations 2.76 and 2.77. This law has explained largest amount of experiments, which are often observed around RT; however, strong deviations from pure Arrhenius behavior are often reported [233–235], possibly indicating the onset of the coherent transport regime.

According to the Friedman–Holstein model [236], the charge carriers in the adiabatic regime can follow the effective motion of the lattice and will possess a high probability of hopping to the adjacent site. The conductivity due to the hopping of N_p small polaron in such regime is [237]

$$\sigma_{ac} = \frac{N_p e^2 d_i^2}{2} \frac{\omega_0}{2\pi k_B T} f(T) \exp\left(-\frac{E_h - t_h}{3k_B T}\right), \qquad (2.78)$$

where
  d_i is the distance between neighboring hopping sites
  $f(T) = 1$ for $t_h \ll k_B T$ and $f(T) = k_B T/t_h$ for $t_h \gg k_B T$
  E_h is the hopping activation energy
  t_h is the hopping integral

In a nonadiabatic regime, the carrier cannot follow the lattice oscillations, and the time required for a carrier to hop from one site to another is large compared to the duration of a coincidence even between two polaron sites. For this case, the *ac* conductivity should be written as [237]

$$\sigma_{ac} = \frac{N_p e^2 d_i^2 t_h}{2h} \left(\frac{\pi}{4k_B T E_h}\right)^{1/2} \exp\left(-\frac{E_h}{3k_B T}\right). \qquad (2.79)$$

To describe electron transfer in amorphous metals, Mott developed the variable range hopping (VRH) model [238]. This theory balances the likelihood of tunneling between random energy electron potential wells with the likelihood of gaining enough thermal energy to move to a nearby site. In this hopping conduction process, each state can have only one electron of one spin direction. If the carrier localization is very strong, the charge will hop to the nearest state with the probability proportional to $\exp(-\Delta E_{ij}/k_B T)$ (see the respective term of Equation 2.9). At less strong localization of charge carriers in a conjugated polymer, this model seems to be more suitable. As the temperature is decreased, fewer states fall within the allowed energy range and the average hopping distance increases. This results in the dependence of both the *dc* and *ac* conductivities of a d-dimensional system on temperature as [239,240]

$$\sigma_{dc}(T) = \sigma_0 \exp\left[-\left(\frac{T_0}{T}\right)^{\frac{1}{d+1}}\right], \qquad (2.80)$$

$$\sigma_{ac}(T) = \frac{1}{3}\pi e^2 k_B T n^2 (\varepsilon_F) \langle L \rangle^5 \omega_e \left[\ln \frac{\omega_0}{\omega_e} \right]^4 = \sigma_0 T, \qquad (2.81)$$

where

$\sigma_0 = 0.39\omega_0 e^2 [n(\varepsilon_F)\langle L \rangle/(k_B T)]^{1/2}$ and $T_0 = 16/k_B n(\varepsilon_F)z\langle L \rangle$ at $d = 1$, $\sigma_0 = \omega_0 e^2$ $[9/8\pi k_B T n^{3/2}(\varepsilon_F)/\langle L \rangle]^{1/2}$ and $T_0 = 18.1/k_B n(\varepsilon_F)\langle L \rangle^3$ at $d = 3$ in equation for *dc* conductivity

z is the number of nearest-neighbor chains

$\langle L \rangle = (L_\parallel L_\perp^2)^{1/3}$, $L_\parallel$, and $L_\perp$ are the average length of charge wave localization function and its projection in parallel and perpendicular directions, respectively

T_0 is the percolation constant or effective energy separation between localized states depending on disorder degree in amorphous regions

The conductivity is essentially determined by the phonon bath and by the distributional disorder of electron states in space and energy respectively above and below T_0. The distance R and average energy W of the charge carriers hopping are $R^{-4} = 8\pi k_B T n(\varepsilon_F)/9\langle L \rangle$ and $W^{-1} = 4\pi R^3 n(\varepsilon_F)/3$, respectively. According to this model, as the temperature falls to zero, the conductivity rises also to zero.

It is important to note that the VRH theory was originally developed for amorphous semiconductors and does not consider specific nature of nonlinear charge carriers in conjugated polymers. For homogeneous systems, the localization length $\langle L \rangle$ is greater than the disordered length scale, for example, greater than the structural coherence length ζ in a polymer that has both crystalline and amorphous regions. Since Mott's power $T^{-1/(d+1)}$ law is usually observed in homogeneously disordered insulating systems, the observation of a similar temperature dependence of conductivity at low temperatures in conjugated polymers can evidence for the presence of homogeneous disordering. For inhomogeneous systems with inhomogeneous doping, phase segregation of doped and undoped regions, partial dedoping, large-scale morphological disorder, etc., should also be taken into consideration.

The semiconductor-like behavior of the conductivity is usually explained by the existence of potential barriers between highly conjugated clusters. These barriers have been suggested to be conjugated defects, segments of undoped or weakly doped polymer chains, interchain or interfibrillar contacts, or other inhomogeneities. The charge carriers hop over or tunnel through such potential barriers. The latter process is temperature independent; therefore, the temperature dependence of the conductivity should arise from other processes influencing the charge transfer between highly conjugated clusters. Sheng's models take into account the charging energy of conjugated clusters [241] or the random thermal fluctuation-induced motion of charge carriers on both sides of the barrier [242].

When the size of the clusters is small enough (typically around 20 nm or less) and if the energy barrier between them is higher than k_BT, charge carrier transfer can be induced only by thermal activation. In this case, it is limited by the intercluster barrier and leads to a temperature-dependent conductivity following Equation 2.80 with $d=1$ [243]. In larger clusters, the fluctuation-induced tunneling of charge carriers between highly conjugated clusters prevails. In this instance the random thermal motion of charge carriers within the conjugated clusters induces a randomly alternating potential across the gap between clusters. According to Sheng's approach, the conductivity is determined by charge carrier transfer through parabolic barrier as [244]

$$\sigma(T) = \sigma_0 \exp\left(-\frac{T_1}{T_2 + T}\right), \tag{2.82}$$

where σ_0, T_1, and T_2 are the constant dependent on the width and height of the tunneling barrier in the polymer.

Pietronero [245] and Kivelson and Heeger [246] have proposed the model considering the scattering of charge carriers on the lattice optical phonons in metal-like clusters embedded into amorphous polymer matrix. In the framework of this model, the total conductivity of such polymer system can be expressed as [245,246]

$$\sigma_{ac}(T) = \frac{Ne^2 d_{1D}^2 M t_0^2 k_B T}{8\pi\hbar^3 \alpha^2}\left[\sinh\left(\frac{E_{ph}}{k_B T}\right)-1\right] = \sigma_0 T\left[\sinh\left(\frac{E_{ph}}{k_B T}\right)-1\right], \tag{2.83}$$

where
  M is the mass of the polymer unit
  t_0 is the transfer integral for the π-electron equal approximately to 2.5–3 eV
  E_{ph} is the energy of the optical phonons
  α is the constant of electron–phonon interaction equal, for example, for *trans*-PA to 4.1×10^8 eV/cm

If spin traps are initiated in polymer matrix, the dynamics of spin charge carriers can be explained in terms of the Hoesterey–Letson formalism modified for amorphous low-dimensional systems containing spin traps with concentration n_t and depth E_t [247]. In the case of low trap concentration limit, one obtains

$$\sigma(T) = \frac{\omega_0 Ne^2 d^2}{k_B T}\left(\frac{R_{ij}}{d}\right)^2 \exp\left(-\frac{2R_{ij}}{r}\right)\exp\left(\frac{E_t}{2k_B T_c}\right)\exp\left[-\frac{E_t}{2k_B T}\left(\frac{E_0}{k_B T}\right)^2\right], \tag{2.84}$$

where

d is the lattice constant

$T_c = E_t/2k_B\ln(n_t)$ is the critical temperature at which the transition from trap-controlled to trap-to-trap hopping transport regimes occurs

E_0 is the width of intrinsic energetic distributions of hopping states in the absence of traps

Charge transfer in conjugated polymers, especially highly doped by heavy atoms with large atomic number, z, can also be realized according to the Elliott mechanism [1,248]

$$\sigma^{-1} = T_{1E}^{-1} = k\Delta g^2/\tau_r = \alpha(\lambda/\Delta E)^2/\tau_r, \tag{2.85}$$

where

T_{1E}^{-1} is the EPR spin–lattice relaxation rate proportional to the EPR spectral broadening $\delta(\Delta B_{pp})$

k is the constant determining experimentally lying near 0.1 for alkali metals [249]

Δg is the shift of g-factor

τ_r is the relaxation time of the electron momentum

λ is the spin–orbit coupling constant

ΔE is the energy splitting between the valence band and the spin–orbit state

Two main concepts of charge transfer in highly doped polyaniline and other conjugated polymers are now postulated. The first of them considers an elemental charge transfer by polaron along and between individual polymer chains even in heavily doped polymer [105,205]. According to the other approach, the doping leads to the increase of number and size of metal-like domains in amorphous phase of a polymer in which the polarons are considered to be localized and the charge is transferred by 3D delocalized electrons [250–252]. The formation of such Q3D crystalline domains in the polymer matrix is confirmed by the increase of its Pauli susceptibility with the polymer doping. In the framework of the latter approach, the electronic and dynamic properties of the polymer are determined mainly by Q3D intradomain and Q1D interdomain charge transfer. As in the case of other conjugated polymers, the energy levels of polarons in the crystalline polymer phase are merged into metal-like band leading to the formation of polaron lattice [253,254]. If diamagnetic bipolarons and/or antiferromagnetic interacting polaron pairs each possessing two elemental charges are formed in heavily doped polymer, it can be considered as a disordered system on the metal-insulator boundary in which the charge is transferred according to the modified Drude model [254]. Such a system is characterized as a Fermi glass, in which the electronic states at the Fermi energy ε_F are localized due to disorder and the charge is transferred by the phonon-assisted Mott VRH between exponentially localized states near the Fermi level ε_F [255].

Thus, the numerous transport mechanisms with different temperature dependencies can be realized in conjugated polymers and their nanocomposites. These mechanisms are directly associated with the evolution of both crystalline and electron structures of the systems. The formation of the states of the polaron- and bipolaron-type charge carriers in conjugated polymer systems leads to a variety in transport phenomena, including the contribution of such mobile nonlinear excitations to the conductivity.

3

Instrumentation and Experimental Setup

3.1 EPR Techniques

Requirements for the EPR experimental technique are mainly determined by the tasks scheduled [256]. If one expects only to analyze if free spins exist in the sample or not, in this case, the simplest routine EPR spectrometer can be used. It can detect the presence of spins with a single line and estimate their number in the sample under study. Currently, the X-band EPR technique is the most developed and prevalent. This is due mainly to the availability starting from the 1940s of an appropriate element base and a relative simplicity of the experiments at this waveband.

The typical X-band EPR spectrometer is drawn schematically in Figure 3.1. The main parts of this device are microwave (MW) power oscillator 1, magnet 9, and MW reflecting cavity (resonator) 11. The EPR spectrum is usually measured keeping constant MW frequency during the sweeping of the magnetic field strength. If the energy of the MW quants equals the energy splitting of the spin states, the resonance condition (2) is fulfilled and MW radiation is absorbed. Typically, the ESR spectrum is given as the first derivative of the absorption peak, resulting from magnetic field modulation necessary for the use of a lock-in amplifier. The field modulation determines, among others, the time resolution of the EPR spectrometer. It is limited to the inverse of the field modulation frequency. Usually, a modulation frequency of 100 kHz is used, which would result in a maximum time resolution of 10 ps. MW power is generated by oscillator 1 previously based on vacuum klystron or backward wave oscillator and then on a solid-state Gunn diode. This part also contains an appropriate automatic frequency control circuit. MW power is splitted and transmitted to the two branches of the MW bridge. The referent part of the bridge consists of the MW attenuator 2 and MW phase shifter 3, whereas the signal measuring part contains more precession MW attenuator 6, one-way MW circulator 7, and MW reflecting cavity (resonator) 11 situated between the magnet poles 9. The sample under study in quartz ampoule is placed at the center of the MW cavity 11, and both the branches are tuned to reach a balanced zero signal at the output of the MW diode 4. Once Equation 2.2 is fulfilled for paramagnetic centers (PC) analyzed, they absorb a little part of MW

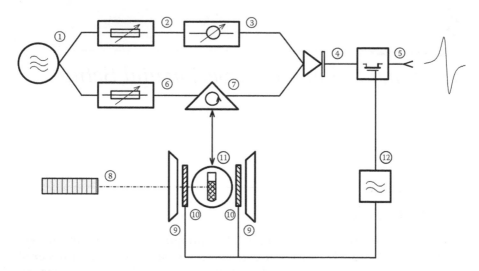

FIGURE 3.1
Sketch of X-band electron paramagnetic resonance spectrometer: 1, microwave (MW) oscillator; 2, MW attenuator; 3, MW phase shifter; 4, MW detector; 5, alternating current (*ac*) phase detector; 6, MW attenuator; 7, MW circulator; 8, light source (laser); 9, magnet pole; 10, *ac* modulation coils; 11, MW cavity with a sample; 12, *ac* oscillator.

energy in the measuring branch and are excited to the higher energy level. This causes an imbalance in the MW bridge and appearance at its output of resonant *ac* signal under a scanning of a sample in an external magnetic field B_0. This signal is transformed into a resulting *dc* EPR response by phase detector 5. Normally, both the cavity and magnet of the spectrometer are provided with a hole through which the PC can be initiated in the sample under its direct irradiation or illumination by appropriated source 8. The more spins are stabilized and/or initiated in the sample, the higher integral signal is registered on the device output. One of the simple exemplary devices, portable X-band EPR spectrometer PS 100X operating at 9.72 GHz, is characterized by a sensitivity of 4×10^{11} spin/mT, a resonance instability of 3×10^{-5} h^{-1}, and a maximal MW power of 150 mW inducing $B_1 = 48$ µT in the center of the cavity with an unloaded quality factor of $Q \approx 5000$ and mode H$_{102}$.

If, however, some spin ensembles with near magnetic parameters are stabilized in a sample, their lines can overlap. The width of individual EPR lines of most organic radicals usually amounts to 0.1–1.0 mT, and *g*-factor value differs from g_e by $(1–10) \times 10^{-4}$. Therefore, the spectral resolution or splitting of lines of different radicals, $\delta B \cong \delta g B_0 / g \cong 10$–100 mT, is smaller than their width at X-band, resulting in the overlapping of their EPR spectra. This complicates the identification of such radicals in solids and the analysis of structural and dynamic properties of microenvironment by using their *g*-factors [98]. This means that in order to attain the satisfactory resolution of free radicals' EPR spectra, the condition $\delta B / B_0 < 2 \times 10^{-5}$ should be valid. If the linewidth of PC do not change with registration frequency, such condition is fulfilled at millimeter waveband EPR. The separation in field of two lines with different g-factors will scale with the field, so if the linewidths remain constant, then the lines will eventually be resolved as the field is increased. Besides, Equation 2.5 shows that the susceptibility of the method also increases at high spin precession frequency ω_e.

The main advantages of the high-frequency/field continuous wave (CW) EPR spectroscopy may be compiled as follows:

1. Increase in spectral resolution due to the higher magnetic field
2. Increase in detection sensitivity for samples of limited quantity due to higher resonator filling factor
3. Increase in orientation selectivity in the investigation of disordered systems
4. Accessibility of spin systems with larger zero-field splitting due to the larger microwave quantum energy
5. Simplification of spectra due to the reduction of second-order effects at high fields

According to Equation 2.2, the registration of PC with $g \cong 2$, for example, at D-band EPR requires MW power source and a strong (near 5 T) permanent magnetic field. Conventional metals may induce a magnetic field of up to ~1.2 T, so respective cryogenic systems (cryostats) based on a superconducting solenoid are most promising for inducing very strong magnetic fields. At present superconducting magnets, providing magnetic fields with the strength up to 30 T and low inhomogeneity, are commercially available and widely used in magnetic spectroscopy. The devices of a submillimeter range with a laser source of polarizing radiation have magnets producing field strength up to 100 T in a pulse regime inside a multisectional superconducting solenoid of a peculiar shape [257]. However, the field of application of this device is strongly restricted because of its low concentration sensitivity, stipulated by a unit quality value of a measuring cavity, and a high cost of its cryogenic equipment.

The first D-band EPR investigation of PC stabilized in different solids showed the growth of sensitivity and resolution of the method at this waveband [122,125,126,258]. It was demonstrated that the study of model, biological, and organic polymer systems at higher registration frequencies allows to increase sufficiently the efficiency of the method, to obtain qualitative new information on these and other objects, and to solve various scientific problems. During this time, the variety of EPR technique, CW, pulse, combined, has been expanded [98]. For example, the Bruker Corporation has developed and started supplying in scientific centers of EPR spectrometers operating at wide (1–263 GHz) wavebands. Besides, different homemade EPR spectrometers were constructed worldwide which operate at MHz [105] up to THz [259] spin precession frequencies that correspond to 10^5–10^{-3} mm wavelengths. The operating frequency was defined in most cases by the investigation purposes and physical properties of the objects under study.

It should be noted that if nuclear magnetic resonance spectroscopy uses pulse techniques as a standard for decades, in EPR most spectrometer systems are CW (i.e., MW illuminates with a fixed MW frequency and sweeps the applied magnetic field). Pulse techniques, which involve MW illuminating the sample with a sequence of short pulses then looking at what it reradiates, offer time-resolved experiments, the ability to resolve closely spaced lines, and are suitable mainly for measuring relaxation. To excite the whole spectrum at once, however, requires very short, very high power pulses and requires the system dead time (the time after the last illuminating pulse before measurement can begin) to be minimized. The system dead time often means that the signal from any broad lines has decayed before measurement starts.

In the 1970s, the first CW multifunctional D-band (2 mm, 140 GHz) EPR5-01 spectrometer with a 100 kHz *ac* modulation was developed and designed for various physical–chemical studies [260]. It still operates in the Russian Institute of Chemical Physics in Chernogolovka for the study of PCs in various condensed media, conjugated polymer systems among them. The basic block scheme of this device is drawn in Figure 3.2. The spectrometer is assembled as a direct amplification circuit with a H_{011} type reflecting cavity, a double MW T-bridge and n-SbIn bolometer operating as an MW detector at temperature of liquid helium (4.2 K). The main part of the spectrometer includes the MW klystron or Gunn diode oscillator *11* with appropriate elements of the waveguide section, a cryostat *1* with a superconducting solenoid *2*, and the MW bolometric n-SbIn sensor *14*. In the center of the cryostat, the tunable MW cavity *4* with a sample *5* encircled by modulating *3* and temperature-sensitive *6* copper coils is placed. A sample filled into a quartz capillary with an external diameter and a length of 0.6 and 7 mm, respectively, is placed into the center of the cavity between two cylindrical pistons. The temperature of the sample can be controlled within 6–380 K region by helium or nitrogen gaseous flow. The quality factor Q of the cavity with an inner diameter of 3.5 mm and an operation length of 1.5 mm lies near 2000. The value of microwave field magnetic component B_1 in the center of the cavity is equal to 20 µT. The magnetic field inhomogeneity in the point of sample arrangement does not exceed 10 µT/mm. The absolute point-sample sensitivity of the spectrometer is 5×10^8 spin/mT at room temperature that is unique for EPR spectroscopy. The latter value is three orders of magnitude lower than N_{min} calculated from Equation 2.5 due to the smaller P and Q_0 values as compared with X-band. The concentration sensitivity for aqueous samples is 6×10^{13} spin/mT cm^3.

In order to obtain the spectra of conjugated polymers at higher frequencies/fields, a homemade EPR spectrometer with a unit quality factor of the sensitive cell [261] can also be used. The signal in this device is registered optically

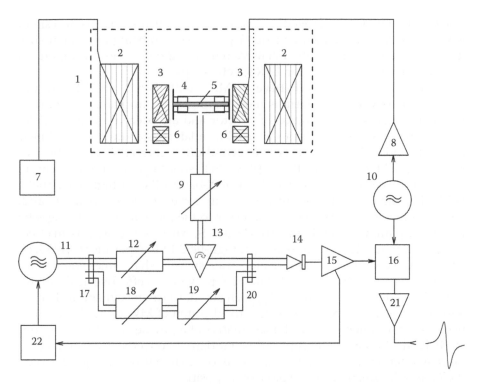

FIGURE 3.2

Sketch of D-band electron paramagnetic resonance spectrometer: 1, helium cryostat; 2, super-conducting solenoid; 3, alternating current (*ac*) modulation coil; 4, microwave (MW) cavity with two cylindrical side tuning and fixed pistons; 5, quartz capillary with a sample; 6, temperature-sensitive coil; 7, solenoid current supply; 8, *ac* modulator amplifier; 9 and 19, MW phase shifter; 10, *ac* oscillator; 11, MW oscillator; 12 and 18, MW attenuators; 13, MW circulator; 14, MW super-low-temperature (4.2 K) barretter; 15 and 21, *ac* preamplifiers; 16, phase detector; 17 and 20, directional MW couplers; 22, MW oscillator power supply with a section of MW frequency auto adjustment. (Modified from Krinichnyi, V.I., *Synth. Metals*, 108(3), 173, 2000. With permission.)

as absorption of a pumped far infrared (FIR) light guide CO_2 laser, providing emission lines at several wavelengths from 570 μm to 2 mm. The magnetic field is induced by a resistive coil system with homogeneity of about 0.3 mT in a sphere of 8 mm diameter. This technique was inter alia used success-fully in multifrequency (9.7–430 GHz) EPR study of PC in both forms of poly-acetylene [262] (see Section 4.1).

3.2 EPR Standards

A quite important problem is to choose the appropriate standards for a precise device tuning, a magnetic field scanning, and *g*-factor scale calibration [256].

The standard must be arranged as close to a sample as possible inside a small MW cavity without reducing its quality; therefore, 6 being of a small size, it must produce a sufficiently intensive narrow signal. Various compounds may be used as a g-standard in EPR spectroscopy.

The powder of MgO with trace of Mn^{2+} with $I = 5/2$, $a = 8.74$ mT, and $g_{eff} = 2.00102$ is widely utilized as mostly simple and available standard in EPR experiments for calibration of g-factor and an external magnetic field. At centimeter wavebands it can be used as a lateral standard situated near a sample under study. However, the second-order correction to its effective resonance field, $\delta B = a^2[(I(I+1) - m^2]/2B_0$ [263], should be taken into account. This value exceeds, for example, 1 mT at X-band EPR. At higher magnetic fields this value becomes smaller, reaching, for example, 65 μT at D-band, and does not contribute an essential error to the measurement of magnetic resonance parameters and can be neglected. So Mn^{2+} can be used as an effective standard for the g-factor scaling at high-field EPR experiments [125,126]. At these wavebands, Mn^{2+} ions demonstrate six very sharp lines equidistanted by hyperfine constant $a_{iso} = 8.74$ mT that provides an effective field calibration over a 40 mT range. Practically, such standard can be either attached by a toluene solution of polystyrene to a fixed piston of MW cavity [125,126] or used as environmental neutral quasi-matrix of a powder-like sample under study [264]. The signal intensity of this standard is conveniently regulated by plunger rotation about its axis, that is, due to the change of the angle between the B_0 and B_1 directions in the place of its position.

2,2-Diphenyl-1-picrylhydrazyl (DPPH) seems to be the other well-known standard due to its extremely intense line. However, its use at high fields [265] has been hampered by the fact that the g-factor depends on the solvent used under DPPH preparation by its recrystallization and, even then, different crystal qualities under the same preparation are sometimes found to have g-factors varying within a 2.0030–2.0043 range. Nevertheless, it is possible to find small (nano) DPPH crystals [266] with a narrow single line and to use them to calibrate other standards.

Some organic conducting ion-radical-based single crystals of lower dimensionality can be also used for calibration of g-factors of PC in different solids [216]. One of them, (fluoranthene)$_2$PF$_6$, is characterized by extremely narrow peak-to-peak EPR linewidth ($\Delta B_{pp} \leq 5$ μT at W- and D-bands) and the value of its g-factor obtained with a high degree of accuracy, $g_{xx} = 2.00226$, $g_{yy} = 2.00258$, and $g_{zz} = 2.00222$ (needle axis). The other ion-radical-based single crystals, (naphthalene)$_2$PF$_6$ and (perylene)$_2$PF$_6$, demonstrate at D-band EPR solitary spectra with $\Delta B_{pp} = 0.18$ mT, $g = 2.00316$ and $\Delta B_{pp} = 0.15$ mT, $g = 2.00321$, respectively. This allows them to be easily recognized and differentiated by adjusting the magnitude of the modulating field in the study of most systems. Besides, fast relaxation of PC in such standards allows using them in combination with the CW saturation method for determination of relaxation parameters of spin charge carriers in conjugated polymers and their nanocomposites. It should be noted, however, that their applicability for such purposes decreases at temperatures much below 200 K due to a phase transition. If the sample is stored for a long time at high temperature and air presence, it can also slowly be deteriorated. A more stable ion-radical salt, (dibenzotetrathiafulvalene)$_3$PtBr$_6$, is characterized by strong EPR line with $\Delta B_{pp} = 0.44$ mT and $g = 2.01628$. The precise device adjustment in the registration of a real χ' or an imaginary χ'' component of paramagnetic susceptibility is obtained by an attainment of the symmetric first and second derivatives of dispersion and absorption standard signals, respectively, in the device output. AC modulation phase can be finely tuned by a minimization of the $\pi/2$-out-of-phase unsaturated signal of a standard with the following phase change by $\pi/2$. In this case the $\pi/2$-out-of-phase signal attenuation is not less than 23 dB. It was shown [216] that this sample being placed in a condensed system can be reoriented in high magnetic field, so then its line is shifted by around 28 mT at D-band EPR. This effect may be successfully used for the study of physical properties of a matrix (see Section 2.4).

Isolated paramagnetic vacancy V_{Si}^- in 3C–SiC that exhibits a narrow, near 50 μT, strong, and stable resonance EPR signal observable within 1.2–300 K temperature region and does not show any aging effect [267] can also be used in some experiments.

1,3-*bis*-Diphenylene-2-phenyl allyl free radicals being embedded into polyethylene are characterized by high spin concentration and, therefore, fast spin relaxation [268]. However, it is characterized by a relatively broad (near 1 mT) EPR spectrum, which may interfere with $g = 2$ spectra. This makes such a system a suitable standard in EPR study at relatively low temperatures.

KC$_{60}$ metallic fullerene crystalline polymer has the advantage of a temperature-independent intensity, linewidth near 0.2 mT, and resonance position within wide temperature region [269]. It was stated that this material is a suitable standard in high temperature (3–300 K) and frequency (up to 225 GHz) ranges. It has temperature-independent magnetic spin susceptibility and only small variations in the linewidth with temperature and spin precession frequency. The material is inert; it does not degrade under atmospheric conditions. It has a high magnetic susceptibility and can thus be used in low sensitivity apparatus. The spin–lattice relaxation rate is fast at low temperatures; thus, saturation problems are not important in CW EPR study of solid-state samples, especially organic polymers. Its usefulness by measuring the susceptibility of a newly synthesized 2D fulleride polymer, Mg$_{4+x}$C$_{60}$, was demonstrated. Another fullerene-based material, P@C$_{60}$ or N@C$_{60}$ (C$_{60}$ fullerene molecule with encapsulated ^{31}P or ^{14}N atom), embedded in C$_{60}$ crystalline matrix is characterized by an extremely narrow (less than 10 μT) EPR spectrum [270] and may be used as a g-factor standard as well. It demonstrates a characteristic ^{31}P doublet with ($a_{iso} = 4.94$ mT) or ^{14}N triplet ($a_{iso} = 0.5665$ mT) spectrum with no additional fine structure splitting at ambient temperatures due to fast rotation of the molecule. At lower temperatures a rotation is frozen and a spin–lattice relaxation becomes longer, but fine structure splitting remains small. It is not sensitive when exposed to air.

Li:LiF is also characterized by a narrow (less than 0.1 mT) solitary EPR spectrum with $g = 2.002293$ [271] and, therefore, may also be successfully used as EPR standard for g-calibrations at millimeter waveband EPR within, for example, a field range of about 28 mT for the BRUKER E680 94 GHz spectrometer, which is well sufficient for typical organic radicals. There is a very good reproducibility of the field measurements on repetitive removal of the Li:LiF sample and reinsertion into the probe head with a typical deviation of the line position in the range of 10 μT. This allows the use of this standard for calibration before and/or after the measurement of interest as long as the reproducibility of the field sweep is sufficient. The major disadvantage is that line shape depends on a sample quality.

As it is shown in the following, main organic conjugated polymers with embedded nanoadducts are characterized by a high concentration of stable spins, are not sensitive to air, and in some cases may be used as a g-factor standard in multifrequency EPR experiments.

4

Magnetic, Relaxation, and Dynamic Parameters of Charge Carriers in Conjugated Polymers

Lattice constants of some conjugated polymers and their nanocomposites are summarized in Table 4.1. They are used for inter alia calculation of dynamic parameters of spin ensembles stabilized in polycrystalline or low-dimensional solids from Equations 2.51 through 2.56 and Equation 2.71.

4.1 Polyacetylene

Many fundamental properties of polyacetylene are determined by the existence of paramagnetic centers (PC) localized and/or delocalized along the polymer chains [292]; therefore, most studies of these compounds have been performed by electron paramagnetic resonance (EPR) method [104,105,293,294]. In the study of conjugated polymers, the EPR and NMR [205] methods are complementary to one another; however, the former allows one to register directly the relaxation and dynamic parameters of spin charge carriers [204].

Thermodynamically more stable *trans*-PA can be obtained by thermal, chemical, or electrochemical treatment of *cis*-form of PA [68,295]. During the thermal isomerization, the concentration of unpaired electrons largely increases from $\sim10^{18}$ spins/g (or one spin per ~44.000 CH units) in *cis*-PA up to $\sim10^{19}$ spin/g (or one spin per 3.000–7.000 CH units) in *trans*-PA [296]. This is accompanied by the linewidth decrease down to 0.03–0.2 mT [104,105]. The latter value depends on the average length of *trans*-chains and demonstrates a linear dependence on concentration of sp^3 defects, n_d [297]. Besides, $\Delta B_{pp} \propto z^{2.3}$ dependence was obtained [298] for PA doped by ions of metals with the atomic number Z. The linewidth of partially stretch-oriented *trans*-PA was registered to be sensitive to the direction of the external magnetic field B_0. This value is equal to 0.48 mT at $B_0 \| c$ orientation (c is the direction of the stretching) and reduces down to 0.33 mT as B_0 direction turns by 90° with respect to c direction [299]. Such a tendency was reproduced also by Mizoguchi et al. [300–302] and Bartl et al. [297].

TABLE 4.1

Lattice Constants of Some Conjugated Polymers and Their Nanocomposites

Sample[a]	a (nm)	b (nm)	c (nm)	References
Cis-PA	0.745	0.440	0.430	[272,273]
Trans-PA	0.399	0.729	0.251	[272,273]
PPP (monoclinic cell)	0.806	0.555	0.430	[274,275]
PPP (pseudo-orthogonal cell)	0.781	0.553	0.420	[274,275]
PPS	0.867	0.561	1.026	[276]
PP	0.820	0.735	0.682	[277,278]
PPV	0.790	0.605	0.658	[234]
PANI–EB	0.765	0.575	1.020	[230]
PANI:HCA	0.705	0.860	0.950	[230]
PANI:SA (pseudo-orthogonal cell)	0.43	0.59	0.96	[253]
PANI:CSA	0.59	1.0	0.72	[279]
PANI:DBSA (orthorhombic cell)	1.178	1.791	0.716	[280]
PANI:TSA	0.440	0.600	1.10	[281]
POT–ES	0.47	0.64	0.99	[282,283]
PMA	0.715	0.800	1.199	[284]
PMNA	0.867	0.590	1.065	[285]
PTTF–CH_3–C_6H_4	—	0.70	1.90	[286]
PTTF–C_2H_5–C_6H_4	—	0.95	2.40	[286]
PTTF–THA	—	0.60	1.20	[286]
PT	0.365	1.25	0.39	[177]
P3HT (theory)	1.67	1.56	0.38	[287]
P3HT	1.663	0.775	0.777	[288,289]
P3OT (theory)	2.05	1.56	0.38	[287]
P3OT	2.03	0.480	0.785	[290]
P3DDT	2.583	0.775	0.777	[288,289]
PCDTBT	—	1.89	2.09	[291]

[a] *cis*-PA, *cis*-polyacetylene; *trans*-PA, *trans*-polyacetylene; PPP, poly(*p*-phenylene); PPS, poly(*p*-phenylene sulfide); PP, polypyrrole; PPV, poly(*p*-phenylene vinylene); PANI, polyaniline; POT, poly(*o*-toluidine); PMA, poly(*o*-methoxyaniline); PMNA, poly(*m*-nitroaniline); PTTF-C_nH_{2n+1}-C_6H_4, poly(tetrathiafulvalene) with methyl (n = 1) and ethyl (n = 2) substitutes and phenyl bridges; PTTF–THA, poly(tetrathiafulvalene) with tetrahydroanthracene bridges; PT, polythiophene; P3HT, poly(3-hexylthiophene); P3OT, poly(3-octylthiophene); P3DDT, poly(3-dodecylthiophene); PCDTBT, poly[*N*-9″-hepta-decanyl-2,7-carbazole-alt-5,5-(4′,7′-di-2-thienyl-2′,1′,3′-benzothiadiazole)].

EPR spectrum of *trans*-PA may be generally considered as the sum of contributions due to highly mobile solitons, whose number increases as the temperature decreases, and fixed impurities probably related to the presence of traces of catalyst molecules or oxygen after the polymerization process. For most cases, no significant change in *g*-factor of doped (*p*- or *n*-type) *trans*-PA was found [104,105]. Being combined with the monotonic variation of a number of PCs upon doping, this result suggests that there is no drastic change in the nature of unpaired electrons, affecting EPR signal.

It was shown that pure *cis*-PA demonstrates no EPR signal that is in agreement with the Su-Schrieffer-Heeger (SSH) model, which does not predict a soliton-like PC in this isomer. However, real *cis*-PA samples contain short *trans*-PA segments (5%–10%) on the chains' ends [68] where solitons can be pinned [303,304], explaining weak and broad (0.7–0.9 mT) line generally observed in EPR spectrum. PCs in *cis*-PA are characterized by isotropic $g = 2.0026$ and by A-tensor with $A_{xx} = -1.16$, $A_{yy} = -2.32$, and $A_{zz} = -3.46$ mT components (x, y, and z axes are directed along a, b, and c crystallographic axes, respectively) [176]. Numerous investigations of the paramagnetic susceptibility of neutral PA showed [296] that both its conformers demonstrate Curie paramagnetism (when inverted temperature dependence, $\chi \propto T^{-1}$, is realized for paramagnetic susceptibility χ at $T < 300$ K). At the same time, Tomkiewicz et al. [305] have shown that the magnetic susceptibility of *cis*-PA does not follow the Curie law at a temperature region of 4–300 K. The reason of such disagreement of the results is still unclear.

Spin–lattice relaxation time of proton nuclei in pristine and AsF_3-doped *trans*-PA was measured by Nechtschein et al. [306] as a function of the nuclear spin precession frequency ω_p. $T_1 \propto \omega_p^{1/2}$ dependence was found for both *trans*-PA samples, which coincides with a characteristic motion spectrum of Q1D spin diffusion, and Q1D diffusion rate of the neutral soliton was found to be $D_{1D} = 6 \times 10^{14}$ rad/s at room temperature (RT) in neutral *trans*-PA and has square temperature dependence. As the valuations shown [306], the diffusion rate can be increased by more than three orders of magnitude at introduction of different dopants into the polymer. It should be noted, however, that the soliton motion influences indirectly the relaxation time of a nuclear spin. As the interaction of a diffusing proton with an immobilized electron spin is also characterized by frequency dependence $T_1 \propto \omega_p^{1/2}$ [307,308], it can lead to an incorrect interpretation of the results obtained by NMR spectroscopy. So Masin et al. [309] showed on the base of kinetics of ^{13}C NMR signal fading that *trans*-PA has not mobile unpaired electrons at all. Ziliox et al. have shown [310], however, that such a conclusion can be true only for the specific samples investigated by Masin et al. [309].

The spin dynamics in *trans*-PA was studied also by direct CW [300,301,311,312] and spin echo [313–315] EPR methods. It was pointed out that soliton in *trans*-PA moves along and between polymer chains with respective diffusing coefficients D_{1D} and D_{3D}. To realize Q1D motion, the spin diffusion in *trans*-PA should be extremely anisotropic because the soliton cannot hop directly from one chain to another. Nevertheless, in real systems, the soliton can diffuse between the chains with D_{3D} coefficient [220,221]. The analysis of $T_{1,2} \propto \omega_e^{1/2}$ and $D_{1D}(T) \propto T^{-2}$ dependencies obtained by the CW EPR method at $\omega_e = 5$–450 MHz frequency region evidences on diffusive spin Q1D motion in *trans*-PA with RT $D_{1D} \leq 10^{13}$ rad/s and anisotropy $A = D_{1D}/D_{3D}$

varies approximately in 10^5–10^8 range [105,293,311]. The analogous dependence for diffusion rate was found also at higher spin precession frequencies, $\omega_e = 9$–14 GHz [314,316]. However, the detailed spin echo and CW EPR measurements of both *trans*-PA and its deuterated analog *trans*-(DH), carried out by Shiren et al. [313], showed a relatively low Q1D spin diffusion with $D_{1D} \sim 10^{11}$ rad/s. This result differs from the SSH theory, which predicted a higher D_{1D} value [170,306] and with $D_{1D} > 10^{15}$ rad/s experimentally obtained from photoconductometry of pristine *trans*-PA [317]. Some data concerning spin dynamics determined by the NMR as well as by the CW and pulse EPR methods are not compatible [313]. In order to explain the discrepancy between the results obtained with complementary magnetic resonance methods, Holczer et al. [318] suggested the existence of two kinds (diffusive and localized) of PC in *trans*-PA. Wang et al. have supposed from the analysis of EPR study [319] that the interchain spin exchange and therefore electron relaxation and soliton diffusion rate depend strongly on the method of synthesis of *trans*-PA. However, it was not confirmed by Mizoguchi et al. [320].

For description of such charge carrier mobility, some theoretical conceptions based on either the Brownian Q1D diffusion of solitons interacting with lattice phonons [321] or solitons scattering on optical and acoustic phonons of *trans*-PA [322] have been proposed. One of them predicts square temperature dependence of the D_{1D} value. In the second case, $D_{1D}(T) \propto T^{-1/2}$ and $D_{1D}(T) \propto T^{1/2}$ dependencies are predicted for optical and acoustic phonons, respectively. It was calculated [323] that the velocity of soliton 1D diffusion near the Fermi level should not exceed $v_F = 3.8 \times 10^{15}$ s^{-1}.

In contrast to traditional semiconductors, in polyacetylene with different doping levels can be realized various charge transport mechanisms, for example, charge activation to a mobility edge [324], the phonon-induced charge tunneling between soliton states [219,325] and Mott variable range hopping (VRH) [326,327], characterizing of different frequency and temperature dependencies $\sigma_{ac}(\omega_e, T)$. Such a variety of electron transport in this polymer is associated with a formation of the nonlinear soliton-like excitations and can be directly connected with the evolution of both crystalline and electron structures of the polymer system.

Therefore, the data concerning the soliton dynamics in *trans*-PA obtained by the same authors and by different methods are very contradictable and sometimes have no simple interpretation. EPR spectroscopy seems to be a more suitable direct method for the investigation of spin composition and dynamics, which, however, is limited by low spectral resolution and high spin–spin exchange at convenient registration frequencies ω_e. This complicates the separate registration of localized and mobile π-radicals with close magnetic parameters in *trans*-PA [318]. There are some other magnetic methods for the study of PA and other conjugated polymers, for example, the method of electrically detected magnetic resonance [89,328], time-domain EPR [329], and electron nuclear double resonance (ENDOR) [293,330]; however, the increase of registration frequency and magnetic field was proved

[201,331,332] to be the more effective mean for the study of fundamental properties of nonlinear excitations in these compounds.

In order to study the nature and magnetic resonance parameters of PC in PA, a series of *cis-* and *trans-*PA films prepared by a Shirakawa method using a Ziegler–Natta catalyst [333,334] was investigated at wide waveband (9.7–430 GHz) EPR [262].

At X-band EPR, *cis-* and *trans-*PA samples are characterized by a single symmetric line (Figure 4.1), with $g = 2.0026$ and ΔB_{pp} equal to 0.7 and 0.22 mT, respectively. A small EPR linewidth broadening (0.05–0.17 mT) is observed at the freezing point of the polymer down to 77 K. This is probably due to a smaller libration motion of different parts of the polymer chains. A linewidth is observed to increase about 0.1 mT when the sample is exposed to oxygen.

At K-band EPR, an insignificant increase of the linewidth of both the PA samples is observed (see Figures 4.1 and 4.2).

At W-band, *cis-*PA shows ΔB_{pp} increase up to 0.84 mT with a small broadening of the high-field peak, whereas *trans-*PA spectrum demonstrates a line with effective $g = 2.00270$, $\Delta B_{pp} = 0.37$ mT and line asymmetry factor A/B near to 1.1. The change in the shape and symmetry of the PA spectra can be originated due to either anisotropy of g-factor or the PA conductivity growth, and the first was analyzed to be the more evident reason for such spectra transformation.

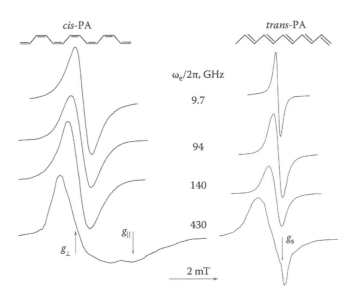

FIGURE 4.1
Electron paramagnetic resonance spectra of *cis-* and *trans-*polyacetylene registered at different electron precession frequencies at room temperature. The g-factors of localized ($g_\perp$ and $g_\parallel$) and mobile (g_s) neutral solitons are shown. (Modified from Krinichnyi, V.I., *Synth. Metals*, 108(3), 173, 2000. With permission.)

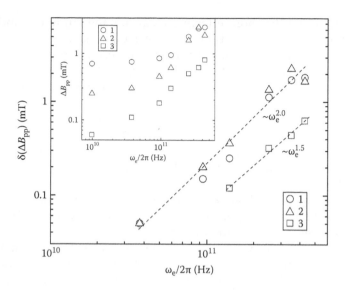

FIGURE 4.2

Logarithmic dependencies of electron paramagnetic resonance linewidth ΔB_{pp} for localized paramagnetic centers in *cis*-(1) and *trans*-(2) polyacetylene samples as well as for delocalized paramagnetic centers in *trans*-polyacetylene (3) versus electron precession frequency at room temperature. The broadening $\delta(\Delta B_{pp})$ of localized (with respect to ΔB_{pp} value defined at 9.7 GHz) and delocalized (with respect to ΔB_{pp} value defined at 94 GHz) paramagnetic centers are also shown as function of ω_e. (Modified from Krinichnyi, V.I., *Synth. Metals*, 108(3), 173, 2000. With permission.)

The increase of the spin precession frequency up to 140 GHz leads to the further manifestation of the g-factor anisotropy of PC in *cis*-PA and the increase of the *trans*-PA line asymmetry up to $A/B = 1.3$. Moreover, at this waveband, ΔB_{pp} of *cis*- and *trans*-PA increases up to 1.1 and 0.5 mT, respectively.

At higher frequencies, the anisotropy of g-factor becomes more evident for PC in *cis*-PA (Figure 4.1). This line shape can obviously be attributed to the localized PC with the following **g** tensor components: $g_\perp = 2.00283$, $g_\parallel = 2.00236$, and isotropic (averaged) $g_{iso} = 1/3(2g_\perp + g_\parallel) = 2.00267$. $g_\perp$ value differs from g_e by $\Delta g = 5 \times 10^{-4}$. In a perturbation theory, such a difference corresponds to an unpaired electron excitation from σ_{c-c} orbital to an antibinding π^* orbital with $\Delta E_{\sigma-\pi^*} = 14$ eV. The latter parameter lies near $\Delta E_{\sigma-\pi^*} = 14.5$ eV calculated for common C–C bond in π-conjugated systems [335]. The other electron transitions with a greater ΔE_{ij} do not influence g-factor. Thus, the spectrum line shape and the agreement between the calculated and measured $\Delta E_{\sigma-\pi^*}$ values support the existence of localized PC with anisotropic g-factor in PA. The line shape of *trans*-PA spectra remains almost unchanged with increasing operation frequency. One can only notice the further increase in both width and asymmetry of the line (Figure 4.1).

The conversion of PA line shape at *cis–trans* isomerization apparently indicates the appearance of mobile solitons in polymer matrix. The proximity of the isotropic *g*-values of the localized and delocalized PCs indicates the averaging of their **g** tensors' main values due to mobility with the rate of [112,336]

$$D_{1D}^0 \geq \frac{(g_\perp - g_e)\mu_B B_0}{\hbar}. \tag{4.1}$$

Thus, two types of PC exist in *trans*-PA, namely, neutral solitons moving along the main polymer axis with $D_{1D}^0 \geq 10^9 \text{rad}/\text{s}$ rate and those pinned on short polymer chains with the relative contribution of 1:18 or 6×10^{-5} (60 ppm) and 1.1×10^{-3} (1100 ppm) spins per carbon atom, respectively. The latter value is two orders of magnitude smaller than that reported by Goldberg et al. [176].

The analysis of the *cis*- and *trans*-PA line shape shows that at $\omega_e/2\pi \geq 140$ GHz, their low-field parts can be described by a Lorentzian function in the center and by a Gaussian one on the wing. At the same time, their high-field parts are Lorentzian. The spin exchange frequency $\nu_{ex} = \omega_{ex}/2\pi$ obtained from the analysis is equal to 3×10^7 and to 1.2×10^8 s^{-1} for the localized PCs in *cis*- and *trans*-PA, respectively. These values are in agreement with spin exchange frequency $\nu_{ex} \geq 10^7$ Hz estimated for *trans*-PA by Holczer et al. [318]. Thus, for the frequencies higher than 16 GHz, the condition $\omega_{ex} < \Delta\omega_{ij}$ holds; hence, spin packets become noninteracting, and ΔB_{pp} value varies according to Equation 2.19.

The dependencies of linewidth and broadening for both localized and delocalized PCs versus electron precession frequency are shown in Figure 4.2. It is seen in Figure 4.2 that the linewidth of PC localized in *cis*- and *trans*-PA changes quadratically with the spin precession frequency, and hence, it can be described by the relationship (2.19). This is an additional evidence for a week interaction of spin packets in these samples. At the same time, the line of mobile PC broadens with frequency as $\Delta B_{pp}^{deloc} \propto \omega_e^{1.5}$. The relation $\Delta B_{pp}^{deloc} \propto (\Delta B_{pp}^{loc})^\alpha$, where $\alpha = 1.3\text{--}1.4$, is valid for both types of PCs in *trans*-PA. It implies that ΔB_{pp}^{delocc} value reflects Q1D spin diffusion in the sample at all electron precession frequencies and changes according to Equation 2.21.

Considering that $\Delta B_{pp}^{deloc} = 1/\left(g_e T_2^{deloc}\right)$ at $\omega_e \to 0$ limit and $T_2(\omega_e \to \infty)/T_2(\omega_e \to 0) = 3.3$ (see Equation 2.52) and by using spin–spin relaxation time for the delocalized PC $T_2^{deloc} = 0.1$ ms at RT [201], it is easy to calculate their zero-field linewidth to be equal to 18 µT. Such a value corresponds to the neutral soliton linewidth (12–38 µT) obtained by Holczer et al. [318].

Figure 4.3a shows the temperature dependencies of normalized spin susceptibility χ of some *cis*-PA samples. These values are well fitted by last two terms of Equation 2.6 with $J_{af} = 0.033\text{--}0.059$ eV for different samples. The analogous consequence of interaction mainly as energy fluctuation of electron polarization of few meV is observed in organic crystal semiconductors [140].

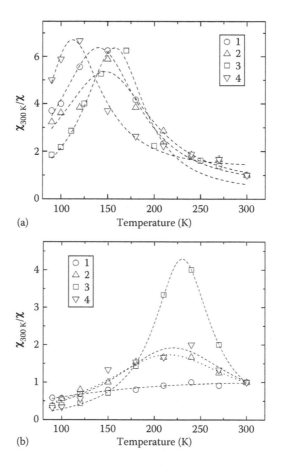

FIGURE 4.3

Temperature dependencies of invert spin susceptibility $\chi^{-1}(T)$ of *cis*-(a) and *trans*-polyacetylene (b) samples with numbers 2 (1), 4 (2), 6 (3), and 5 (4) (see Table 4.2) with respect to those measured at room temperature. The dashed lines show dependencies calculated using Equation 2.6 fitted with appropriate experimental data of *cis*-polyacetylene with J_{af} equal to 0.052 (1), 0.054 (2), 0.059 (3), and 0.033 (4) eV (a) and *trans*-polyacetylene with J_{af} equal to 0.069 (1), 0.068 (2), 0.081 (3), and 0.074 (4) eV (b).

As it is seen from Figure 4.3a, the interaction of the spin magnetic moment dominates in paramagnetic susceptibility only at high temperatures. At temperatures lower than some characteristic temperature $T_c \approx 110$–160 K, the localized spins determine paramagnetic susceptibility, following the Curie law (third term of Equation 2.6). The contributions of these processes of unpaired electrons' magnetic moment orientation are different for the *cis*-PA samples investigated.

The susceptibility of PC in *trans*-PA samples is also characterized by an anomalous temperature dependence (Figure 4.3b). As in the case of *cis*-PA, the main contribution to χ of *trans*-PA at high-temperature region gives last two terms of Equation 2.6. Higher J_{af} values obtained for these samples (0.068–0.081 eV) can be explained by the increase of both rigidity of the polymer chains and by its packing density during *cis–trans* isomerization. It is probably the reason for the critical temperature shift up to $T_c \approx 210$–230 K.

In contrast to *trans*-PA, *cis*-PA is characterized by more sharp primary branch of the $\chi(T)$ curve at $T < T_c$. This effect should be caused by the significant increase in number of neutral solitons with amplitude A leading to the decrease in distance between spins rss and a growth of interaction of these charge carriers with the probability $W_R \propto A\exp(-2Ar_{ss})$ [337]. Besides, the defrosting of soliton Q1D diffusion leads to an additional increase of probability of interaction between solitons, $W_{ss} \propto T^{-2}$ (see Table 4.2). The discrete levels of neighboring solitons localized at midgap are narrowed due to overlapping of unpaired electrons wave functions and are transformed to soliton band of limited weight. As in the case of *cis*-PA, the contribution of terms of Equation 2.6 is determined by different properties and history of *trans*-isomer.

Cis–trans isomerization leads to the increase in concentration and spin–lattice relaxation rate of both types of PCs in *trans*-PA samples. For thick *cis*-PA samples with a smaller packing density, such isomerization becomes easier and yields a greater amount of *trans*-PA isomer with longer and more rigid π-conjugated chains. This is confirmed by the increase in activation energy of chain libration together with the increase of energy transfer from spin reservoir to the lattice [201]. It is important to note that this isomerization leads only to a small change in J_{af}. This can be possibly explained by the fact that *cis–trans* isomerization corresponds not to the noticeable structural changes but only to the increase of π-conjugation and spin–phonon interaction.

The investigation of *trans*-PA samples at D-band EPR has shown [201] that the spectrum shape and the ratio of concentration of mobile and localized solitons do not change with the *trans*-PA doping by iodine vapor when their σ_{dc} value increases up to ~10 S/m. This fact confirms the assumption proposed by Nechtschein et al. [170] on the existence of both mobile and fixed solitons on short conjugated chains, which become charged and diamagnetic under doping. Thus, during *cis–trans* isomerization of pristine PA, the concentration of pinned solitons increases remarkably and mobile PCs appear. This process leads to the increase in dc conductivity by the same orders of magnitude, probably *via* motion of the delocalized PCs. The difference in $\Delta\omega_{ij}$ and ν_{ex} values for the centers of both types leads to a sharp narrowing in its low-frequency EPR spectrum (e.g., by 4–5 times at $\nu_{ex} \approx 10^{10}$ Hz) under the isomerization. This differs from the opinion still existing that such a change of spectrum linewidth occurs only due to highly mobile neutral solitons [71,104,105,306,321].

The decrease in the rate of spin–spin exchange in high fields can be due to the effect of fast passage in both PA isomers (Figure 4.4). It was shown

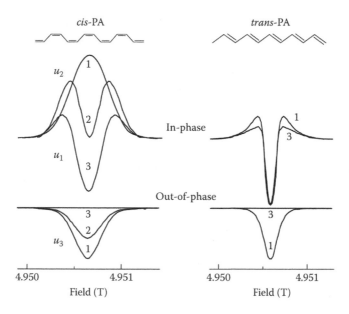

FIGURE 4.4

In-phase and $\pi/2$-out-of-phase components of the first derivative of dispersion signal of *cis*- and *trans*-polyacetylene registered at D-band electron paramagnetic resonance at $B_1 = 20$ µT (1), $20 > B_1 > 1$ µT (2), and $B_1 = 1$ µT (3). (Modified from Krinichnyi, V.I., *Synth. Metals*, 108(3), 173, 2000. With permission.)

[200,201,338] that $\omega_m T_1 > 1$ and $\omega_m T_1 < 1$ hold for *cis*- and *trans*-PA, respectively, so then their dispersion spectrum $U(\omega_e)$ is defined, respectively, by u_2, u_3 and u_1, u_3 terms of Equation 2.35. This enables one to calculate independently T_1 and T_2 values for various PA samples by using respective Equations 2.36 through 2.39.

The temperature dependencies of spin–lattice and spin–spin relaxation times of PCs in some *cis*-PA films synthesized by different methods and *trans*-PA ones obtained by isomerization of appropriate *cis*-PA samples are presented in Table 4.2 [200,201]. One can see that T_1 value of both PA isomers is a function decreasing monotonically with temperature, whereas T_2 value demonstrates the different temperature dependencies. Note that relaxation parameters presented in Table 4.2 for *trans*-PA samples are effective values of localized and mobile PCs. So determining experimentally effective relaxation times and concentrations of localized and mobile PCs, one can estimate separately the relaxation parameters for both types of PCs in *trans*-PA.

The spin–lattice relaxation time may be written as $T_1 = An^{-1}\omega_e^\alpha T^\beta$, where A is a constant, α is equal to 3 and -0.5 for *cis*-PA and *trans*-PA, respectively, and β varies from -1.5 to -3.5 as a function of sample thickness [200,201]. This relation indicates mainly the two-phonon Raman relaxation process in *cis*-PA and the more complicated spin–lattice interaction in *trans*-PA.

TABLE 4.2

Spin–Lattice and Spin–Spin Relaxation Times Obtained as $T_{1,2} = AT^\alpha$ for *cis*-Polyacetylene and *trans*-Polyacetylene Samples of Various Thickness Synthesized by Different Methods and Determined from Their Saturated D-Band Electron Paramagnetic Resonance Spectra by Using Equations 2.36 through 2.39

| No. | *Cis*-Polyacetylene | | | | *Trans*-Polyacetylene | | | |
| | T_1 (s) | | T_2 (s) | | T_1 (s) | | T_2 (s) | |
	A	α	A	α	A	α	A	α
1	4.2×10^{-2}	−1.6	1.8×10^{-9}	1.2	2.7	−2.6	1.0×10^{-7}	0.5
1 [a]	0.37	−2.0	7.7×10^{-9}	1.0	—	—	—	—
2	6.1×10^{-3}	−1.4	1.5×10^{-7}	0.5	0.1	−2.2	7.2×10^{-8}	0.3
2 [a]	0.77	−2.3	1.0×10^{-7}	0.5	—	—	—	—
3	1.4	−2.3	9.5×10^{-8}	0.5	2.0×10^{-3}	−1.7	1.3×10^{-5}	−0.9
3 [a]	290	−3.3	1.7×10^{-8}	0.8	—	—	—	—
3 [b]	52	−2.7	1.2×10^{-8}	0.8	—	—	—	—
3 [a,b]	6.5	−3.6	4.2×10^{-9}	1.0	—	—	—	—
3 [c]	—	—	—	—	62	−3.5	2.1	−3.0
4	0.65	−2.1	9.6×10^{-9}	0.9	4.1×10^{-3}	−1.5	2.9×10^{-6}	−1.0
4 [a]	10	−2.6	2.8×10^{-9}	1.1	—	—	—	—
5	27	−2.5	2.4×10^{-7}	0.3	4.0×10^{-4}	−1.2	9.1×10^{-6}	−0.7
5 [d]	—	—	—	—	8.3×10^{-4}	−1.3	5.0×10^{-6}	−0.6
6	3.1×10^3	−3.5	3.4×10^{-8}	0.7	1.7×10^{-4}	−1.1	1.0×10^{-6}	−0.8
7	1.6×10^3	−2.7	4.2×10^{-9}	1.0	1.1×10^{-2}	−1.9	9.1×10^{-5}	−1.2
8	8.3×10^2	−2.6	9.1×10^{-9}	0.9	2.8×10^{-4}	−1.0	2.2×10^{-5}	−0.7
8 [c]	83	−2.7	3.6×10^{-9}	1.1	—	—	—	—

Source: Data from Krinichnyi, V.I., *Synth. Metals*, 108(3), 173, 2000. With permission.

Notes: The measurements were taken [a] at the air presence, [b] after a half-year storage in an inert atmosphere, [c] after a doping by I_3 vapor, [d] after an annealing in an inert atmosphere. The samples of different thickness were synthesized by different methods.

The dependence mentioned earlier for the latter sample is probably due to the mixture of 1D Raman modulation and 3D spin–lattice interaction of the immobilized spins with total probability $W_R \propto \omega_e^{-2}T^2 + \omega_e^2 T$ [339] and also to the diffusive modulation of spin–lattice interaction by Q1D motion of delocalized centers with the probability $p_d \propto \omega_e^{-1/2}$ [162].

Furthermore, a mobile neutral soliton interacts with N protons of a polymer chain. Therefore, applying $\Delta\omega_G \propto N^{1/2}\rho(n)$ dependence [156] for the Gaussian part of soliton linewidth, one can evaluate the relative width of an unpaired electron delocalization in *trans*-PA, provided that $\Sigma\rho(n) = 1$ and $\rho(n) = N^{-1}$. The analysis of the line shape of mobile solitons also yields the frequency ω_l and activation energy of small-scale librations of polymer chains [201].

A slight doping of *trans*-PA sample by iodine vapor when its conductivity increases up to 10 S/m leads to a four fold reduction in total spin concentration and approximately a 10-fold decrease of spin–lattice relaxation time (see Table 4.2). A smaller change (by a factor of two) in the latter value occurs in the presence of oxygen in PA matrix [200]. Taking also into consideration $T_1 \propto n^{-\alpha}$ concentration dependence, where α varies from 0.7 to 1.0 in 330–90 K temperature range, one can postulate both J_3 and O_2 molecules to be the traps for the delocalized PC and that the introduction of these molecules leads to the decrease of the density of polymer chains packing in PA.

It should be noted that T_1 and T_2 values are the important parameters of PA, characterizing its structural and conductive properties [200,201]. Indeed, half-year storage of as-prepared *cis*-PA under inert atmosphere results in sufficient increase in its T_1 (Figure 4.5) [340]. The analogous change in T_1 is typical for *cis*-PA irradiated by the electron beam with 1 MGy dose. However, T_1 value for this sample irradiated by the electron beam with 0.50–0.75 MGy dose is practically constant during the same period (Figure 4.5). After a longer storage of the initial and 1 MGy *e*-irradiated samples T_1 decreases to some extent, that can be attributed to a partial both the degradation and *cis–trans* isomerization of *cis*-PA. This effect shows the possibility of stabilization and even the improvement of electronic dynamics characteristics of *cis*-PA *e*-irradiated with the optimal dose.

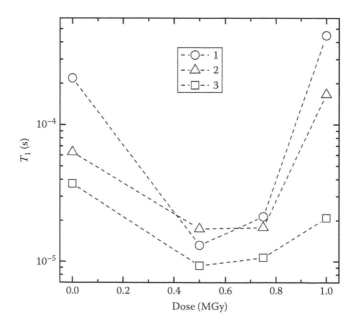

FIGURE 4.5
Dependencies of spin–lattice relaxation time ($T = 120$ K) on *e*-irradiation dose for the initial (3) *cis*-polyacetylene sample and the same sample stored for 6 (1) and 12 (2) months. (Modified from Krinichnyi, V.I., *Synth. Metals*, 108(3), 173, 2000. With permission.)

The relaxation and dynamic processes in an initial and in partially stretch-oriented samples were also studied [341,342]. It was shown that the sp^3 defects and immobile solitons in the initial and stretch-oriented *cis*-PA exhibit weakly axial-symmetric EPR spectra with the aforementioned *g*-factors and linewidth $\Delta B_{pp}^{in} = 1.23$ mT and $\Delta B_{pp}^{str} = 1.45$ mT, respectively, at RT. For the latter one, ΔB_{pp}^{str} value and the other magnetic parameters remain invariable under sample rotation with respect to the direction of an external magnetic field B_0. Under rotation, the linewidth of stretch-oriented *trans*-PA varies nonmonotonically from 0.60 to 0.68 mT at RT.

The temperature dependencies of T_1 and T_2 values for the initial and stretch-oriented *trans*-PA samples are shown in Figure 4.6 as functions of the angle ψ between the external magnetic field B_0 and the stretching directions. Both the $T_1(T)$ and $T_2(T)$ dependencies of *trans*-PA with chaotic chains have a week sensitivity to angle ψ. On the other hand, these values of oriented *trans*-PA are functions changing with ψ. This can be explained by the motion depinning for a part of neutral solitons in *trans*-PA film.

The electron spin relaxation rates, T_1^{-1} and T_2^{-1}, are defined mainly by dipole interaction and partly by HFI between delocalized and fixed spins through Q1D diffusion along macromolecules. Therefore, these values can be determined from Equations 2.51 through 2.58 with appropriate concentrations of localized and mobile PCs per a carbon unit, n_1 and n_2, respectively. In order to compare these data with the soliton theory and with the results already reported [204,316], it was assumed that in *trans*-PA soliton Q1D diffuses along the polymer chain with diffusion coefficient D_{1D} and 3D hops between polymer chains with D_{3D} rate. Since the localized PC predominate in *trans*-PA, the contribution of libration of their chain segments to relaxation mechanism should also be taken into account. The parameters of librations of the chain segments remain almost unchanged at *cis*–*trans* isomerization of PA [201]. In order to neglect these librations, the corresponding relaxation rate of *cis*-PA was subtracted from that of *trans*-PA.

Figure 4.7 displays the temperature dependencies of D_{1D} and D_{3D} values for the *trans*-PA with chaotic and partly stretch-oriented chains. These parameters were calculated by using the lattice constants presented in Table 4.1. Figure 4.7 shows that, as in the case of relaxation times, both the Q1D and Q3D diffusion rates of the soliton are sensitive to the orientation of the polymer chains in an external magnetic field due to Q1D soliton motion. As it is seen from the figure, the phase of $D_{1D}(\psi)$ function for an oriented sample is opposite to that of $D_{3D}(\psi)$ one. Since main *c*-axes of the chains in chaotic *trans*-PA remain arbitrary, these values are averaged over angle ψ. Moreover, the averaged D_{1D} value is well described by the equation $\langle D_{1D} \rangle = D_{1D}^{\perp} \cos^2\psi + D_{1D}^{\parallel} \sin^2\psi$, where $D_{1D}^{\perp}$ and $D_{1D}^{\parallel}$ are the extrema of $D_{1D}(\psi)$ function. Note that a similar function describes an effective spin diffusion in system of low dimensionality [343,344]. Thus, $D_{1D}^{\perp} \gg D_{1D}^{\parallel}$ inequality displays spin delocalization over N_s soliton sites. Taking into account that the soliton

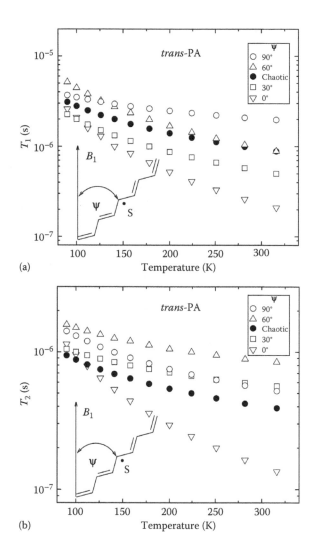

(a)

(b)

FIGURE 4.6
Temperature dependencies of effective spin–lattice T_1 (a) and spin–spin T_2 (b) relaxation times for paramagnetic centers in the initial (3) *trans*-polyacetylene number 3 (see Table 4.2) and that stretch oriented with an orientation degree $A = 0.07$ oriented by lattice *c*-axis with respect to an external magnetic field by ψ equal to 90° (1), 60° (2), 30° (4), and 0° (5). (From Schlick, S.: *Advanced ESR Methods in Polymer Research*. 307–338. 2006. Copyright Wiley-VCH Verlag GmbH & Co. KGaA. Reproduced with permission.)

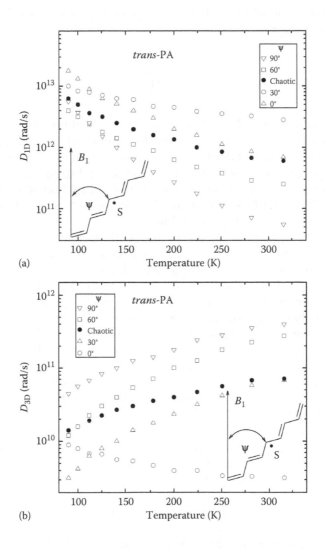

FIGURE 4.7

Temperature dependencies of intrachain D_{1D} (a) and interchain D_{3D} (b) diffusion rates for soliton in the initial (3) *trans*-polyacetylene number 3 (see Table 4.2) and that stretch oriented with an orientation degree $A = 0.07$ oriented by lattice *c*-axis with respect to an external magnetic field by ψ equal to 90° (1), 60° (2), 30° (4), and 0° (5) calculated from Equations 2.51 through 2.54 with $P_0 = 4.3 \times 10^{52} \sin \psi$, $P_1 = 4.8 \times 10^{51}(1 - \cos^4 \psi)$, $P_2 = 4.8 \times 10^{51}(1 + 6\cos^2 \psi + \cos^4 \psi)$, $P_0^{\parallel} = 1.6 \times 10^{52}$, $P_1^{\parallel} = 2.7 \times 10^{51}$, and $P_2^{\parallel} = 1.1 \times 10^{52}$ cm^{-6}, to Equations 2.57 and 2.58 with $Q = 2.34$ mT and $(\langle I_z \rangle - \langle I_0 \rangle)/(\langle S_z - S_0 \rangle) = 0.078$. (From Mizoguchi, K. et al., *Solid State Commun.*, 50(3), 213, 1984; Schlick, S.: *Advanced ESR Methods in Polymer Research.* 307–338. 2006. Copyright Wiley-VCH Verlag GmbH & Co. KGaA. Reproduced with permission.)

hopping is limited by the interchain lattice constant and that the value of the average square of its diffusive hopping step along c-axis is $\langle N_s^2 c^2 \rangle$, the soliton width is easily expressed as

$$N_s^2 = \frac{D_{1D}^{\perp}}{D_{1D}^{\parallel}}. \tag{4.2}$$

Thus, the data obtained [342] confirm the realization of Q1D diffusion motion of solitons in *trans*-PA with the rate significantly exceeding D_{1D}^0 value defined earlier. Besides, it derives from the analysis of the data obtained that relationship (2.19) holds for the linewidth in the whole frequency range used. This relationship characterizes the narrowing of EPR line of a semiconductor, when Q1D spin diffusion motion arises, being the cause of the averaging of **g** tensor components of mobile soliton.

However, the principal evidence for the realization of Q1D spin motion in *trans*-PA is the sensitivity of D_{1D} and D_{3D} values to the orientation of a part of polymer chains in an external magnetic field (Figure 4.7).

The analysis of the data presented in Figure 4.7 gives $N_s(T) \propto T^{1.0}$ temperature dependence for the soliton width and $2N_s = 14.8$ polymer units at RT. This value is in good agreement with the SSH theory [70,71] and also with that derived from magnetic resonance experiments [170]. Extrapolation to the low-temperature range allows one to suppose the lowest $T_0 \approx 60$ K, where the soliton width starts to increase.

It is important to note that at such temperatures, the bend in the experimental $D_{1D}(T)$ function and the difference between the experimental and theoretical results occur [170,313].

The spin dynamics anisotropy D_{1D}/D_{1D} was determined to be almost temperature independent and equal to 30 for chaotic (initial) and to 45 for stretch-oriented *trans*-PA samples [342]. Note that if the intra- and interchain spin diffusion rates of the chaotic *trans*-PA presented in Table 4.2 may be written as $D_{1D} = AT^\alpha$ and $D_{3D} = BT^\beta$, the decrease of α values from -2 to -5 is accompanied by the increase of β value from 0.4 to 7 and by the decrease of RT anisotropy from 10 to 10^4.

Thus, the experimental data evidences that soliton Q1D diffusion in *trans*-PA with the rate, which exceeds significantly minimum rate D_{1D}^0 calculated earlier from Equation 4.1. This conclusion is also confirmed by the averaging of **g** tensor components of mobile PC at fulfillment of Equation 4.1 and by the *trans*-PA EPR spectral line narrowing according to Equation 2.19 at a wide frequency range. However, the more explicit evidence for the soliton Q1D diffusion is the sensitivity of D_{1D} and D_{3D} values to the orientation of a polymer in an external magnetic field. In this case, the dynamics of solitons can be described by the modified well-known Burgers–Korteweg–de Vries equation for Q1D motion of solitary waves in nonlinear medium [337]

$$\frac{\partial u}{\partial t} + \left(D_{1D}^0 + \varepsilon u\right)\frac{\partial u}{\partial x} - \kappa\frac{\partial^2 u}{\partial x^2} + \beta\frac{\partial^3 u}{\partial x^3} = 0, \tag{4.3}$$

where ε, κ, and β are, respectively, the parameters of the medium nonlinearity, dissipation, and *reactive* dispersion. In a dissipative system with a small nonlinearity ($\beta \approx 0$), the formation of a mobile front (kink) with $u_2 - u_1$ bounce is possible. The stationary solution of Equation 4.3 for such a quasiparticle yields

$$u(x,t) = \frac{1}{2}(u_1 + u_2) - A\tanh\left[\frac{A\varepsilon(x - D_{1D}t)}{\kappa}\right], \tag{4.4}$$

where $A = (u_2 - u_1)/2$, $D_{1D} = D_{1D}^0 + \varepsilon(u_1 + u_2)/2$, and $N = \kappa/(2A\varepsilon)$ are the amplitude, velocity, and width of the kink, respectively. If the dissipation of the system is neglected ($\kappa \approx 0$), the other quasiparticles, solitons, can be stabilized in such a system, and for a soliton multitude, Equation 4.3 has another integrated solution

$$u(x,t) = A\operatorname{sech}^2\left[\frac{(x - D_{1D}t)}{N}\right], \tag{4.5}$$

where $D_{1D} = D_{1D}^0 + A\varepsilon/3$ and $N^2 = 3\beta/(A\varepsilon) = \beta/\left(D_{1D} - D_{1D}^0\right)$. The dependencies analogous Equations 4.4 and 4.5 were used for description of non-linear lattice deformations of *trans*-PA and its electronic states [71].

From functional dependencies $D_{1D}(T) \propto T^{-n}$ and $N_s(T) \propto T^{n/2}$ ($n \approx 2$) obtained experimentally [342], one obtained the relation $N_s^2 \propto D_{1D}^{-1}$ for *trans*-PA. It means that if the condition $D_{1D} \ll D_{1D}^F = 2.4 \times 10^{16}$ rad/s holds, the soliton motion in the present sample can indeed be described by Equation 4.3 with a stationary solution shown in Equation 4.5 that is universal for nonlinear systems. Undoubtedly, different *trans*-PA samples can be characterized by different sets of ε, κ, and β constants in Equation 4.3 because the character of a motion of quasiparticles in these polymers can somewhat deviate from that described earlier.

RT *ac* conductivity of the doped *trans*-PA obtained from Equation 2.71 should be near to 0.1–1 S/m even if all solitons were participating in charge transfer, in spite of increasing disorder and coulombic pinning. This value is the same orders of magnitude smaller than that usually achieved for highly doped *trans*-PA. Moreover, the slight doping of *trans*-PA causes a deceleration of Q1D and an acceleration of Q3D spin diffusion [341]. Hence, high conductivity of *trans*-PA cannot be achieved only with intrachain soliton motion.

The temperature dependencies of *ac* ($\omega_e/2\pi = 1.4 \times 10^{11}$ Hz) conductivities, σ_{1D} and σ_{3D}, calculated from Equation 2.71 for one of the slightly doped

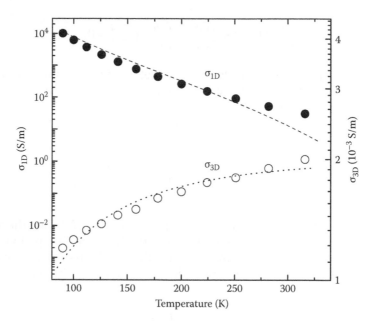

FIGURE 4.8

The temperature dependencies of conductivity σ_{1D} and σ_{3D} due to soliton motion in slightly doped *trans*-PA number 3 (see Table 4.2). The dependencies calculated using Equation 2.73 with $\sigma_0 = 2.8 \times 10^{-11}$ S s/K m, $k_3 = 6.6 \times 10^{14}$ s K$^{14.2}$, and $n = 13.2$ (dashed line) and $\sigma_0 = 7.5 \times 10^{-19}$ S s/K m, $k_3 = 4.1 \times 10^2$ s K$^{14.0}$, and $n = 13.0$ (dotted line) are shown as well. (Modified from Krinichnyi, V.I., *Synth. Metals*, 108(3), 173, 2000. With permission.)

trans-PA are presented in Figure 4.8. Spin dynamics in slightly doped *trans*-PA sample, as in the case of other PA samples [326], should be described in terms of the Kivelson theory [219]. For the comparison, the theoretical functions $\sigma_{ac}(T)$ calculated by using Equation 2.73 with different fitting parameters by using the method described earlier [219] are also shown in Figure 4.8. In fact, both the $\sigma_{1D}(T)$ and $\sigma_{3D}(T)$ dependencies seem to be fitted by Equation 2.73, however, at γ_0 less than theoretical one. This confirms the applicability of the Kivelson theory for charge transport by soliton in *trans*-PA.

Hence, the charge transfer in pristine *trans*-PA can be described in the following manner. In *cis*-PA, the solitons trapped on short-chain segments are the dominated spins. Electron hopping between soliton states is governed mainly by librations of polymer chains, so the probability of such process and hence σ_{dc} are too small. During *cis*–*trans* isomerization of PA, the length of π-conjugated chains increases and the number of PCs grows; however, only a few percents of these charge carriers become mobile. Nevertheless, such soliton depinning causes a drastic increase in probability of the tunneling electron transfer between the soliton states in *trans*-PA and thus in its conductivity up to $\sigma_{dc} \sim 10^{-3}$ S/m. It should be stressed that because the solitons play an auxiliary role, the described charge transport mechanism may be correct, however, only for pristine and slightly doped *trans*-PA. At higher doping levels, the electron hopping between soliton states is unlikely to be the dominant charge transport mechanism, and the conductivity is determined mainly by other parallel processes. Thus, charge transfer mechanism and dynamic processes in PA strongly depend on the structure, conformation, packing density, and length of conjugated chains of this polymer.

PA-like domains can be formed under treatment of some conventional polymers. The treatment of, for example, polyvinyl chloride by fuming sulfuric acid leads to the appearance in its dingy matrix regions [39]. So modified material exhibited at RT a Lorentzian weak EPR singlet with $\Delta B_{pp} = 0.54$ mT and $g = 2.0031$, characteristic of π-electron systems. The linewidth and signal intensity changed weakly, while the temperature decreased down to $T = 77$ K. Magnetic parameters of this material appeared to be close to those of solitons formed in small amount of *trans*-PA usually presented in *cis*-PA [104]. Thus, the change in electronic and paramagnetic properties of the film can be explained by dehydrochlorination during its oleum treatment with the formation of *trans*-polyacetylene regions with unpaired electrons delocalized along neutral solitons, $[-CHCl-CH_2-]_m \rightarrow [-CHCl-CH_2-]_m-[CH=CH \dot{=} CH=CH-]_n$. It was found that so-prepared material becomes a highly sensitive and selective polymer sensor for water molecules. This effect was explained in the following manner. Being weak electron acceptors, water molecules can partially accept the electron density from such solitary charge carriers [104], on the surface or in the bulk of the film, thus providing p-type conductivity of the sample. When the water molecules diffuse into the polymer bulk, they form bridge-type hydrogen bonds between the conjugated chains. As a result, the solitons acquire a positive charge in the *trans*-polyacetylene fragments. Moreover, solitons that may occur in the spreading water associates [345] may also participate in charge transfer between polymer chains. Thus, the macroconductivity of the sample contacting water vapor increases considerably as a result of both intra- and interchain charge transfer. Other molecules are not able to form such associations that determine the selectivity of the sensor to water molecules. Such feature can be used in highly efficient molecular sensors.

4.2 Poly(*bis*-Alkylthioacetylene)

The derivative of *trans*-PA, poly(*bis*-alkylthioacetylene) (PATAC) [346] shown in Figure 1.2, is also an insulator in a neutral form. From the ^{13}C NMR study [347], the conclusion was made that PATAC has sp^2/sp^3-hybridized carbon atom ratio typical for polyacetylene; however, in contrast with the latter, pristine polymer has a more twisted backbone. The *dc* conductivity of PATAC increases under chemical doping from $\sigma_{dc} \approx 10^{-12}$ S/m up to $\sigma_{dc} \approx 10^{-8} - 10^{-2}$ S/m depending on the kind and/or concentration of an anion introduced into the polymer in a liquid or gas phase. The conductivity of PATAC irradiated by argon laser increases up to $\sigma_{dc} \approx 10^3 - 2 \times 10^4$ S/m depending on the absorbed dose [348]. Hall coefficient measurement of the laser-modified PATAC [348] has shown that the charge carriers are of *p*-type and their mobility depends on the temperature as $\mu \propto T^n$ with $0.25 \leq n \leq 0.33$ at $80 \leq T \leq 300$ K. The mobility obtained at RT was close to $\mu = 0.1 - 8$ cm^2/V s that is close to $\mu = 2$ cm^2/V s obtained for *trans*-PA [349].

The X-band EPR study [348] has shown that π-like PCs with different mobility exist in laser-modified PATAC; however, the conductivity of treated polymer is mainly determined by the dynamics of diamagnetic bipolarons. It was proposed that as the temperature increases, the bipolaron mobility decreases and its concentration in laser-modified PATAC increases, so then these processes should lead to the extremely low (close 10^{-3} K^{-1}) temperature coefficient of the PATAC *dc* conductivity. At the optimal polymer treatment by laser, the concentration, *g*-factor, and peak-to-peak linewidth of PCs change at RT, respectively, from $N \approx 2.7 \times 10^{14}$ cm^{-3}, $g = 2.0056$, and $\Delta B_{pp} = 0.72$ mT to $N \approx 4.7 \times 10^{17}$ cm^{-3}, $g = 2.0039$, and $\Delta B_{pp} = 0.65$ mT.

An initial, insulating, powder-like PATAC-1 sample and that irradiated by an argon ion laser at $\lambda = 488$ nm with doses of 5, 20, and 80 J/cm^3, PATAC-2, PATAC-3, and PATAC-4, respectively, were studied at both the X- and D-bands EPR at RT [350,351].

At X-band, the PC in the PATAC-2 sample demonstrates a slightly asymmetric Lorentzian spectrum with a weak component at low fields (Figure 4.9). The intensity of the latter component decreases during the PATAC laser modification and/or as the temperature increases. The spectrum simulation has shown that such line asymmetry arises rather due to the anisotropy of *g*-factor. This supposition was confirmed by a more detailed study of the samples at higher registration frequency. D-band EPR spectra of the PATAC-2–PATAC-4 samples are also presented in Figure 4.9. Higher spectral resolution allowed to show the existing PATAC of two types of PCs: polarons localized on the short π-conjugated polymer chain $P_{loc}^{+\cdot}$ with $g_{xx} = 2.04331$, $g_{yy} = 2.00902$, $g_{zz} = 2.00243$, and linewidth $\Delta B_{pp} = 6.1$ mT (D-band) and polarons moving along the main π-conjugated polymer chain $P_{mob}^{+\cdot}$ with $g_{xx} = 2.00551$, $g_{yy} = 2.00380$, $g_{zz} = 2.00232$ ($2g_{\parallel} = g_{xx} + g_{yy}$, $g_{\perp} = g_{zz}$), and $\Delta B_{pp} = 2.7$ mT (D-band). Simulated spectra of these charge carriers are shown in Figure 4.9 as well. The principal *x*-axis of a polymer chain is chosen parallel to its longest molecular *c*-axis, the *y*-axis lies in the C–C–C plane, and the *z*-axis is perpendicular to *x*- and *y*-axes.

Since the *g*-factor of the PATAC samples is considerably higher than that of PA [352,353] and most poly(*p*-phenylene) (PPP)-like conjugated polymers, $g \cong 2.003$ [104,105,118,122,354], one can conclude that the unpaired electron in

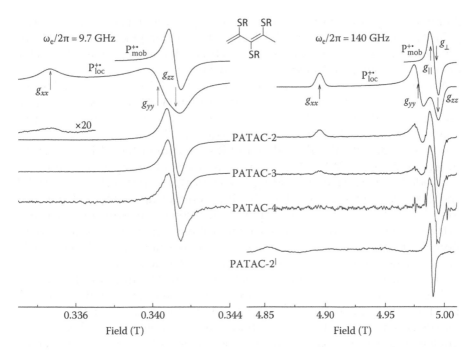

FIGURE 4.9

The absorption X-band and D-band electron paramagnetic resonance spectra of the laser-modified poly(bis-alkylthioacetylene) (PATAC) samples registered at 100 K. At the top, the spectra of radical $P^{+\bullet}_{loc}$ calculated with $g_{xx} = 2.0433$, $g_{yy} = 2.00902$, and $g_{zz} = 2.00243$ and radical $P^{+\bullet}_{mob}$ calculated with $g_{xx} = 2.00551$, $g_{yy} = 2.00380$, and $g_{zz} = 2.00232$ are shown. An initial, PATAC-1 polymer with R = −CH$_3$ is also shown schematically. (From Krinichnyi, V.I. et al., *Appl. Magn. Reson.*, 23, 1, 2002. With permission.)

PATAC interacts with sulfur atoms. This is typical for other sulfur-containing compounds, for example, poly(tetrathiafulvalenes) (PTTF) [118,354–356] and benzotrithioles [357–359], in which sulfur atoms are involved into the conjugation. Taking into account that the overlapping integral I^p_{c-c} in such organic π-systems depends on the torsion (dihedral) angle θ (the angle between p-orbits of neighboring C atoms) as $I^p_{c-c} \propto \cos\theta$ [335,360], the shift of g-factor from the g-factor for the free electron in Equation 2.3 should be multiplied by the factor $\lambda_s(1 - \cos\theta)/(1 + k_1 \cos\theta)$, where $\lambda_s = 0.047$ eV [134] is the spin–orbit interaction constant for sulfur and k_1 is a constant. The g-factor of PC in sulfur-containing solids in which electrons are localized mainly on the sulfur atom lies in the region of $2.014 \leq g_{iso} \leq 2.020$ [357–359,361]. In tetrathiafulvalene (TTF) derivatives, an unpaired electron is delocalized on 12 or more carbon atoms and four sulfur atoms leading to the decrease of both $\rho_s(0)$ and g_{iso} values [118]. An additional fast spin motion takes place in the PTTF [118,122,359] leading to a further decrease in the $\rho_s(0)$ value and therefore in g_{iso} down to 2.007–2.014 depending on the structure and effective polarity in PTTF samples [118,122]. Due to the smaller g-factor in PATAC, one can expect a higher spin delocalization in this polymer as compared with the organic semiconjugated solids mentioned earlier.

Assuming $\Delta E_{n\pi^*} \approx 2.6$ eV, typical for benzotrithioles and PTTF [358,361], then the g-factor components of PC in the initial PATAC were obtained from Equation 2.3 to be $\rho_s(0) \approx 1.1$ and $\Delta E_{n-\sigma^*} = 15.6$ eV. This means that in the initial polymer, the spin is localized within one monomer unit.

Figure 4.9 depicts also the D-band EPR spectrum of the PATAC-2 sample stored for 2 years (PATAC-2ᶦ). From the analysis of the spectrum, one can conclude the increase of g_{xx} and g_{yy} values of $P_{loc}^{+\cdot}$, respectively, up to 2.0451 and 2.00982. The RT linewidth of this PC increases up to $\Delta B_{pp} = 12.9$ mT as well. This means the increase of spin localization on the sulfur nucleus due to the shortness of polymer chains in amorphous regions during the PATAC storage. On the other hand, such destruction does not lead to the change in the magnetic parameters and concentration of mobile polarons.

Supposing the spin delocalization onto approximately five units [75] also in PATAC, then the $\rho_s(0)$ value determined earlier should decrease down to 0.22. This fits very well with the g-factor measured for the PATAC-2 sample. Further spin delocalization during laser modification of the polymer with higher laser irradiation dose can be accompanied by the decrease of the θ value. This should lead to an additional acceleration of spin diffusion along the polymer chains. The $\rho_s(0)$ and θ values calculated from Equation 2.3 for PC on polaron in the sample PATAC-4 are approximately equal to 0.032° and 21.3°, respectively. The latter value is close to the change in θ ($\Delta\theta = 22°–23°$) at transition from benzoid to quinoid form in PPP [362], from emeraldine base to emeraldine salt form of polyaniline (PANI-EB and PANI-ES, respectively) [182], and from PTTF with phenyl bridges to PTTF with tetrahydroanthracene ones [118] (see Section 4.5). This change in the g-factor supports the assumption made by Roth et al. [348] that laser irradiation leads to a more planar structure of the polymer chains and therefore to a higher spin delocalization and to a higher conductivity of PATAC.

The concentration ratio $\left[P_{mob}^{+\cdot} \right] / \left[P_{loc}^{+\cdot} \right] \approx 1:2$, obtained for the insulating sample PATAC-1, increases during its modification by laser up to 2:1, 4:1, and 10:1 for the PATAC-2, PATAC-3, and PATAC-4 samples, respectively. This means that such a treatment of conjugated polymer leads to increase of the concentration of mobile polarons by the maximum factor of 30 in highly irradiated polymer. On the other hand, the concentration of charge carriers determined from Hall and dc conductivity studies changes by more than 15 orders of magnitude reaching $N \sim 10^{19}$ cm^{-3} in a relatively strong laser-irradiated polymer [348]. This means that the charge, as in the case of some other conjugated polymers, is predominantly transferred by paramagnetic polarons in an initial and slightly modified PATAC. The planarity of chains in highly irradiated polymers increases, so then most polarons collapse into diamagnetic bipolarons becoming the dominant charge carriers in the system.

Spin susceptibility χ of PATAC depends not only on the laser irradiation dose absorbed but also on the temperature. Figure 4.10 demonstrates the temperature dependencies for the $1/\chi$ value of the PATAC samples. The χ value decreases as the temperature decreases from maximum down to a critical temperature $T_c \approx 200$–220 K and then starts to increase at lower temperatures. The observed increase of the magnetic susceptibility at temperatures lower than T_c can result from the formation of clusters with collective localized Curie spins. At higher temperatures, as in the case of PANIs [185, 363] and poly(3-dodecylthiophenes) [364–367], the spin susceptibility of PATAC seems also to include a contribution due to singlet–triplet equilibrium described by the last term of Equation 2.6. The data presented in Figure 4.10 evidence that $\chi(T)$ dependencies of the PATAC samples are fitted well by Equation 2.6 with $J_{af} \approx 0.05$–0.07 eV.

It is obvious that the processes mentioned earlier should be accompanied by a reversible collapse of polarons into bipolarons and dissociation of bipolarons into polarons. However, Stafström et al. [368,369] have shown that the bipolaron state is not the favorable state in main conjugated polymers.

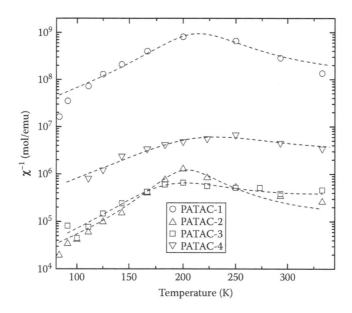

FIGURE 4.10

Temperature dependence of spin susceptibility of the poly(bis-alkylthioacetylene) (PATAC) samples. The appropriate dependencies calculated using Equation 2.6 with $C = 2.4 \times 10^{-6}$ emu K/mol, $c_{af} = 7.2 \times 10^{-4}$ emu K/mol, and $J_{af} = 0.069$ eV (PATAC-1); $C = 4.6 \times 10^{-3}$ emu K/mol, $c_{af} = 0.48$ emu K/mol, and $J_{af} = 0.050$ eV (PATAC-2); $C = 3.7 \times 10^{-3}$ emu K/mol, $c_{af} = 0.26$ emu K/mol, and $J_{af} = 0.042$ eV (PATAC-3); and $C = 2.4 \times 10^{-4}$ emu K/mol, $c_{af} = 2.1 \times 10^{-2}$ emu K/mol, and $J_{af} = 0.054$ eV (PATAC-4) are shown by the dashed lines as well. (Modified from Krinichnyi, V.I., *Synth. Metals*, 108(3), 173, 2000. With permission.)

In order to verify if the observed minimum of the spin susceptibility reflects a structural phase transition in the polymer matrix at T_c, the molecular dynamics in this system was studied. Because the intrinsic spin concentration in the as-prepared PATAC-1 sample is too low, the spin probe EPR method seems more suitable for such investigation of this PATAC sample and its solution in chloroform. Other PATAC samples are not solvable in organic solvents, so they cannot be studied by this method.

Figure 4.11 displays X-band EPR spectra of the stable nitroxide radical introduced into the powderlike PATAC-1 sample and, for comparison, in its solution in chloroform as a spin probe obtained at different temperatures. Such spin probe being introduced into the PATAC bulk is characterized by the isotropic and anisotropic constants of the spin–orbit interaction, $a_{iso} = 1.52$ mT and $A_{zz} = 3.25$ mT, respectively. In its solution, these magnetic parameters of the spin probe appear to increase up to $a_{iso} = 1.59$ mT and $A_{zz} = 3.54$ mT due to the increase of polarity of the radical microenvironment [125,126]. The shape of the spectra evidences that an active fragment

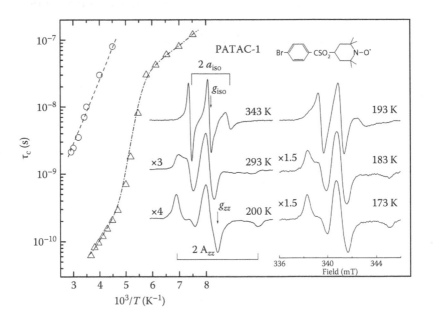

FIGURE 4.11
X-band electron paramagnetic resonance spectra of the nitroxide spin probe in the solid-state unmodified PATAC-1 sample (left) and in its solution in chloroform (right) registered at different temperatures (inset) and the Arrhenius dependence of correlation time τ_c of the spin probe motion in these samples, circle and triangle points, respectively, evaluated in the framework of the model of the nitroxide radical hopping on arbitrary angles. The dependencies calculated using Equation 2.60 with $\tau_c^0 = 3.44 \times 10^{-12}$ s and $E_a = 0.192$ eV (dashed line), $\tau_c^0 = 3.18 \times 10^{-13}$ s and $E_a = 0.124$ eV (high-temperature region), and $\tau_c^0 = 4.03 \times 10^{-10}$ s and $E_a = 0.065$ eV (low-temperature region) (dot-dashed line) are shown as well. The measured magnetic parameters are shown.

of the spin probe moves with a relatively high frequency. This means that the average size of the pockets where the probes are situated in the polymer is considerably large compared with an effective diameter of the nitroxide radical. So we can conclude that the packing order of polymer is relatively low. The analysis of spectra of the nitroxide spin probe has shown that spin probe moves in solid and solved PATAC-1 samples according to the model of thermally activated spin hopping on small arbitrary angles with the correlation time described by Equation 2.60.

Figure 4.11 presents also the Arrhenius dependence for the correlation time of the spin probe motion in both the powderlike PATAC-1 sample and its solution in chloroform. From these dependencies, the activation energies of the probe rotation were calculated to be 0.19 eV for the solid polymer and 0.12 eV (high temperatures) and 0.065 eV (low temperatures) for the solved polymer. In the latter case, the change in the activation energy was attributed to the phase transition in the system at $T \approx 180$ K.

Because the PATAC-1 sample shows no effect in the correlation time due to a phase transition, the results of spin probe studies allow one to conclude that the nonlinear temperature dependence of the spin susceptibility of PATAC samples cannot be explained by a structural phase transition but by a strong ferri- or ferromagnetic interaction in the polymer matrix.

In order to study the spin dynamics in the laser-modified PATAC samples, the temperature dependencies of linewidth ΔB_{pp} of mobile polarons $P_{mob}^{\cdot+}$ in these samples should be analyzed. These dependencies measured at X- and D-band EPR are shown in Figure 4.12. It is seen from the figure that the linewidth measured at X-band EPR is weakly dependent on temperature. However, this value becomes more temperature sensitive at higher spin precession frequency due to the decrease of an interaction between spin packets. The D-band linewidth of the PATAC-1 sample increases as the temperature decreases. At the same time, this value of all laser-modified samples demonstrates the extremal temperature dependence with the critical temperature T_c close to 200–220 K. These functions are similar to $\chi(T)$ presented in Figure 4.10; therefore, such behavior can be associated with the polaron–bipolaron transition at T_c and can probably not reflect the change of the mobility of charge carriers in the laser-treated samples. At the storage, the linewidth of the PATAC-2 sample becomes linearly dependent on the temperature, whereas T_c of the PATAC-3 sample shifts to a lower temperature region (Figure 4.12c). The RT linewidth of the PATAC-2, PATAC-3, and PATAC-4 samples was measured at D-band EPR to be 1.45, 1.34, and 1.31 mT, respectively. So the linewidth of PC $P_{mob}^{\cdot+}$ increases by approximately four times at the transition from the X-band to the D-band mainly due to the effect of spin precession frequency on spin–spin relaxation.

As in the case of other conjugated polymers, saturation effects were registered in dispersion spectra of the PATAC samples (Figure 4.13). The inequality $\omega_m T_1 > 1$ is realized for all PATAC samples; therefore, their in-phase and

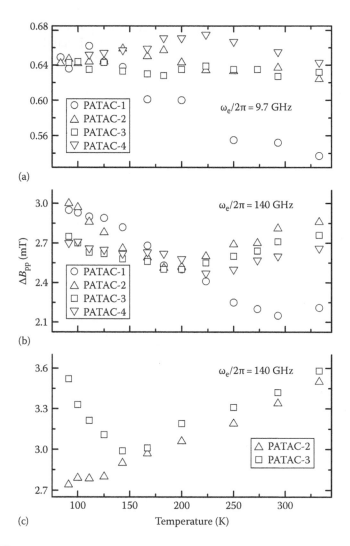

(a)

(b)

(c)

FIGURE 4.12
Temperature dependencies of the linewidth of mobile polarons R_2 in the poly(bis-alkylthio-acetylene) (PATAC) samples determined from their X-band (a) and D-band (b) electron paramagnetic resonance absorption spectra. This value of some PATAC samples stored for 2 years is also shown in (c) as a function of temperature. (From Krinichnyi, V.I. et al., *Appl. Magn. Reson.*, 23, 1, 2002. With permission.)

$\pi/2$-out-of-phase components of the dispersion signal are determined mainly by the last two terms of Equation 2.35. The monomers of the PATAC contain sulfur nuclear stipulating for the anisotropy of g-factor. This allows investigating anisotropic torsion librations of the pinned polarons near the main x-axis of the polymer chains.

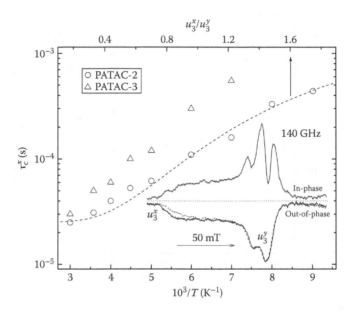

FIGURE 4.13
Arrhenius dependencies for correlation times of super slow librations near the polymer main x-axis of polarons $P_{loc}^{+\cdot}$ localized on polymer chain segments in the PATAC-2 and PATAC-3 samples evaluated from their saturation transfer electron paramagnetic resonance (ST-EPR) spectra. The dependence calculated using Equation 2.65 is shown by the dashed line. Inset: in-phase and $\pi/2$-out-of-phase D-band dispersion spectra of the PATAC-2 sample registered at 280 K (solid lines) and 143 K (dashed line).

Figure 4.13 presents the correlation times of such macromolecular librations in the PATAC-2 and PATAC-3 samples determined from their ST-EPR spectra from Equation 2.65 in the framework of the activation motion. These parameters were determined for these samples to be $\tau_c^x = 6.3 \times 10^{-6} \exp(0.043 \text{ eV}/k_B T)$ and $\tau_c^x = 3.1 \times 10^{-6} \exp(0.062 \text{ eV}/k_B T)$ s, respectively. The increase in the activation energy of the polymer chain librations evidences the strong dependence of the superslow macromolecular dynamics of both the pinned spins and polymer segments on the polymer treatment level. The higher the laser irradiation doses, the more rigid the polymer matrix becomes and the higher the energy for the polymer chain motion required (Figure 4.13).

Figure 4.14 demonstrates the temperature dependence of effective spin–lattice and spin–spin relaxation times of the laser-modified PATAC-2 sample determined from Equations 2.36 and 2.37. Relaxation parameters calculated for the same sample stored for 2 years are presented in Figure 4.14 as well. It is seen from the figure that both relaxation times of polarons initiated in the sample decrease considerably during the storage. Besides, this process leads to the extremal temperature dependence with $T_c \approx 160$ K at which the semiconductor–metal transition occurs. The analogous phenomenon was registered

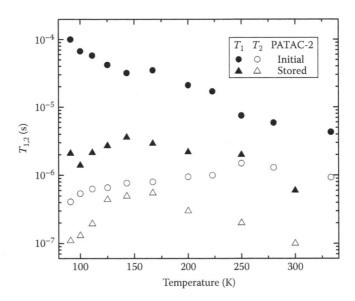

FIGURE 4.14
Temperature dependence of the effective spin–lattice T_1 and spin–spin T_2 relaxation times for as-modified PATAC-2 sample and that stored for 2 years. (From Krinichnyi, V.I. et al., *Appl. Magn. Reson.*, 23, 1, 2002. With permission.)

in the study of PANI [182,185], poly(*p*-phenylene vinylene) [370], poly(3,4-ethylene-dioxy-thiophene) [371], and other conjugated polymers [372].

The involvement of spectral components in the motional exchange and therefore their shift to the spectrum center at the laser modification of PATAC apparently indicates the appearance of polarons moving along the PATAC-2 polymer chains with the rate (see Equation 4.1) $D_{1D}^0 \geq 2 \times 10^{10}$ rad/s. As in the case of other conjugated polymers with heteroatoms [182,356,373], $\Delta B_{pp} \propto T_2^{-1} \propto \omega_e^{-1/2}$ relations are realized for modified PATAC samples (see Figure 4.12); thus, the spin charge dynamics in PATAC can also be determined using the method described earlier.

Figure 4.15 exhibits temperature dependencies of the D_{1D} and D_{3D} values calculated for as-modified and stored for 2 years PATAC-2 sample from Equations 2.45, 2.46, 2.51, and 2.52 under assumption that the spins are situated near sites of cubic lattice. It is seen from Figure 4.15 that the rate of the spin intrachain diffusion in the samples remains invariable as the temperature decreases down to $T_c \cong 200$ K due probably to the compensation effect of the T_1 increase, and the n decrease, however, starts to increase as the temperature decreases further. At the same time, D_{3D} decreases monotonically with the temperature decrease of the as-modified PATAC-2 sample. With sample storage, this value has a weak fall in temperature (Figure 4.15).

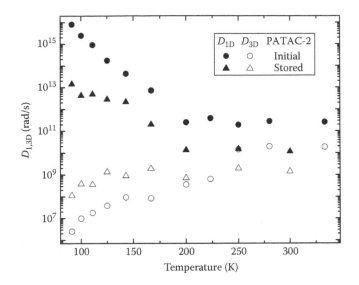

FIGURE 4.15
Temperature dependencies of the intrachain D_{1D} and interchain D_{3D} spin diffusion of mobile polarons in as-modified PATAC-2 sample and that stored for 2 years, which are determined using Equations 2.46, 2.51, and 2.52. (From Krinichnyi, V.I. et al., *Appl. Magn. Reson.*, 23, 1, 2002. With permission.)

Figure 4.16 shows the contribution of both Q1D and Q3D polaron motions into the *ac* conductivity of the as-modified and stored PATAC-2 sample determined from Equation (2.71) with c_{1D} = 0.251 nm for *trans*-PA [273] and $c_{3D} \approx r_0$ = 0.357 nm. Analogously to the spin Q1D diffusion rates, σ_{1D} of these polymers consists of the temperature-independent part at $T \geq T_c$ and temperature-dependent part at $T \leq T_c$. The latter term can be interpreted according the model of the charge carrier scattering on the lattice optical phonons proposed for metal-like clusters in conjugated polymers [245,246]. As can be seen in Figure 4.16, the $\sigma_{1D}(T)$ dependence for the PATAC-2 sample is fairly well fitted by Equation 2.83 with E_{ph} = 0.18 eV. This value is close to that (0.13 eV) obtained for PANI doped by hydrochloric acid [184] but exceeds by a factor 3–4 the activation energy of macromolecular librations in PATAC samples determined from their ST-EPR spectra (see Figure 4.13). The conductivity due to Q1D spin motion in the stored PATAC-2 sample is also fitted by Equation 2.83 with E_{ph} = 0.15 eV. From the σ_0 values obtained, $N = 4.2 \times 10^{18}$ cm^{-3} and $d_{1D} = c = 0.251$ nm for *trans*-PA [273], one gets $\alpha = 7.7 \times 10^9$ eV/cm and $\alpha = 4.1 \times 10^9$ eV/cm for these samples, respectively. The latter values lie close to $\alpha = (4 - 20) \times 10^9$ eV/cm determined for highly doped PANI [184]; however, they are higher than that calculated for *trans*-PA.

As in the case of σ_{1D}, the σ_{3D} conductivity of the PATAC samples in the high-temperature region weakly depends on temperature (Figure 4.16).

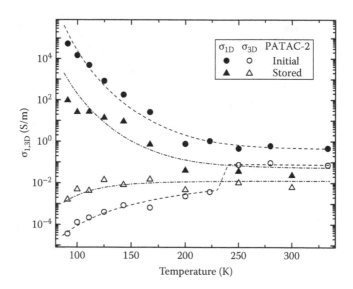

FIGURE 4.16

Temperature dependencies of alternating current (140 GHz) conductivities σ_{1D} and σ_{3D} calculated for the as-modified PATAC-2 sample and that stored for 2 years using Equation 2.71 with $N = 4.2 \times 10^{18}$ cm^{-3}. The dependencies calculated using Equation 2.83 with $\sigma_0 = 2.9 \times 10^{-7}$ S K/m and $E_{ph} = 0.18$ eV (dashed line) and with $\sigma_0 = 8.3 \times 10^{-8}$ S K/m and $E_{ph} = 0.15$ eV (dash-dotted line) are shown. The dependencies calculated using Equation 2.77 with $\sigma_0 = 1.7 \times 10^{-2}$ S K/m and $E_a = 0.051$ eV (dash-dotted line) and with $\sigma_0 = 3.3 \times 10^{-16}$ S s/m K ($T < 230$ K), $\sigma_0 = 5.2 \times 10^{-15}$ S s/m K ($T > 230$ K), and $E_a = 0.055$ eV (dashed line) are shown as well. (From Krinichnyi, V.I. et al., *Appl. Magn. Reson.*, 23, 1, 2002. With permission.)

At low temperatures, the conductivity due to the interchain spin diffusion can be interpreted in the framework of the activation of spin charge carriers to the extended states. Figure 4.16 evidences that σ_{3D} experimental data of the as-modified and stored PATAC-2 sample are fitted well by Equation 2.77 with $E_a = 0.061$ eV and $E_a = 0.051$ eV, respectively. These values are close to the typical activation energy of the interchain spin hopping in some other compounds of low dimensionality. The decrease in E_a evidences the decrease of crystallinity during PATAC storage. The break in the $\sigma_{3D}(T)$ curve could be attributed to a change in the conformation of the system at T_c as it occurs in other conjugated polymers [118,122,354]. The activation energy necessary for the charge carrier hopping between polymer chains is close to those obtained earlier from the PATAC paramagnetic susceptibility and polymer chain librations data. This fact leads to the conclusion about the interference of these processes in the polymer.

From the average RT mobility $\mu \cong 0.5$ cm^2/V s obtained by the Hall study for charge carriers in highly irradiated polymer [348], one can evaluate an intrinsic conductivity of the treated PATAC to be close to 3.4×10^2 S/m. However, $\sigma_{1D} = 0.48$ S/m and $\sigma_{3D} = 8.5 \times 10^{-2}$ S/m obtained from EPR data are much lower. This means that the intrinsic conductivity in highly conjugated PATAC is determined mainly by dynamics of spinless bipolarons with concentration exceeding the number of mobile polarons by approximately two orders of magnitude.

So multifrequency EPR spectroscopy allowed us to postulate the existence of two types of PCs in laser-modified PATAC—polarons moving along polymer chains in crystalline domains and polarons localized on polymer chains in amorphous regions. Assuming that the polaron is covered by electron and excited phonon clouds, one can propose that both spin relaxation and charge transfer should be accompanied by the phonon dispersion. The mobility of the polarons depends strongly on their interaction with other PCs and with the lattice phonons. The charge transfer integral and therefore the intrinsic conductivity of the sample are modulated by macromolecular dynamics, and this dynamics reflects the effective crystallinity of PATAC with metal-like domains. The strong spin–spin interaction at high temperatures leads to the collapse of bipolarons into polaron pairs. The number of bipolarons exceeds the number of polarons at least by two orders of magnitude, so the total conductivity of PATAC is determined mainly by dynamics of diamagnetic charge carriers. Magnetic resonance, relaxation, and dynamic parameters of PATAC are shown to change during its storage that can be explained by degradation of the polymer.

4.3 Poly(p-Phenylene)

PPP and its nanocomposites are considered as suitable material for constructing various molecular devices [374,375]. This is mainly to its simple synthesis and high conductivity. For example, *dc* conductivity of PPP film synthesized electrochemically by direct anodic oxidation of *p*-terphenyl oligomers was reached to be with conductivity of 1.5 S/m [376]. A number of unpaired electrons observed in other simple conjugated polymers, PPP, strongly depends on the technique used for the polymerization and is varied within 10^{17}–10^{19} cm^{-3} [104]. As in the case of *trans*-PA, the RT linewidth of PPP increases at doping with the increase of atomic number of an alkali metal dopant due to a strong interaction between molecule of a dopant and an unpaired electron. Carbonized regions can be formed in PPP under its annealing. Such polymer demonstrates a Korringa spin relaxation mechanism, typical of disordered metals, and a VRH mechanism for carriers localized within weakly conjugated carbonized regions [377]. The charge in PPP is transferred mainly by polarons at low doping level and bipolaron at metallic state [378], similar to that, as it is observed in analogous highly conjugated polymers [74,379]. The lattice constants of PPP are also presented in Table 4.1.

A series of PPP film-like samples of nearly 10 μm thick synthesized electrochemically on a platinum electrode in a BuPyCl–AlCl$_3$ melt were studied at

X- and D-bands EPR: as-prepared and evacuated PPP–Cl$_3^-$ film (PPP-1), the same film after its storage for 4 days (PPP-2), which is exposed for a few seconds to air oxygen (PPP-3); after Cl$_3^-$ dopant removal from the PPP-1 sample (PPP-4); and after BF$_4^-$ redoping of the PPP-1 film (PPP-5) [178,380].

At X-band EPR, PPP-1–PPP-3 samples demonstrate well-pronounced asymmetric single Dysonian spectrum with effective g = 2.0029 (Figure 4.17a). The analogous PPP line shape was also registered at the study of highly lithium-doped PPP [381]. The line asymmetry factor A/B is changed depending on the sample modification. As dopant Cl$_3^-$ is removed, that is, at the transition from PPP-1 film to PPP-4 one, the spectrum mentioned earlier transforms to a two-component one with $g_\perp$ = 2.0034 and $g_\parallel$ = 2.0020 (Figure 4.17b). Such a transition is accompanied by a line broadening and by a drastic decrease in the concentration of spins and charge carriers (Table 4.3). It should be noted that in earlier studies of different PPP samples and other π-conjugated polymers at X-band EPR, there were no such axially symmetric EPR spectra registered [104,379]. With PPP doped by BF$_4^-$ anions (PPP-5), the spectrum shape retains; however, a slight decrease in the concentration of PCs (Table 4.3) and a change in the sign of its dependence on temperature are observed. Using the difference $g_\perp - g_\parallel$ =

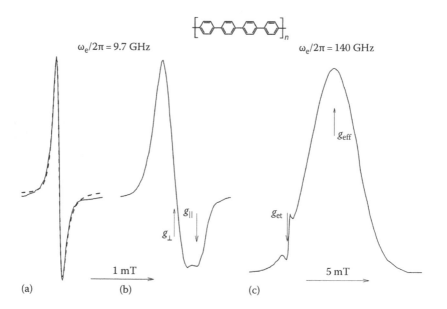

FIGURE 4.17
Typical in-phase modulation X-band electron paramagnetic resonance (EPR) absorption spectrum of PPP-1–PPP-3 (a) and PPP-4–PPP-5 (b) samples. (c) Typical in-phase modulation D-band dispersion spectrum EPR of PPP-4 and PPP-5 films (see the text) registered at 300 K. The spectrum calculated using Equations 2.27, 2.30, and 2.31 with D/A = 0.39 and ΔB_{pp} = 0.109 mT is shown at the left by the dashed line. In the latter spectrum, a narrow line is attributed to the lateral standard, single crystal (DBTTF)$_3$PtBr$_6$ with g = 2.00411. (Modified from Krinichnyi, V.I., *Synth. Metals*, 108(3), 173, 2000. With permission.)

TABLE 4.3

Room Temperature Concentration of Paramagnetic Centers N, Conductivity at Direct σ_{dc} and Alternating $(\omega_e/2\pi = 140\ \text{GHz})\ \sigma_{ac}$ Current, the Line Asymmetry Parameter A/B, Linewidth ΔB_{pp} Calculated from Equations 2.27, 2.30, and 2.31, Spin–Lattice T_1 and Spin–Spin T_2 Relaxation Times, In-Chain Spin Diffusion Rate D_{1D} of Poly(p-Phenylene) Samples of ca. 10 µm Thickness

Parameter	PPP-1	PPP-2	PPP-3	PPP-4	PPP-5
N (cm^{-3})	1.1×10^{19}	3.0×10^{19}	2.0×10^{19}	3.9×10^{17}	5.9×10^{16}
	$(1.9 \times 10^{19})^a$		$(7.1 \times 10^{19})^a$	$(7.7 \times 10^{17})^a$	$(5.8 \times 10^{15})^a$
σ_{dc} (S/m)	10^3–10^4	—	—	10^{-6}	1.0
σ_{ac} (S/m)	1.5×10^5	1.8×10^5	5.3×10^4	—	—
	$(1.1 \times 10^5)^a$				
A/B	$2.3\ (1.9)^a$	2.5	1.4	—	—
ΔB_{pp} (mT)	0.092 (0.182)	0.12	0.22	0.37^b	0.47^b
T_1 (s)	4.1×10^{-7}	5.2×10^{-7}	2.1×10^{-7}	10^{-4}	10^{-4}
T_2 (s)	5.2×10^{-8}	4.0×10^{-8}	2.2×10^{-8}	1.3×10^{-8}	1.0×10^{-8}
D_{1D} (rad/s)	3.4×10^4	3.9×10^5	2.7×10^4	3.4×10^8	8.1×10^5

Source: Data from Krinichnyi, V.I., *Synth. Metals*, 108(3), 173, 2000. With permission.
[a] Measured at 77 K.
[b] The width of high-field spectral component.

1.4×10^{-3}, the minimum excitation energy of unpaired electron $\Delta E_{\sigma\pi^*}$ is calculated from Equation 2.3 to be equal to 5.2 eV. This energy is close to the first ionization potential of polycyclic aromatic hydrocarbons [335]. Consequently, PC can be localized in PPP-4 and PPP-5 samples near cross-linkages that appear as polycyclic hydrocarbons as it was predicted earlier [382].

At D-band EPR, the second bell-like component of Equation 2.35 with $g_{eff} = 2.00319$ attributed to the manifestation of the fast passage of inhomogeneously broadened line is registered in EPR spectrum of the two latter films (Figure 4.17c). This effect allows us to evaluate spin–spin relaxation time, $T_1 \approx 10^{-4}$ s, for PCs in dedoped and redoped samples.

From the analysis of the X- and D-band EPR line shape, the rate of spin packet exchange in neutral and redoped films, $\nu_{ex} = 4 \times 10^7$ s^{-1}, was estimated [380]. This value increases up to 1.8×10^8 s^{-1} for doped film due to the increase in PC concentration and mobility. Isotropic g-factor, $g_{iso} = 1/3$ $(g_{\parallel} + 2g_{\perp})$, of a neutral sample is equal to g-factor of Cl_3^--doped one. This fact shows that **g** tensor components of PCs are averaged due to their Q1D spin diffusion with the rate (see Equation 4.1) $D_{1D}^0 \geq 4.3 \times 10^7$ rad/s in doped PPP(Cl_3^-) sample. However, the RT rates of Q1D spin diffusion in the samples calculated from relaxation times are considerably smaller (Table 4.3) due possibly to low ordering of polymer matrixes. The conductivities of the samples calculated from Equations 2.27, 2.30, and 2.31 are also presented in Table 4.3. The temperature dependence of σ_{ac} for PPP-1 film is determined from $A/B(T)$ one to be $\sigma_{ac} \propto T^{0.23}$. Such a dependence seems to indicate the existence of some conductivity mechanisms in doped PPP sample, namely, Mott VRH [239] and isoenergetic tunneling [383] of charge carriers. A strong decrease of PC concentration, the rates of spin exchange, and spin–spin relaxation processes at the sample dedoping (at transition from PPP-1 to PPP-4) are the evidence for the collapse of most polarons into bipolarons, whose Q1D diffusion causes electron relaxation of the whole spin system.

The intrachain spin diffusion coefficients D_{1D} calculated from Equations 2.45, 2.46, and 2.51 as well as conductivity σ_{1D} due to such spin diffusion calculated from Equation 2.71 are presented in Table 4.3 as well. It is seen that the values determined seem too small for the appearance of a Dysonian component in the PPP spectra.

Thus, EPR data allow the conclusion that in highly doped PPP-1, the charge is transferred mainly by mobile bipolarons and only small part of mobile polarons take part in this process. In PPP-5, these polarons also collapse into spinless bipolarons being the predominant charge carriers in this conjugated polymer.

At the electrochemical substitution of Cl_3^- anion by BF_4^- one, the location of the latter may differ from that of the dopant in the initial PPP sample. However, the morphology of the BF_4^--redoped PPP may be close to that of a neutral PPP-4 film.

As it was established by Goldenberg et al. [178,380], the PPP film synthesized in the BuPyCl-AlCl$_3$ melt is characterized by smaller number of benzoid monomers and by greater number of quinoid units. It leads to a more ordered structure and planar conformation of the polymer that in turn prevents the collapse of the spin charge carriers to bipolaron in highly doped polymer. The anions are removed at dedoping, so then the packing density of the polymer chains grows. It can prevent an intrafibrillar implantation of BF_4^- anions and lead to the localization of dopant molecules in the intrafibrillar free volume of the polymer matrix. The change of charge transfer mechanism at the replace of dopants in PPP should be a result of such a conformational transition.

4.4 Polypyrrole

Polypyrrole is another conjugated polymer widely investigated as perspective molecular system for plastic electronics [384]. The degree of local order varies for polypyrrole depending upon the preparation method, with the degree of crystallinity varying from nearly completely disordered up to ~50% crystalline [385]. In contrast to PANI, the local order in the disordered regions of PP does not resemble that in the ordered regions.

Neutral PP exhibits a complex X-band EPR spectrum with a superposition of a narrow (0.04 mT) and a wide (0.28 mT) lines with $g \approx 2.0026$ [104,386], typical for radicals in polyene and aromatic π-systems [384]. The intensities of both lines correspond to one spin per a few hundred monomer units. Paramagnetic susceptibility of the narrow line is thermally activated, while χ parameter of the broad one has a Curie behavior. These features, together with the temperature dependencies of the linewidth, were interpreted in terms of coexistence of two types of PCs with different relaxation parameters in neutral PP. Doped PP sample exhibits only a strong and narrow (~0.03 mT) X-band EPR spectrum with $g = 2.0028$, which follows the Curie law from 300 to 30 K [104]. However, it was revealed that magnetic susceptibility of PC in doped PP is characterized by a Curie-like susceptibility and a weak Pauli contribution; however, the latter does not contribute to the conductivity mechanism [387,388]. It was obtained that the charge in PP is transferred according to the VRH model [389]. The linewidth of PC stabilized in partly stretch-oriented PP was observed [390] to be the functions of temperature and the angle between the static magnetic field and the stretched direction. The linewidth of PC becomes higher as the oxygen molecules penetrate into PP bulk due to the exchange and dipole–dipole interactions of polarons with oxygen biradicals [391]. Spin relaxation of PP was shown [388,392] to depend on its doping level in relation to several physical properties such as the ratio of the Pauli susceptibility to the total one, the metallic behavior in the conductivity, and the polaron band observed in the optical spectrum. Most of the preceding results imply that EPR signal does not arise from the same species, which carries the charges, because of the absence of correlations of the susceptibility with concentration of charge carriers and the linewidth with the carrier mobility. This was interpreted in favor of the spinless bipolaron formation upon PP doping [393,394]. Thus, EPR signal of doped PP is attributed mainly to neutral radicals and therefore reports little about the intrinsic conjugated processes.

In this case, the method of spin probe described earlier seems to be more effective for the study of structural and electronic properties of PP sample. Only a few papers reported the study of conjugated polymers by using spin label and probe at X-band [210,395–397]. It is explained by the fact that main conjugated polymers in conjugated state are insoluble in convenient solvents that prevent an introduction of a stable radical into their bulk. A low spectral resolution at low-frequency wavebands did not allow the registration of all components of **g** and **A** tensors and therefore the separate determination of the magnetic susceptibility of both spin label and PC on the polymer chain and the measurement of the dipole–dipole interaction between different PCs. PP modified during electrochemical synthesis by nitroxide doping anion, as a label covalently joined to the pyrrole cycle, was studied by Winter et al. [210]. However, in spite of a high concentration of a spin probe introduced into PP, effective X-band EPR spectrum of such sample did not contain lines of the probe.

D-band EPR method of spin probe becomes more effective at investigation of PP synthesized electrochemically on a platinum electrode in an aqueous solution of 0.2 M pyrrole and 0.02 M 2,2,6,6-tetramethyl-1-oxypiperid-4-ylacetic acid [398].

The absorption X- and D-band EPR spectra of a nitroxide radical, introduced as a probe and a counterion into PP and as a probe into a frozen

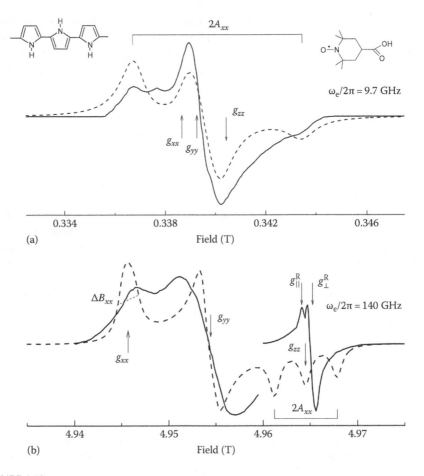

FIGURE 4.18

The X-band (a) and D-band (b) absorption spectra electron paramagnetic resonance of 4-carboxy-2,2,6,6-tetramethyl-1-oxypiperidiloxyl nitroxide radical introduced into frozen (120 K) toluene (dotted line) and conductive polypyrrole (solid line) as a spin probe. The anisotropic spectrum of localized paramagnetic centers marked by the symbol R and taken at a smaller amplification is also shown in the lower part of the figure. The measured magnetic parameters of the probe and radical R are shown. (From Schlick, S.: *Advanced ESR Methods in Polymer Research*. 307–338. 2006. Copyright Wiley-VCH Verlag GmbH & Co. KGaA. Reproduced with permission.)

nonpolar model system, are shown in Figure 4.18. One can see that at X-band EPR, the lines of nitroxide radical rotating with correlation time $\tau_c > 10^{-7}$ Hz overlap with the single line of PC (R) stabilized in PP (Figure 4.18a). Such an overlapping stipulated by a low spectral resolution hinders the separate determination of magnetic resonance parameters of the probe and radical R in PP together with the dipole–dipole broadening of its spectral components.

As it was expected, the spectra of both model and modified polymer systems become more informative at D-band EPR (Figure 4.18b). At this waveband, all canonic components of EPR spectra of the probe in PP and toluene are completely resolved so then all the values of **g** and **A** tensors can be measured directly. Nevertheless, the asymmetric spectrum of radicals R with magnetic parameters $g_\parallel^R = 2.00380$, $g_\perp^R = 2.00235$, and $\Delta B_{pp} = 0.57$ mT is registered on the z-component of the probe spectrum. In nonpolar toluene, the probe is characterized by the following magnetic resonance parameters: $g_{xx} = 2.00987$, $g_{yy} = 2.00637$, $g_{zz} = 2.00233$, $A_{xx} = A_{yy} = 0.60$ mT, and $A_{zz} = 3.31$ mT. The difference $\Delta g = g_\parallel^R - g_\perp^R = 1.45 \times 10^{-3}$ corresponds to an excited electron configuration in R with $\Delta E_{\sigma\pi^*} = 5.1$ eV lying near to an energy of electron excitation in neutral PPP. In conjugated PP, the g_{xx} value of the probe decreases down to 2.00906 and the broadening of its x- and y-components, $\delta(\Delta B_{pp})$, is 4 mT (Figure 4.18b). In addition, the shape of the probe spectrum shows the localization of PC R on the polymer pocket of 1 nm size, that is, the charge is transferred by spinless bipolarons in PP, as it was proposed in the case of PPP:BF_4 and PT:BF_4.

The fragments with a considerable dipole moment are *a priori* absent in neutral PP. Besides, the dipole–dipole interactions between the radicals can be neglected due to low concentration of the probe and PC localized on the chain. Therefore, the change mentioned earlier in the probe magnetic resonance parameters taking place at transition from model nonpolar system to the conjugated polymer matrix may be caused by coulombic interaction of the probe active fragment with the extended spinless bipolarons, each carrying double elemental charge. The effective electric dipole moment of such charge carriers moving near the probe was determined from the shift of g_{xx} component to be equal to $\mu_v = 2.3$ D. The shift of **g** tensor component g_{xx} of the probe may be calculated within the frames of the electrostatic interaction of the probe and bipolaron dipoles by using the following approach.

The potential of electric field induced by bipolaron in the place of the probe localization is determined as [399]

$$E_d = \frac{k_B T (x \coth x - 1)}{\mu_u}, \tag{4.6}$$

where
$x = 2\mu_u \mu_v (\pi \varepsilon \varepsilon_0 k_B T r^3)^{-1}$
μ_u is the dipole moment of the probe
ε and ε_0 are the dielectric constants for PP and vacuum, respectively
r is the distance between an active fragment of the radical and bipolaron

By using the dependence of the growth of an isotropic hyperfine constant of the probe under microenvironment electrostatic field, $\Delta a = 7.3 e r_{NO} t_{cc}^{-1}$ (here r_{NO} is the distance between N and O atoms of the probe active fragment, t_{cc} is the resonant overlapping integral of C=C bond) and the relation $dg_{xx}/dA_{zz} = 2.3 \times 10^{-2}$ mT^{-1} for hexamerous unit ring nitroxide radical [125,126], one can write $\Delta g_{xx} = 6 \times 10^{-3} e r_{NO} k_B T (x \coth x - 1)/(t_{cc} \mu_u)$. By using $\mu_u = 2.7$ D [400], $\mu_v = 2.3$ D, and $r_{NO} = 0.13$ nm [135,136], the value of $r = 0.92$ nm is obtained.

The rate of spin–spin relaxation that stipulates the radical spectrum broadening $T_2^{-1} = T_{2(D)}^{-1} + T_{2(0)}^{-1}$ consists of the relaxation rate of the radical noninteracting with the environment $T_{2(0)}^{-1}$ and the growth in the relaxation rate due to dipole–dipole interactions $T_{2(D)}^{-1} = \gamma_e \delta(\Delta B_{x,y})$. The characteristic time τ_c of such an interaction can be calculated from the broadening of the spectral lines using Equation 2.52 with $J(\omega_e) = 2\tau_c / (1 + \omega_e^2 \tau_c^2)$. The inequality $\omega_e \tau_c \gg 1$ is valid for most condensed systems of high viscosity, so then averaging the lattice sum over angles, $\sum \sum (1 - 3\cos^2 \vartheta)^2 r_1^{-3} r_2^{-3} = 6.8 r^{-6}$ [156], and using $\gamma_e \delta(\Delta B_{pp}) = 7 \times 10^8$ s^{-1} and $r = 0.92$ nm calculated earlier, one can determine $T_{2(D)}^{-1} = 3\langle \omega^2 \rangle \tau_c$ or $\tau_c = 8.1 \times 10^{-11}$ s. This value lies near to the polaron interchain hopping time, $\tau_{3D} \cong 1.1 \times 10^{-10}$ s, estimated for lightly doped PP [388]. Taking into account that the average time between the translating jumps of charge carriers is defined by the diffusion coefficient D and by the average jump distance equal to a product of lattice constant d_{1D} on half-width of charge carrier $N_p/2$, $\tau_c = 1.5\langle d_{1D}^2 N_p^2 \rangle / D$, and by using then $D = 5 \times 10^{-7}$ m^2/s typical for conjugated polymers, one can determine $\langle d_{1D} N_p \rangle = 3$ nm equal approximately to four pyrrole rings. This value lies near to a width of the polaron in both polypyrrole and PANI but, however, is smaller considerably than N_p obtained for polydithiophene [75].

Thus, the shape of the probe spectrum reports about a very slow motion of the probe due probably to an enough high pack density of polymer chains in PP. The interaction between spinless charge carriers with an active fragment of the probe results in the redistribution of the spin density between N and O nuclei in the probe and therefore in the change of its magnetic resonance parameters. This makes it possible to determine the distance between the radical and the chain along which the charge is transferred together with a typical bipolaron length in doped conjugated polymers. The method allows also to evaluate characteristic size of a cavity in which the probe is localized and, therefore, the morphology of the sample under study.

4.5 Poly(tetrathiafulvalene)

In recent decades, the electron donor TTF and its derivatives have been a subject of chemical and physical studies, due to the fact that many compounds of this group can form electrically conjugated charge transfer salts [1]. In order to design conjugated TTF-based polymers more suitable for molecular electronics, different PTTF samples were synthesized in which TTF units with hydrogen, methyl, and ethyl side substitutes R are linked *via* phenyl bridges (PTTF–R–C$_6$H$_4$) and *via* tetrahydroanthracene bridges (PTTF–THA) (Figure 1.2). PTTF–R–C$_6$H$_4$ and PTTF–THA were synthesized from the corresponding 4,4$^|$-*p*-phenylene(di-1,3-dithiolium-perchlorates) by deprotonation with tertiary aliphatic amines [401,402] and 4,5,10,11-tetrahydroantra[1,2-*d*;5,6-*d*$^|$]

bis(1,3-dithioliumperchlorate) [403], respectively. These conjugated polymers were shown [404] to be suitable, for example, for design of chemical sensors.

Iodine-doped PTTF is a *p*-semiconductor with highest *dc* conductivity on the order of $\sigma_{dc} \approx 0.1 - 0.01$ S/m depending on the structure of monomer unit [405]. The temperature dependence of *dc* conductivity was obtained to be explained in terms of a VRH and a thermally activated hopping at low- and high-temperature ranges, respectively [406]. EPR and Mössbauer measurements of doped PTTF indicate a polaron–bipolaron charge transfer mechanism depending on the doping level and the temperature [355,407,408].

PTTF–CH$_3$–C$_6$H$_4$, PTTF–C$_2$H$_5$–C$_6$H$_4$, and PTTF–THA samples were studied at X-band EPR [355,405]. EPR spectrum of the simplest PTTF–H–C$_6$H$_4$ sample was analyzed to be a superposition of a strongly asymmetric spectrum of immobilized PC with $g_{xx} = 2.0147$, $g_{yy} = 2.0067$, and $g_{zz} = 2.0028$ and a symmetric spectrum caused by mobile polarons with $g = 2.0071$. A relatively high value of **g** tensor evidences for the interaction of an unpaired electron with sulfur atom having large spin–orbit coupling constant. Roth et al. have shown [355,405] that the spin–lattice relaxation time T_1 of an undoped PTTF–C$_2$H$_5$–C$_6$H$_4$ sample depends on the temperature as $T_1 \propto T^{-\alpha}$ with α equal to two within $100 < T < 150$ K temperature range and to one at higher temperatures. The addition of a dopant causes the change of a line shape of the PTTF–C$_2$H$_5$–C$_6$H$_4$ sample due to the appearance of a larger number of mobile PCs. Such a change in the magnetic and relaxation parameters was attributed to the conversion of the bipolarons into the paramagnetic polarons induced by doping by nanoadducts and/or temperature increase. In neutral and slightly doped polymers, the charge is transferred by small polarons [408]. However, it is difficult to carry out at X-band EPR the detailed investigation of doped PTTF samples with low concentration of the immobilized PC for the further analysis of spin effect in electron relaxation and dynamics.

The nature, composition, and dynamics of PC in initial and iodine-doped PTTF samples mentioned earlier were studied by multifrequency EPR method more completely [118,373,409].

Multifrequency EPR spectra of PTTF–CH$_3$–C$_6$H$_4$ sample with different I-doping levels are presented in Figure 4.19. They allow one to determine more correctly all terms of the anisotropic **g** tensor and to separate the lines attributed to polarons with different mobility. Computer simulation shows that the anisotropic EPR spectrum of, for example, PTTF–CH$_3$–C$_6$H$_4$ consists of localized PC $P_{loc}^{+\cdot}$ with slowly temperature-dependent magnetic parameters $g_{xx} = 2.01189$, $g_{yy} = 2.00544$, and $g_{zz} = 2.00185$ and more mobile PCs $P_{mob}^{+\cdot}$ with $g_{xx} = 2.00928$, $g_{yy} = 2.00632$, and $g_{zz} = 2.00210$. The analogous spectrum of $P_{loc}^{+\cdot}$ in PTTF–C$_2$H$_5$–C$_6$H$_4$ is characterized by the magnetic parameters $g_{xx} = 2.01424$, $g_{yy} = 2.00651$, and $g_{zz} = 2.00235$, whereas PCs with nearly symmetric spectrum are registered at $g^P = 2.00706$. The canonic components of **g** tensor of PC localized in PTTF–THA are $g_{xx} = 2.01292$, $g_{yy} = 2.00620$, and $g_{zz} = 2.00251$, whereas more mobile PCs with weakly asymmetric spectrum

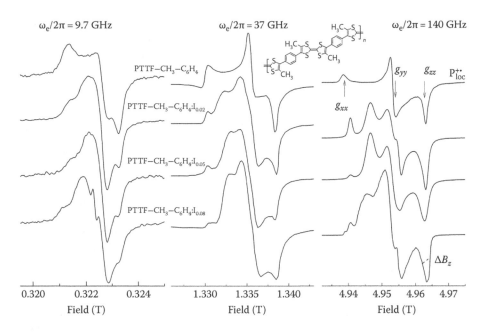

FIGURE 4.19

Typical in-phase X-band, K-band, and D-band absorption spectra electron paramagnetic resonance of the initial and I-doped PTTF–CH$_3$–C$_6$H$_4$ samples registered at room temperature. The measured magnetic parameters are shown.

are characterized by the parameters $g_{||}^P = 2.00961$ and $g_{\perp}^P = 2.00585$. These g-factors exceed the corresponding magnetic parameters of the polarons in PATAC (see Section 4.2) evidencing of the larger interaction of an unpaired electron with sulfur nuclear in PTTF. The ratio of concentrations of the localized and mobile PCs is 20:1 in neutral PTTF–CH$_3$–C$_6$H$_4$, 1:1.8 in PTTF–C$_2$H$_5$–C$_6$H$_4$, and 3:1 in neutral PTTF–THA.

As g^P value is close to the average g-factor of immobilized polarons, two types of PCs with approximately equal magnetic parameters exist in PTTF, namely, polarons moving along the polymer main axis with minimum rate (see Equation 4.1) of $D_{1D}^0 \geq 3 \times 10^{10}$ rad/s and polarons pinned on traps and/or on short polymer chains. The comparatively large iodine ions soften the polymer matrix at the doping of a polymer, so then the mobility of its chains increases. It seems just a reason for the growth in a polymer of a number of delocalized polarons (Figure 4.19). The main terms of **g** tensor of some PC in PTTF-C$_2$H$_5$-C$_6$H$_4$ are averaged completely due to their mobility, whereas such an averaging takes place only partially in the case of other PTTF samples. This fact can be explained by a different structure and conformation of the polymers' matrix.

So the high spectral resolution at D-band EPR allows to determine separately all components of spectra of the polarons with different mobility and then to analyze their temperature dependence. Figure 4.20 shows the linewidth of the localized and mobile polarons changes in the PTTF–CH$_3$–C$_6$H$_4$ as a function of I-doping level y and temperature. It is seen from the figure that the polymer doping leads to the decrease in the RT linewidth of immobilized polarons and to the increase of this parameter of delocalized polarons. The linewidth of the localized polarons monotonically increases with the temperature decrease in whole temperature region (Figure 4.20). On the other hand, the linewidth of the mobile polarons in doped samples first decreases at the temperature decrease down to $T_c \sim 230$ K and then starts to increase at the further polymer cooling as in the case of localized polarons. The decrease of the spectrum linewidth of the mobile polarons in PTTF samples with the temperature decrease down to T_c indicates the mobility of polarons to become more intensive. Such a change in ΔB_{pp}^{mob} is analogous to the line narrowing of spin charge carriers in classic metals. Further broadening of the mobile polaron line at $T \leq T_c$ should be explained by a localization of part of the P$_{mob}^{+}$ PC. In order to verify this supposition, an additional ^{1}H NMR study of the PTTF–CH$_3$–C$_6$H$_4$ samples was carried out.

Figure 4.20b demonstrates typical ^{1}H NMR spectrum of slightly doped PTTF–CH$_3$–C$_6$H$_4$ samples registered at RT which consists of two narrow and one broad lines attributed to protons of different CH$_3$ groups in insulating and conjugated phase of the samples, respectively. The linewidth of these nucleus change with the temperature as it is shown in the figure. The linewidth of both types of protons in an insulating phase does not change visibly at $T \geq T_c \approx 240$ K and starts to increase as the temperature of the sample becomes lower than T_c. At the same time, the line of protons in conjugated polymer phase first narrows down at the temperature decrease down to T_c and, as in the case of the protons in insulating polymer phase, broadens at lower temperatures (Figure 4.20). This fact confirms additionally the supposition mentioned earlier made on the existence of polarons with different mobility in crystalline and amorphous polymer phases and on the localization of part of mobile polarons at temperatures lower than T_c. Polaron motion along the polymer chain induces not only an interaction between electron spins but also an interaction of electron and proton spins expressed by the last term of Equation 2.1. Such interactions depend also on the spin precession frequency. RT linewidth ΔB_{pp} of the EPR spectral components of polarons immobilized in, for example, PTTF–C$_2$H$_5$–C$_6$H$_4$ increases from 0.28 to 0.38 and then to 3.9 mT, while electron spin precession frequency $\omega_e/2\pi$ increases from 9.5 to 37 and then to 140 GHz, respectively. On the other hand, linewidth of mobile polarons increases from 1.02 to 1.15 and then to 17.5 mT, respectively, at such a transition. The width of all NMR lines also increases by a factor of 1.5–2 with the increase of $\omega_p/2\pi$ from 300 to 400 MHz. The fact that the mobile PC has a broader line than the pinned ones can be explained by the higher probability of its interaction with other electron and nuclear spins and with the dopant ions due to mobility. This feature is typical for

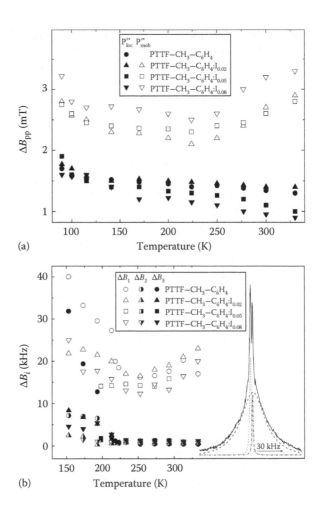

FIGURE 4.20

(a) Temperature dependence of the linewidth of localized $P_{loc}^{+\cdot}$ and mobile $P_{mob}^{+\cdot}$ polarons in the initial and I-doped PTTF–CH$_3$–C$_6$H$_4$ samples determined from their D-band electron paramagnetic resonance spectra. (b) ^{1}H NMR spectrum 300 MHz (solid line) containing two narrow lines with a width ΔB_2 and ΔB_3 attributed to protons in CH$_3$ groups being in different conformations in polytetrathiafulvalene (PTTF) insulating phase (dotted dash-dotted lines) and a broad enveloping line with a width ΔB_1 of these groups in PTTF conjugated phase (dashed line) (inset) as well as the change in width of these lines, ΔB_2, ΔB_3, and ΔB_1, respectively, with temperature.

conjugated polymers [118], however, disagrees with that obtained earlier for such system at X-band EPR [355,405].

The concentration of PC in PTTF–CH$_3$–C$_6$H$_4$ slightly increases from 2×10^{17} cm^{-3} up to 3×10^{17} cm^{-3} at the polymer doping. This parameter also changes slightly from 3×10^{18} cm^{-3} up to 4×10^{18} cm^{-3} at the doping of PTTF–THA sample. The PC concentration in doped PTTF–C$_2$H$_5$–C$_6$H$_4$ is 5×10^{17} cm^{-3}.

Paramagnetic susceptibility of the initial PTTF samples and that obtained for the samples with doping levels $y \leq 0.08$ and also $0.2 \leq y \leq 2.0$ [405] is presented in Figure 4.21 as function of temperature. This value changes monotonically with the temperature at low doping level. The analysis of the data obtained has shown that the susceptibility of the initial and slightly iodine-doped PTTF–CH_3–C_6H_4 samples with $y = 0.02$, 0.05, and 0.08 is determined mainly by the second term of Equation 2.6 with $C = 2.3 \times 10^{-4}$, 3.8×10^{-4}, 3.4×10^{-4}, and 5.9×10^{-4} emu K/mol, respectively. This parameter of the initial and slightly iodine-doped PTTF–THA samples with $y = 0.08$ and 0.12 is also governed by the Curie term of Equation 2.6 with $C = 1.10 \times 10^{-2}$, 3.1×10^{-3}, and 4.8×10^{-3} emu K/mol, respectively. As the doping level increases, the temperature dependence of susceptibility becomes more complex (Figure 4.21). Spin susceptibility of highly doped polymers decreases as the temperature decreases from RT down to $T_c \approx 150$ K and then increases at $T \leq T_c$ (Figure 4.21). This means that starting from some critical concentration spins interact and therefore form quasi-pairs or triplets, as in the case of PA. As a result of such interaction, the last term of Equation 2.6 prevails in spin susceptibility, especially at $T \geq T_c$. Indeed, Figure 4.21 shows that experimental data obtained for high-doped PTTF–CH_3–C_6H_4 polymers are well fitted by last two terms of Equation 2.6. The Curie constant is nearly equal for all samples,

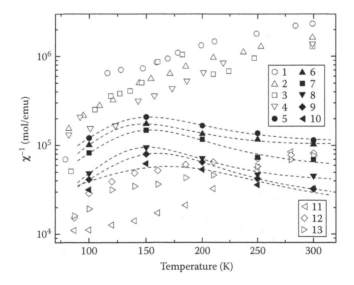

FIGURE 4.21
Inverted effective magnetic susceptibility of polarons in the PTTF–CH_3–C_6H_4–I_y samples with the doping levels (y) equal to 0.00 (1), 0.02 (2), 0.05 (3) 0.08 (4), 0.2 (5), 0.4 (6), 0.6 (7), 0.8 (8), 1.0 (9), and 2.0 (10) and in the PTTF–THA samples with the doping levels (y) equal to 0.00 (11), 0.08 (12), and 0.12 (13) as function of temperature. Dashed lines show the dependencies calculated using Equation 2.6 with respective parameters presented in Table 4.4.

whereas the J_{ex} value increases from 0.038 up to 0.057 with the doping level increase from 0.2 up to 2.0 (Figure 4.21).

The linewidth of PC follows Equation 2.19 with the ω_e increase in 37 ≤ $\omega_e/2\pi$ ≤ 140 GHz range that is evidence of a weak interaction between spin packets in this polymer. This provokes MW saturation of PC in PTTF at comparatively small B_1 values at D-band EPR and therefore the manifestation of the fast passage effect in both the in-phase and π/2-out-of-phase components of its dispersion signal (Figure 4.22). $\omega_m T_1 > 1$ inequality is realized for undoped PTTF samples, so that the first derivation of its dispersion signal

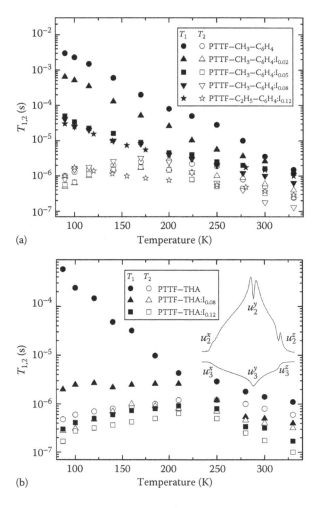

(a)

(b)

FIGURE 4.22

Temperature dependence of effective spin–lattice T_1 and spin–spin T_2 relaxation times of polarons in the initial and I-doped PTTF–CH$_3$–C$_6$H$_4$ (a) and PTTF–THA (b) polymers determined from their in-phase (top) and π/2-out-of-phase (down) D-band dispersion spectra (inset).

is determined mainly by the u_2 and u_3 terms of Equation 2.35. On the other hand, $\omega_m T_1 < 1$ condition is actual for slightly doped PTTF samples; therefore, their dispersion signals are determined by u_1 and u_3 terms of the equation mentioned earlier.

The u_i^x, u_i^y, and u_i^z terms of the dispersion signal U in Equation 2.35 are attributed to PCs with a strongly asymmetric distribution of spin density that are differently oriented in an external magnetic field. The simulation of the PTTF dispersion spectra allowed to identify them as a superposition of a predominant asymmetric spectrum with $g_{xx} = 2.01189$, $g_{yy} = 2.00564$, $g_{zz} = 2.00185$ (undoped PTTF–CH$_3$–C$_6$H$_4$), $g_{xx} = 2.01356$, $g_{yy} = 2.00603$, $g_{zz} = 2.00215$ (PTTF–C$_2$H$_5$–C$_6$H$_4$), and $g_{xx} = 2.01188$, $g_{yy} = 2.00571$, $g_{zz} = 2.00231$ (undoped PTTF–THA) of immobile polarons and a spectrum, attributed to mobile polarons (Figure 4.22).

The temperature dependencies of effective relaxation times of PC in PTTF samples determined from their D-band EPR spectra using Equations 2.36 through 2.39 are summarized in Figure 4.22. T_1 value of PC in all PTTF samples with phenyl bridges and in undoped PTTF with THA bridges was shown to change monotonically with temperature as $T^{-\alpha}$ with exponent α lying near 3 and 5 for mobile and pinned polarons, respectively. This value determined for PTTF at D-band EPR is larger than that measured for immobile radicals by using spin echo technique at X-band EPR [405]. A difference between T_1^{mob} and T_1^{loc} can be caused, for example, by a strong interaction between different PCs.

The macromolecular motion in PTTF, similarly to ordinary ones [410], is *a priori* strongly anisotropic carrying out with the correlation time $\tau_c \geq 10^7$ s^{-1}. Such dynamics was studied at D-band EPR [373,409] by using the ST-EPR method [114] based on the saturation effect mentioned earlier.

As it is seen from Figure 4.23, the heating of PTTF leads to the growth of its $K_{mov} = u_3^x / u_3^y$ parameter as the result of anisotropic librational reorientations of the pinned polarons near the main x-axis of the polymer chains. The Arrhenius dependencies of correlation times of superslow molecular motion in different PTTF samples calculated from Equation 2.65 are shown in Figure 4.23. The maximum value of τ_c is ca. 10^{-4} s at $T = 75$ K for PTTF–CH$_3$–C$_6$H$_4$ when $K_{mov} = 0.07$. Macromolecular librations in the PTTF samples were analyzed to carry out within activation law according to Equation 2.60, and the preexponential factor τ_{c0}^x and activation energy E_a of macromolecular librations of such dynamics calculated within this approach are summarized in Tables 4.4 and 4.5. It was concluded from the data obtained that the slight doping of PTTF matrices doubles the activation energy of librations of their macromolecules. E_a values obtained at D-band EPR are comparable with that determined at lower registration frequency for interchain charge transfer in doped PTTF [355,405] that indicates the interaction of pinned and mobile polarons in this polymer matrix.

In order to compare experimental results with the polaron theory, Q1D diffusion motion of mobile PC along and between PTTF's chains with the diffusion coefficients D_{1D} and D_{3D}, respectively, was assumed.

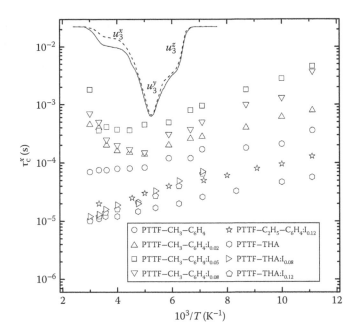

FIGURE 4.23

Arrhenius dependencies of correlation time τ_c^x of x-anisotropic libration of polarons local-ized on chains in the initial and I-doped polytetrathiafulvalene samples evaluated from their saturation transfer electron paramagnetic resonance (ST-EPR) spectra. Typical ST-EPR spectra registered at low and high temperatures are shown in the inset by the dashed and solid lines, respectively.

The temperature dependencies of effective D_{1D} and D_{3D} calculated for PC in different PTTF samples by using Equations 2.45, 2.46, 2.51, and 2.52, and the data presented in Figure 4.22 are shown in Figure 4.24. Assuming that the spin delocalization over the polaron in PTTF occupies approximately five monomer units [75], the maximum value of D_{1D} does not exceed 2×10^{12} rad/s for PTTF samples at RT. This value is at least two orders of magnitude lower than that determined earlier by low-frequency magnetic resonance methods for polarons in polypyrrole [411] and PANI [75] but higher than D_{1D}^0 evaluated earlier. RT anisotropy of spin dynamics in PTTF is $A = D_{1D}/D_{3D} \geq 10$. Figure 4.24 shows that D_{1D} value increases at RT by at least one to two orders of magnitude at transition from PTTF–C_2H_5–C_6H_4 to PTTF–THA and to PTTF–CH_3–C_6H_4 samples due probably to more planar chain conforma-tion in PTTF–CH_3–C_6H_4 sample. Indeed, the variation of g_{xx} value is $\Delta g_{xx} = 1.32 \times 10^{-3}$ at such a transition. Assuming that the overlapping integral t_{cc} of macromolecules depends on dihedral angle θ (*i.e.*, the angle between p-orbits of neighboring C atoms) as $t_{cc} \propto \cos \theta$ and that spin density on sulfur atom ρ_s depends on $\rho_s \propto \sin \theta$ [335], one can calculate from Equation 2.3 the dif-ference $\Delta \theta$ at such a transition to be $\Delta \theta = 22°$. Note that analogous change in

TABLE 4.4

C, c_{af} and J_{af} Values Determined from Equation 2.6, τ_{c0}^{x} and E_a Determined from Equation 2.60, σ_{01}, α, and E_a Determined from Equation 2.77, σ_{02}, T_0 Determined from Equation 2.80 with $d = 3$, for Initial and Iodine-Doped PTTF–CH$_3$–C$_6$H$_4$:I$_y$ Samples with Different Doping Levels y

Parameter	PTTF–CH$_3$–C$_6$H$_4$:I$_y$									
y	0.00	0.02	0.05	0.08	0.2	0.4	0.6	0.8	1.0	2.0
C (10^{-3} emu K/mol)	–mu K/mol	2.5	2.7	5.3	4.7	2.6	—	—	—	—
c_{af} (emu K/mol)	—	—	—	—	0.20	0.24	0.40	0.64	0.92	1.10
J_{af} (eV)	—	—	—	—	0.038	0.035	0.042	0.039	0.046	0.057
τ_{c0}^{x} (s)	$\tau 2.5 \times 10^{-5}$	4.1×10^{-5}	6.7×10^{-5}	2.6×10^{-5}	-0657/mol	—	—	—	—	—
E_a (eV)	0.019	0.023	0.033	0.036	—	—	—	—	—	—
σ_{01} (S/K m s)	6.3×10^{-20}	1.6×10^{-16}	5.4×10^{-14}	1.1×10^{-11}	-1136/mol	—	—	—	—	—
α	0.16	1.1	2.1	3.2	—	—	—	—	—	—
E_a (eV)	0.11	0.16	0.12	0.14	—	—	—	—	—	—
σ_{02} (10^{12} S/K$^{0.5}$ m)	–0.50k K/mol	1.5	–0.50k K/	—	—	—	—	—	—	—
T_0 (10^8 K)	—	—	—	—	6.2	6.1	—	—	6.0	—

TABLE 4.5

τ_{c0}^x and E_a Determined from Equation 2.60, σ_{01}, α, and E_a, Determined from Equation 2.77, σ_{02} and E_{ph} Determined from Equation 2.83 for Initial and Iodine-Doped PTTF Samples with Different Doping Levels y

Parameter	PTTF–C$_2$H$_5$–C$_6$H$_4$:I$_y$		PTTF–THA:I$_y$	
y	0.12	0.00	0.08	0.12
τ_{c0}^x (s)	9.8×10^{-6}	5.2×10^{-6}	3.1×10^{-6}	2.4×10^{-6}
E_a (eV)	0.021	0.019	0.038	0.041
σ_{01} (S/K m s)	2.7×10^{-15}	1.9×10^{-10}	1.6×10^{-17}	1.8×10^{-17}
α	0.0	2.8	0.0	0.0
E_a (eV)	0.052	0.19	0.09	0.07
σ_{02} (S/m)	4.2×10^{-5}	—	2.0×10^{-3}	4.1×10^{-3}
E_{ph} (eV)	0.11	—	0.025	0.024

θ takes place at transition from benzoid to quinoid form of PPP [72] and at transition from emeraldine base to salt form of PANI [181,185] and treatment of PATAC (see Section 4.2).

Effective *ac* and *dc* conductivity of PTTF depends on motion of charge carriers inside crystallites and between them, respectively. Let us consider both these processes.

Figure 4.25a depicts the temperature dependencies of σ_{dc} measured for an initial and iodine-doped PTTF–CH$_3$–C$_6$H$_4$ with different doping level [406]. Amorphous phase prevails in an undoped sample, so then the charge transport should be determined by polarons diffusing through both the crystal and amorphous phases of the polymer, respectively, in framework of the activation and Q1D VRH charge transport. This is the reason why the temperature dependence of σ_{dc} is well fitted by Equation 2.76 with E_a = 0.31 eV and by Equation 2.80 with d = 1 and T_0 = 510 K (Figure 4.25a). As the number of dopant molecules increases, the VRH charge transport mechanism becomes as predominant and the dimensionality of the polymer system d increases from 1 up to 3. The percolation parameter T_0 is near 6×10^8 K for doped PTTF samples that is near to that earlier obtained for PTTF–CH$_3$–C$_6$H$_4$ [286]. One can then calculate from Equation 2.80 the density of states $n(\varepsilon_F)$ at the Fermi level ε_F, the charge hopping distance R, and the phonon frequency $\nu_{ph} = \omega_{ph}/2\pi$ to be, respectively, 1.1×10^{-4} states/eV, 16.4 nm, and 5×10^{12} Hz in doped PTTF.

AC conductivity of the PTTF sample depends not only on its doping level y but also on its structure and the spin precession frequency ω_e (Figure 4.25b). The value of σ_{ac} varies with the latter parameter as $\sigma_{ac} \propto \omega_e^{1/2}$. The transition from –CH$_3$ to –H substitute leads to the increase in the *ac* conductivity due possibly to more ordered morphology of PTTF–H–C$_6$H$_4$ conjugated polymer. This conductivity, however, does not reflect directly charge process due to

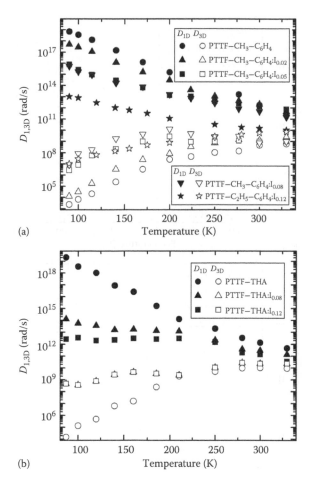

FIGURE 4.24
Temperature dependencies of the effective rates of polaron intrachain D_{1D} and interchain D_{3D} diffusion in the polytetrathiafulvalene samples with phenyl (a) and tetrahydroanthracene (b) bridges determined from Equations 2.46, 2.51, and 2.52 and the data presented in Figure 4.22.

polaron motion. In order to determine more detailed charge transport process, the σ_{1D} and σ_{3D} should be analyzed.

The σ_{1D} and σ_{3D} values of the PTTF samples calculated using Equation 2.71 are presented in Figure 4.26 as function of temperature. It is seen that the σ_{1D} value of an initial and doped PTTF–CH$_3$–C$_6$H$_4$ samples strongly depends on temperature following $\propto T^{-\beta}$ law, where $\beta = 10$–14. Undoped PTTF–THA demonstrates stronger temperature dependence with $\beta = 16$. Such dependencies cannot be described by tunnel charge transfer mechanisms. Indeed, $\sigma_{ac}(T)$ dependencies calculated in framework of charge isoenergetic hopping between polaron and bipolaron states from Equation 2.73 and shown by doted lines do not fit $\sigma_{1D}(T)$ ones experimentally obtained for all PTTF

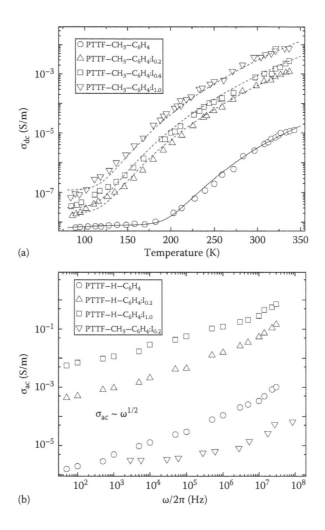

FIGURE 4.25

(a) Direct current conductivity of the initial and iodine-doped PTTF–CH$_3$–C$_6$H$_4$ samples as function of temperature and doping level y. The dependence calculated as combination of Equation 2.76 with $\sigma_0 = 0.62$ S/m and $E_a = 0.31$ eV and of Equation 2.80 with $\sigma_0 = 8.9 \times 10^{-7}$ S K$^{0.5}$/m, $d = 1$, and $T_0 = 510$ K is shown by the solid line. Dashed lines show the functions calculated using Equation 2.80 with respective parameters summarized in Table 4.4. (b) Dependence of alternating current conductivity of some polytetrathiafulvalene samples on the registration frequency. (From Gruber, H., Roth, H. K., Patzsch, J., and Fanghanel, E.: Electrical-Properties of Poly(tetrathiafulvalenes), 99.1990. Copyright Wiley-VCH Verlag GmbH & Co. KGaA. Reproduced with permission.)

samples. Therefore, another mechanism should be proposed for interchain charge transport mechanism in PTTF. The most acceptable charge dynamic process can be suggested in terms of the Kivelson–Heeger theory of polaron interaction with the lattice phonons. The data presented in Figure 4.26 evidence that *ac* conductivity of one PTTF–C$_2$H$_5$–C$_6$H$_4$ and both PTTF–THA

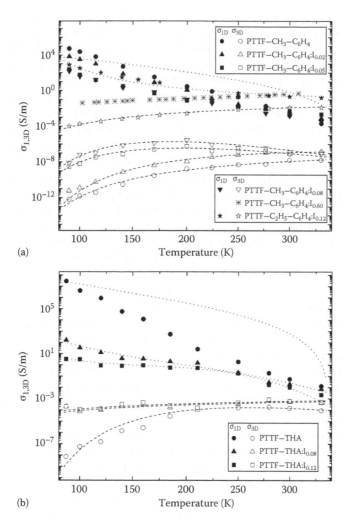

(a)

(b)

FIGURE 4.26

Temperature dependence of alternating current (*ac*) (140 GHz) conductivity of the initial and I-doped polytetrathiafulvalene samples due to the intrachain and interchain spin motion calculated using Equation 2.45 using the data presented in Figure 4.24. For the comparability, by the * symbol, the *ac* conductivity obtained for iodine-doped PTTF–CH$_3$–C$_6$H$_4$ with $y = 0.6$ by the convenient conductometry method is shown. (With kind permission from Springer Science+Business Media: *Electronic Properties of Polymers*, Small polarons in polymeric polytetrathiafulvalenes (PTTF), 107, 1992, 121–124, Patzsch, J. and Gruber, H.) Top dotted lines represent temperature dependencies calculated using Equation 2.73 with $\sigma_0 = 2.1 \times 10^{-10}$ S s K/m, $k_3 = 7.1 \times 10^{12}$ s K$^{9.4}$, and $n = 8.4$ (a) and $\sigma_0 = 6.0 \times 10^{-7}$ S s K/m, $k_3 = 1.4 \times 10^{14}$ s K$^{10.1}$, and $n = 9.1$ (b). The theoretical functions $\sigma_{1D}(T)$ calculated using Equation 2.83 with $\sigma_0 = 4.2 \times 10^{-5}$ S/m and $E_{ph} = 0.11$ eV (lowermost dotted line) (a) and $\sigma_0 = 2.0 \times 10^{-3}$ S/m and $E_{ph} = 0.025$ eV (middle dotted line) and $\sigma_0 = 4.1 \times 10^{-3}$ S/m and $E_{ph} = 0.024$ eV (lowermost dotted line) (b) are shown. The dependencies $\sigma_{3D}(T)$ calculated using Equation 2.77 with $\kappa = 1 - \alpha k_B T/E_a$ and respective parameters summarized in Tables 4.4 and 4.5 are also shown by the dashed lines.

iodine-doped samples can indeed be fitted by Equation 2.83. The energy of the lattice phonons determined for these polymers are summarized in Table 4.5. The σ_{3D} value of an initial and slightly doped PTTF–CH$_3$–C$_6$H$_4$ monotonically decreases with the temperature decrease (Figure 4.26). On the other hand, this value of undoped PTTF–THA and PTTF–CH$_3$–C$_6$H$_4$ with $y \geq 0.05$ somewhat increases at the sample cooling from RT down to $T_c \approx$ 170–200 K and then decreases on further cooling. Such dependence can be suggested within the thermally activated interchain polaron hopping in conduction band tails with activation energy E_a. Temperature dependencies of conductivity due to such spin motion calculated from Equation 2.77 are shown by lines in Figure 4.26. The values of α and E_a obtained from the fitting of experimental data of different PTTF samples are presented in Tables 4.4 and 4.5. It is seen from the data presented that the ratio E_a/α decreases from 0.690 obtained for undoped sample down to 0.145 and to 0.057 and to 0.044 with the increase of its doping level y up to 0.02 and then to 0.05 and to 0.08. For comparison, the temperature dependence of σ_{ac} of highly iodine-doped PTTF–CH$_3$–C$_6$H$_4$ sample determined by conductometric method [408] is also presented in Figure 4.26. This flat curve evidences for closeness of both the conductivity terms in the polymer.

It is seen from Figure 4.26 that σ_{3D} value increases at the replacement of phenyl bridges by tetrahydroanthracene ones in PTTF. This fact allows us to conclude that this term of *ac* conductivity and therefore the anisotropy of charge carrier transfer depend on the polymer structure. Indeed, the structure of a polymer becomes more planar at such transition, so then the polymer chains are situated closer in PTTF–THA and the probability of interchain transfer increases due to the increase of appropriate transfer integral. This obviously leads to the decrease in the anisotropy of charge transport in this polymer.

The Fermi velocity v_F was determined for PTTF samples to be near to 1.9×10^7 cm/s [118]. So the mean free path l_i of a charge was determined to be $l_i = v_{1D}c_{1D}^2\bar{v}_F^{-1} = 10^{-2} - 10^{-4}$ nm for the PTTF samples. The l_i is less than lattice constant *a*; therefore, the charge transfer is incoherent in this polymer, which cannot be considered as quasi-one-dimensional metal. For such case, the interchain charge transfer integral can be determined as [166] $t_\perp = v_F\hbar/2a = 0.05$ eV. This value lies near activation energy of chain librations and also near normalized activation energy of interchain polaron hopping in doped PTTF–CH$_3$–C$_6$H$_4$ and PTTF–THA samples (Table 4.4). These facts allow us to conclude that the conductivity of PTTF is determined mainly by interchain phonon-assisted hopping of polarons, which, in one's turn, is stimulated by macromolecular dynamics. This evidences for the interaction of molecular and charge dynamics in PTTF. It confirms also the supposition [412] that the fluctuations of lattice oscillations, for example, librations among them, can modulate the electron interchain transfer integral in conjugated compounds.

The rise of libron–exciton interactions at the polymer doping evidences for the formation of a complex quasiparticle, namely, molecular–lattice polaron [140] in doped PTTF. According to this phenomenological model, molecular polaron is additionally covered by lattice polarization so then its mobility becomes sum of mobilities of molecular and lattice polarons. The energy of formation of such molecular–lattice polaron E_p in PTTF was determined to be 0.19 eV [373]. This value is smaller than that (0.1 eV) obtained for other disordered conjugated polymers [222]. Therefore, the characteristic time, necessary for polarization of both atomic and molecular orbits of polymer, can be determined as $\tau_p \approx \hbar E_p = 3.5 \times 10^{-15}$ s. This value is more than two orders of magnitude smaller than intra- and interchain hopping times for charge carriers in PTTF (Figure 4.24). This leads to the conclusion that the time τ_h required for the hopping of charge carriers in PTTF sufficiently exceeds the polarization time for charge carriers' microenvironment in the polymer, that is, $\tau_h \gg \tau_p$. This inequality is a necessary and sufficient condition for electronic polarization of polymer chains by a charge carrier.

Thus, both the Q1D and Q3D spin dynamics are realized in PTTF affecting the charge-transfer process. Q3D polaron diffusion dominates in the polymer conductivity; however, Q1D diffusion of spin and spinless charge carriers also plays an important role in effective conductivity of the polymer.

4.6 Polyaniline

Within the class of conjugated polymers, PANI (Figure 1.2) is of particular interest because of its excellent stability under ambient conditions and well perspective of utilization in molecular electronics [413,414]. The PANI family is known for its remarkable insulator-conductor transition as a function of doping (protonation or oxidation) level y [415]. Depending on the doping level y, it can be in leucoemeraldine (LE), pernigraniline (PN), emeraldine base (EB), or emeraldine salt (ES) forms. PANI–LE is fully reduced state. PANI–PN is fully oxidized state with imine links instead of amine links. These two forms are poor conductors, even when doped with an acid. PANI–EB is a neutral form, whereas PANI–ES is a p-type semiconductor with hole charge carriers [416]. It is semicrystalline, heterogeneous system with a crystalline (ordered) region embedded into an amorphous (disordered) matrix [230]. When PANI is doped with an acid, intermediate bipolaron and more stable polaron structures are formed. Such charge carriers transfer elemental charges during their intrachain and interchain diffusion as well as hopping inside and between well-ordered crystallites [417]. In polaron structure, a cation radical of one nitrogen acts as a hole that can transfer an elemental positive charge. The electron from the adjacent nitrogen (neutral) jumps to this hole and it becomes electrically neutral initiating motion of the holes. However, in bipolaron structure, this type of movement is not possible since two holes are adjacently located. This polymer differs from other organic conjugated polymers in several important aspects. In contrast with these polymers, it has no charge conjugation symmetry. Besides, both carbon rings and nitrogen atoms are involved in the conjugation. The phenyl rings of PANI can rotate or flip, significantly altering the nature of electron–phonon interaction. So an additional mobility of macromolecular units can modulate sufficient electron–phonon interactions and, therefore, lead to more complex mechanism of electron transfer in PANI [418,419]. These results are somewhat different in magnetic and charge transport properties of PANI compared with other conjugated polymers. It was shown theoretically and experimentally [420] that charges in PANI, as in the case of other conjugated polymers, are transferred by polarons moving along individual polymer chains. At low y, a hopping charge transfer between polarons and bipolarons predominates in PANI. In modified polymer, a number of such charge carriers increase and their energy levels merge and form metal-like band structure, so-called polaron lattice [253,417]. Stronger spin–orbit and spin–lattice interactions of the polarons diffusing along the chains are also characteristic of PANI. Upon protonation of PANI–EB or oxidation of PANI–LE insulating forms of PANI, their conductivity increases by more than 10 orders of magnitude, whereas a number of electrons on the polymer chains remain constant in the ES form of PANI [421]. Such a doping is accompanied by appearance of the Pauli susceptibility [422,423], characteristic for classic metals, due to the formation of high-conductive completely protonated or oxidated clusters with the characteristic size about 5 nm in amorphous polymer. While doping of PANI–EB films with the sulfonated dendrimers gives dc macroconductivity up to ca. 10^3 S/m, the hydrogensulfated fullerenol-doped materials show metallic characteristics with RT σ_{dc} as high as 10^4 S/m [424] that is about six orders of magnitude higher than the typical value for fullerene-nanomodified conjugated polymers. In some cases, diamagnetic bipolarons [417] and/or antiferromagnetic interacting polaron pairs [425] each possessing two elemental charges can also be formed in heavily doped polymer. The effective crystallinity of polymer increases up to ~50%–60%. The lattice constants of the PANI–EB and PANI–ES forms are presented in Table 4.1.

Crystallinity and, therefore, conjugated properties of PANI essentially depend on structure of a dopant introduced. The mechanism of charge transfer in heavily doped PANI is also depend sufficiently on the nature of a dopant as well as on the method of the polymer synthesis. For instance, the Fermi level in PANI doped with sulfuric (PANI:SA) or hydrochloric (PANI:HCA) acid lies in the region of localized states, therefore, is considered as a Fermi glass with localized electronic states [279], whereas the Fermi level energy ε_F of PANI highly doped with camphorsulfonic acid (PANI:CSA) lies in the region of extended states governing metal behavior of the latter near the metal–insulator boundary [426,427]. On the other hand, optical (0.06–6 eV) reflectance measurements of, for example, in PANI:CSA [254,279,428,429] suggest that this polymer is a disordered Drude-like metal near the metal–insulator boundary due to improved homogeneity and reduced degree of structural disorder. From optical measurements, it was determined that the effective charge carrier mass $m^* \approx 2m_e$, the mean free path $l^* \approx 0.7$ nm, and the density of states at the Fermi level $n(\varepsilon_F) \approx 1$ state per eV per two ring repeat units [279,428]. Studies of the effect of doping level on both the electronic transport and film morphology of PANI:CSA have shown a direct correlation between the degree of crystallinity (induced by hydrogen bonding with the CSA counter ion) and the metallic electronic properties [430–432]. This leads to the improvement of crystallinity and metallic conductivity of the polymer in the series PANI:HCA → PANI:SA → PANI:CSA at comparable modification levels. However, this deduction is not always conformed to results obtained at PANI study by other methods and other authors [3].

Typical *dc* conductivity is frequently a result of various electronic transport processes and is ca. 10^3–10^5 S/m for disoriented and oriented PANI–ES [61,165,166,415,421,426,433–437]. For example, this value has been determined for PANI:CSA to vary in the range $\sigma_{dc} \approx (1.0$–$3.5) \times 10^4$ S/m at RT [437,438]. The variety of conjugated properties makes difficult to investigate completely and correctly by usual experimental methods true charge dynamics along a polymer chain, which can be masked by interchain, interglobular, and other charge-transfer processes. The inhomogeneity of distribution of the counterion molecules results in an additional complexity of experimental data interpretation.

The electronic structure of PANI–ES has been described theoretically by the metallic polaron lattice model [417,439] with a finite $n(\varepsilon_F)$ value [440]. An analysis of experimental data on the temperature dependencies of *dc* conductivity, thermoelectric power, and Pauli-like susceptibility allowed MacDiarmid and Epstein et al. [166,250,251,433,441,442] to declare that PANI–EB is a completely amorphous insulator in which Q3D granular metal-like domains of characteristic size of 5 nm are formed during its doping and transformation into PANI–ES. A more detailed study of the complex MW dielectric constant, paramagnetic resonance linewidth, and electric field dependence of conductivity of PANI–ES [165,166,230,253,433,443] allowed them to conclude that both chaotic and oriented PANI–ES consist of some parallel chains strongly coupled into *metallic bundles* between which Q1D VRH charge transfer occurs and in which 3D electron delocalization takes place. The intrinsic 3D *ac* conductivity of the domains was evaluated using Drude model [444] as $\sigma_{ac} \cong 10^9$ S/m at 6.5 GHz [445], which was very close to the value expected by Kivelson and Heeger for the metal-like clusters in highly doped Naarmann *trans*-PA [246]. However, this value of the sample does not exceed the σ_{ac} mentioned earlier. It means that other processes, which make difficult its determination, mask the true process of electron transfer by usual experimental methods.

The polaronic charge carriers in PANI are characterized by electron spin $S = \frac{1}{2}$, so then the direct EPR spectroscopy is also widely used for the study of relaxation and dynamic properties of charge carriers in this system [42,105,118,122,184,446]. The oxidation or protonation of PANI–EB leads to the monotonic increase in PC concentration accompanied with strong line narrowing [447,448]. Lapkowski et al. [448] and MacDiarmid and Epstein [449] showed the initial creation of Curie spins in EB, indicating a polaron formation, followed by a conversion into Pauli spins, which indicates the formation of the polaron lattice in highly conductive PANI–ES [450,451]. The method enables to determine electron relaxation and diffusion of spin charge carriers even in PANI with chaotically oriented chains in scale from several macromolecular units [105,122]. These parameters are important to understand how relaxation and transport properties of spin charge carriers depend on the structure and dynamics of their microenvironment (lattice, anion, etc.). It should be noted that diffusion of electron spin effects nuclear relaxation of protons in PANI, so, in principle, the NMR spectroscopy can additionally be used for the study of electron spin dynamics in PANI [205]. Such investigations were mainly carried out for highly doped PANI:HCA [452–456]. It was noted [105], however, that the data on PANI proton relaxation experimentally obtained by NMR method [452–454] reflect an electron spin dynamics indirectly and consequently cannot give a correct enough conception of charge transfer in polymer. On the other hand, EPR method allows directly and more precisely determining spin composition, relaxation and dynamics of polarons in PANI and its nanocomposites.

The method allows to determine both relaxation times T_1 and T_2 as well as the diffusion of spin charge carriers along and between chains even in PANI with chaotically oriented chains in scale from several macromolecular units [105,122]. Both relaxation times of polarons govern their peak-to-peak EPR linewidth ΔB_{pp}, whereas the dependencies $T_1 \propto \omega_p^{1/2}$ and $\Delta B_{pp} \propto \omega_e^{1/2}$ obtained by comparatively low-frequency NMR and EPR methods, respectively, for highly protonated PANI–ES were interpreted in terms of their Q1D diffusion and Q3D hopping in polymer backbone. D_{1D} value was obtained to be, respectively, near to 10^{14} and 10^{12} rad/s and weakly depend on the doping level, while D_{3D} value strongly depends on y and correlated with both dc and ac conductivities of PANI:HCA [452]. The RT anisotropy of such motion $A = D_{1D}/D_{3D}$ varies from 10^4 in PANI–EB down to 10 in PANI–ES. This fact was interpreted in favor of existence of single high-conductive chains even in highly protonated PANI, between which Q1D charge transfer is realized [457]. Such an interpretation differs from the alternate model of formation of Q3D metal-like clusters in amorphous phase of the polymer [250,251]. Besides, the diffusion constants were determined from EPR linewidth that may reflect different processes carried out in PANI. Indeed, at registration frequencies less than 10 GHz, the lines of multicomponent spectra or spectra of different radicals with close magnetic resonance parameters overlap due mainly to low spectral resolution. So the linewidth of PANI at these frequencies generally represents a superposition of various contributions of localized and delocalized PCs.

The presence in the PANI of oxygen molecules can also affect its magnetic resonance parameters. In this case, the change in linewidth of organic systems is normally explained by dipole–dipole interaction of polarons with spin $S = ½$ with the oxygen molecules possessing sum spin $S = 1$. It was found

[458–462] that oxygen can reversibly broad EPR spectrum of PANI without remarkable change of its conductivity. Initial pumping of the sample leads to more promising effect of spin–spin interaction on spin relaxation [462–465]. However, Kang et al. have shown [466] that the contact of PANI:HCA with air leads to a reversible decrease in the intensity and an increase in the width of the EPR spectrum of PANI with a simultaneous decrease in its conductivity. Such change in the polymer properties was explained by the decrease of the polaron mobility at its interaction with air. The opposite effect, however, was registered in the study of polypyrrole [467] and PANI:HCA [153]. In the latter case, the diffusion of the oxygen into the polymer bulk was proposed to lead to a reversible increase in the polymer linewidth and conductivity due to acceleration of a polaron motion along the polymer chain.

As in the case of convenient conductors and other metal-like conjugated polymer nanocomposites, some highly doped PANI samples demonstrate EPR spectra with Dyson contribution as a result of interaction of MW field with both the spin and spinless charge carriers [179,180]. This additionally results in ambiguous interpretation of the data obtained on electron relaxation and dynamics and also on mechanism of charge transport in such systems.

In this part, the data of the multifrequency EPR study of the magnetic and charge transport properties of PANI–EB and PANI–ES doped with SA, HCA, CSA, and 2-acrylamido-2-methyl-1-propanesulfonic acid (AMPSA) and *para*-toluenesulfonic acid (TSA) up to $y \approx 0.60$ are summarized [42,181–187,264,354,356,446,468–474].

Figure 4.27 shows the X- and D-band EPR spectra of an initial PANI–EB sample and PANI slightly doped with different numbers of sulfuric acid molecules. At the X-band, PANI–EB demonstrates a Lorentzian three-component EPR signal consisting of asymmetric and symmetric spectra of PC that should be attributed, respectively, to localized ($P_{loc}^{\cdot +}$) and delocalized ($P_{mob}^{\cdot +}$) PCs. PC $P_{mob}^{\cdot +}$ keeps line symmetry at higher doping levels. At the D-band, the PANI EPR spectra became Gaussian and broader compared with the X-band ones (Figure 4.27), as it is typical for PC in other conjugated polymers. At this waveband, delocalized PCs demonstrate an asymmetric EPR spectrum at all doping levels. The analysis of EPR spectra obtained at both wavebands showed that the line asymmetry of $P_{mob}^{\cdot +}$ PC in undoped and slightly doped PANI samples can be attributed to anisotropy of the g-factor that becomes more evident at the D-band EPR. The linewidth of these PCs weakly depends on the temperature.

Therefore, the $P_{loc}^{\cdot +}$ with strongly asymmetric EPR spectrum can be attributed to a - (Ph – N$\overset{\bullet}{\text{H}}$– Ph)- radical with $g_{xx} = 2.006032$, $g_{yy} = 2.003815$, $g_{zz} = 2.002390$, $A_{xx} = A_{yy} = 0.45$ mT, and $A_{zz} = 3.02$ mT, localized on a short polymer chain. The magnetic parameters of this radical differ weakly from those of the Ph – NH– Ph radical [135,136], probably because of a smaller delocalization of an unpaired electron on the nitrogen atom ($\rho_N^\pi = 0.39$) and of the more planar

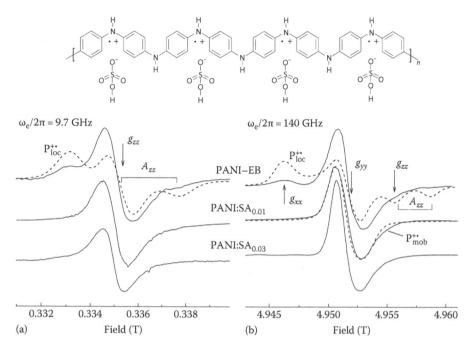

FIGURE 4.27
Typical room temperature X-band (a) and D-band (b) electron paramagnetic resonance absorption spectra of PANI–EB and this sample slightly doped by sulfuric acid. The absorption spectra calculated with $g_{xx} = 2.006032$, $g_{yy} = 2.003815$, $g_{zz} = 2.002390$, $A_{xx} = A_{yy} = 0.45$ mT, $A_{zz} = 3.02$ mT ($P_{loc}^{+\bullet}$), and $g_\perp = 2.004394$ and $g_\parallel = 2.003763$ ($P_{mob}^{+\bullet}$) are shown by the dashed lines. At the top, the PANI:SA nanocomposite is shown schematically. (From Krinichnyi, V.I., *Appl. Phys. Rev.*, 1(2), 021305/01, 2014. With permission.)

morphology of the latter. Assuming a McConnell proportionality constant for the hyperfine interaction of the spin with nitrogen nucleus $Q = 2.37$ mT [135,136], a spin density on the heteroatom nucleus of $\rho_N(0) = (A_{xx} + A_{yy} + A_{zz})/(3Q)$ can be estimated to be 0.55. At the same time, another radical $P_{mob}^{+\bullet}$ is formed in the system with $g_\perp = 2.004394$ and $g_\parallel = 2.003763$ that can be attributed to PC $P_{loc}^{+\bullet}$ delocalized on more polymer units of a longer chain. Indeed, the model spectra presented in Figure 4.27 well fit both the PCs with different mobility. The lowest excited states of the localized PC were determined from Equation 2.3 at $\rho_N^\pi = 0.56$ to be $\Delta E_{n\pi^*} = 2.9$ eV and $\Delta E\sigma_{\pi^*} = 7.1$ eV [475].

It was found [468] that PANI is sensitive to the water molecules, as it was found for short *trans*-PA regions in polyvinyl chloride treated by fuming sulfuric acid [39] (Section 4.1). The saturation of PANI–EB with water vapor broadens its spectrum from 0.25 up to 0.90 mT; however, its doping with sulfuric acid narrows the line down to 0.15 mT; the saturation of the doped sample broadens its signal up to 0.52 mT. X-band EPR spectrum of polarons in this sample is characterized by a symmetric Lorentzian solitary line with $g_{iso} = 2.0031$. At D-band EPR, its spectrum becomes Gaussian and the anisotropy of polaron's g-factor becomes more evident. At this registration frequency, the initial and modified samples reveal typical spectra of polarons with axially symmetrical distribution of unpaired electron in which effective relaxation time lies near 3×10^{-7} s. The saturation of the initial PANI sample with the water vapor causes the broadening of D-band EPR spectral components from 0.46 up to 0.59 mT at unchanged $g_\perp = 2.00301$ and $g_\parallel = 2.00249$. The doping is found not to lead to a noticeable change in its magnetic parameters. However, the exposure of doped PANI samples to an air resulted in the narrowing of the individual EPR components from 0.41 mT down to 0.32 mT and shift resulting spectrum to $g_\perp = 2.00288$ and $g_\parallel = 2.00271$. This change in the spectrum shape indicates a significant rearrangement of microenvironment of unpaired electrons localized on chains caused by the diffusion of water molecules into the polymer matrix bulk. Since the energy of the excited configuration of macromolecular systems is inversely proportional to the g-factor shift, $\Delta E \propto (g_\perp - g_\parallel)^{-1}$ (see Section 2.1), the change mentioned earlier in the EPR spectrum shape may be explained by the growth of the ordering of the doped and water-vapor-saturated PANI, in accordance with the analogous conclusion made from the analysis of its x-ray phase data [468].

Let us show how the main properties of PANI depend on the number and structure of counterions introduced into its bulk.

Radical $P_{loc}^{+\cdot}$ stabilized in PANI:HCA also demonstrates the strongly anisotropic spectrum with the canonic components $g_{xx} = 2.00522$, $g_{yy} = 2.00401$, and $g_{zz} = 2.00228$ of g tensor, and hyperfine coupling constant $A_{zz} = 2.27$ mT (Figure 4.27). PC R_2 is registered at $g_\perp = 2.00463$ and $g_\parallel = 2.00223$.

It was shown earlier [125,126] that g_{xx} and A_{zz} values of nitroxide radicals localized in a polymer are sensitive to changes in the radical microenvironmental properties, for example, polarity and dynamics. The shift of the PC $P_{mob}^{+\cdot}$ spectral x-component to higher fields with y and/or a temperature increase may be interpreted not only by the growth of the polarity of the radical microenvironment but also by the acceleration of the radical dynamics near its main molecular x-axis. The effective g-factors of both PCs are near to one another, that is, $\langle g_{loc} \rangle = 1/3(g_{xx} + g_{yy} + g_{zz}) \approx \langle g_{mob} \rangle = 1/3(2g_\perp + g_\parallel)$. This indicates that the mobility of a fraction of radicals R_1 along the polymer chain increases with the polymer doping. Such a depinning of the mobility results in an exchange between the spectral components of the PC and, hence, a decrease in the anisotropy of its EPR spectrum. In other words, radical $P_{loc}^{+\cdot}$ transforms into radical $P_{mob}^{+\cdot}$, which can be considered as a polaron diffusing along a polymer chain with a diffusion rate (see Equation 4.1) $D_{1D}^0 \geq 6.5 \times 10^8$ rad/s.

As in the case of other conjugated polymers, at the D-band in both in-phase and $\pi/2$-out-of-phase components of the dispersion EPR signal of neutral and slightly doped PANI, the bell-like contribution with Gaussian spin packet distributions due to the adiabatically fast passage of the saturated spin packets by a modulating magnetic field is registered (see inset in Figure 4.28b). This effect was not observed earlier in studies of PANI at lower registration frequencies [104,105]. It can be used for the determination of relaxation and dynamic parameters of these PCs.

Relaxation times of an initial and slightly doped PANI:SA and PANI:HCA samples calculated from Equations 2.36 through 2.39 are shown in Figure 4.28 as functions of temperature. The figure demonstrates that the increase in the

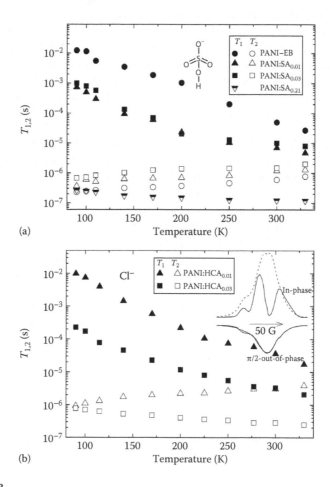

(a)

(b)

FIGURE 4.28

Temperature dependencies of spin–lattice T_1 and spin–spin T_2 relaxation times of polarons in emeraldine base form of polyaniline (PANI) and PANI–EB and in emeraldine base form of PANI, doped with (a) sulfuric (PANI:SA) and (b) hydrochloric (PANI:HCA) acids with different doping levels y calculated using Equations 2.36 through 2.39. In the inset, solid and dashed lines show in-phase and $\pi/2$-out-of-phase dispersion D-band electron paramagnetic resonance spectra of PANI–EB registered at 300 and 200 K, respectively. (From Krinichnyi, V.I., *Appl. Phys. Rev.*, 1(2), 021305/01, 2014. With permission.)

doping level of the polymer leads to shortening of the effective relaxation times of PC, which can be due to an intensification of the spin exchange with the lattice and with other spins stabilized on neighboring polymer chains of highly conjugated domains. It should be noted that at high temperatures, spin relaxation in the polymer is mainly determined by the Raman interaction of the charge carries with lattice optical phonons. The probability and rate of such a process are dependent on the concentration n of the PC localized, for example, in ionic crystals ($W_R \propto T_1^{-1} \propto n^2 T^7$) and in π-conjugated polymers ($W_R \propto T_1^{-1} \propto n T^2$) [339]. The available data suggest that the T_1^{-1} values of PC in slightly (up to y ≤ 0.03) doped PANI are described by a dependence of the type $T_1^{-1} \propto n T^{-k}$, where $k = 3$–4. The k exponent decreases with the y increase. This indicates the appearance of an additional channel of the energy transfer from the spin ensemble to the lattice at the polymer doping, as it occurs in conventional metals. At a medium degree of oxidation y ≥ 0.21, the electronic relaxation times become comparable and only slightly temperature dependent because of an intense spin–spin exchange in metal-like domains of higher effective dimensionalities. Providing that $T_1 = T_2$ for the PANI sample with $y = 0.21$, an effective rate of Q1D and Q2D spin motions in this polymer can be evaluated as well. It seems that both the rates calculated in the frameworks of such spin diffusions should be near to one another.

The shape of the π/2-out-of-phase dispersion signal of PC localized in PANI–EB changes with the temperature (see inset in Figure 4.28b) indicating the defrosting of anisotropic macromolecular librations in this polymer. The correlation of such motions was determined from Equation 2.65 with $\tau_{c0}^x = 5.4 \times 10^{-8}$ s and $\alpha = 4.8$ to be $\tau_c^x = 3.5 \times 10^{-5} \exp(0.015 \text{ eV}/k_B T)$. Similar dependencies were also obtained for slightly doped samples. The activation energy of the polymer chain librations lies near to that determined for PANI:HCA [182]. Upper limit for correlation time, calculated using Equation 2.61 with $\vartheta = 45°$, $B_1 = 0.01$ mT, g_{xx} and g_{zz} values measured for $P_{loc}^{+\cdot}$ PC, is equal to 1.3×10^{-4} s and corresponds to $u_3^x/u_3^y = 0.22$ (see Equation 2.61) at 125 K.

The relaxation times of electron and proton spins in PANI should vary depending on the spin precession frequency as $T_{1,2} \propto n^{-1} \omega_{p,e}^{1/2}$ [105,205]. This is the case for PANI:SA and PANI:HCA; therefore, the experimental data obtained for these polymers can be explained by a modulation of electronic relaxation by Q1D diffusion of $P_{mob}^{+\cdot}$ radicals along the polymer chain and by Q3D hopping of these centers between chains with the diffusion coefficients D_{1D} and D_{3D}, respectively.

Figure 4.29 illustrates the temperature dependencies of dynamic parameters D_{1D} and D_{3D} calculated for both types of PCs in several PANI samples from the data presented in Figure 4.28 using Equations 2.44, 2.51, and 2.52 at $L \cong 5$ [75]. It seems to be justified that the anisotropy of the spin dynamics is maximum in the initial PANI sample and decreases as y increases.

Investigating highly doped PANI at $\omega_e/2\pi \le 10$ GHz, it was concluded [452] that RT anisotropy of the spin dynamics in such polymers remains very high

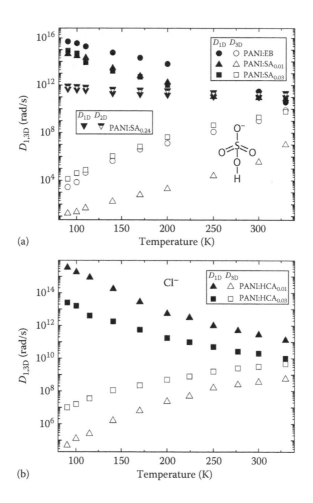

(a)

(b)

FIGURE 4.29

Temperature dependencies of coefficients of the intrachain and interchain polaron diffusion in undoped, PANI–EB and in PANI:SA (a) and PANI:HCA (b) samples with different doping levels calculated using Equations 2.46, 2.51, and 2.52. Counterions SA and HCA are shown schematically. (From Krinichnyi, V.I., *Appl. Phys. Rev.*, 1(2), 021305/01, 2014. With permission.)

even at $y = 0.6$. However, our experimental data indicate that this parameter is high only in PANI:HCA and PANI:SA with $y < 0.21$. Such a discrepancy is likely due to the limitations appearing at the study of spin dynamics at low precession frequencies. At $y \geq 0.21$, the system dimensionality seems to grow in PANI–ES, and at high temperatures the spin motion tends to become almost isotropic. The increase in the dimensionality at the polymer doping is accompanied by a decrease in the number of electron traps, which reduces the probability of electron scattering by the lattice phonons and results in the virtually isotropic spin motion and relatively slight temperature dependencies of both the electronic relaxation and diffusion rates of PC, as is the case for amorphous inorganic semiconductors [62,239].

By assuming that the diffusion coefficients D of spin and diamagnetic charge carriers have the same values, one can obtain from Equation 2.71 RT $\sigma_{1D} = 10$ S/m and $\sigma_{3D} = 1 \times 10^{-3}$–0.5 S/m for the PANI:HCA sample with $0 < y < 0.03$. At $D_{1D} = D_{3D}$, these values were determined to be $\sigma_{1D} = (5\text{–}18) \times 10^3$ S/m and $\sigma_{3D} = (3\text{–}10) \times 10^3$ S/m. Thus, the conclusion can be drawn that σ_{3D} grows more strongly with y that is the evidence for the growth of a number and a size of 3D quasi-metal domains in PANI–ES.

Let us consider the charge transfer mechanisms in the initial and slightly doped PANI samples. The fact that the spin–lattice relaxation time of PANI is strongly dependent on the temperature (see Figure 4.28) means that, in accordance with the energy conservation law, electron hops should be accompanied by the absorption or emission of a minimum number of lattice phonons. Multiphonon processes become predominant in neutral PANI because of a strong spin–lattice interaction. For this reason, an electronic dynamics process occurring in the polymer should be considered in the framework of Kivelson's formalism [219–221] of isoenergetic electron transfer between the polymer chains involving optical phonons, because spin and spinless charge carriers probably exist even in an undoped polymer (see Figure 4.27). Figure 4.30 shows that the experimental data for σ_{1D} of the initial PANI sample is fitted well by Equation 2.73 with $\sigma_0 = 2.7 \times 10^{-8}$ S K s/m, $k_3 = 3.1 \times 10^{12}$ s K$^{9.5}$, and $n = 8.5$. In contrast to undoped *trans*-PA, some quantities of charged carriers exist even in the initial PANI–EB sample, so the Kivelson mechanism mentioned earlier can determine its conductivity. Such an approach is not evident for PANI doped up to $0.01 \leq y \leq 0.03$ with less strong temperature dependence. The model of charge carrier scattering on optical phonons of the lattice of metal-like domains described earlier seems to be more convenient for the explanation of the behavior of their conductivity.

The concentration of mobile spins in PANI:HCA$_{0.01}$ sample is $y_p = 6.1 \times 10^{-5}$ per one benzoid ring. Taking into account that each bipolaron possesses dual charge, $y_{bp} = 1.2 \times 10^{-3}$ and $\langle y \rangle = 2.3 \times 10^{-2}$ can be obtained. The concentration of impurity is $N_i = 2.0 \times 10^{19}$ cm^{-3}, so then the separation between them $r_i = (4\pi N_i/3)^{-1/3} = 2.28$ nm is obtained for this polymer. The prefactor γ_0 in Equations 2.72 and 2.73 is evaluated from the $\sigma_{dc}(T)$ dependence to be 3.5×10^{19} s^{-1}. Assuming spin delocalization over five polaron sites along the polymer chain [75], $\xi_{\|} = 1.19$ nm is obtained as well. Using $2\Delta_0 = 3.8$ eV [476], typical for π-conjugated polymers $E_p \approx 0.1$ eV [222], $D_{3D} = 3.6 \times 10^8$ rad/s determined for PANI:HCA$_{0.01}$, $\xi = 0.20$ nm, $\xi_{\perp} = 0.079$ nm and $t_{\perp} = 7.1 \times 10^{-3}$ eV are obtained for this sample from Equations 2.74 and 2.75, respectively. The similar procedure gives $\langle y \rangle = 7.9 \times 10^{-2}$, $\gamma_0 = 2.1 \times 10^{17}$ s^{-1}, $\xi_{\perp} = 0.087$ nm, and $\xi = 0.21$ nm for PANI:HCA$_{0.03}$ sample with $y_p = 1.1 \times 10^{-3}$ and $y_{bp} = 1.2 \times 10^{-2}$.

As shown in Figure 4.30, the $\sigma_{1D}(T)$ dependence for the PANI:SA$_{0.01}$ and PANI:SA$_{0.03}$ samples is fairly well fitted using Equation 2.83 with $E_{ph} = 0.12$ and 0.11 eV, respectively. These values are near to energy (0.19 eV) of the

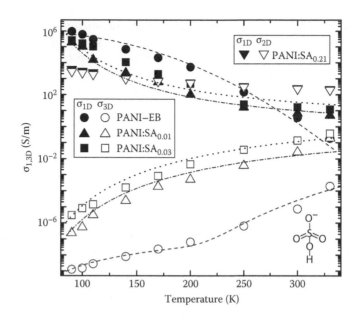

FIGURE 4.30
Temperature dependency of the alternating current conductivity due to polaron motion along (σ_{1D}, filled symbols) and between (σ_{3D}, open symbols) polymer chains in the PANI–EB and slightly doped PANI:SA samples as well as an effective rate of spin diffusion in PANI:SA$_{0.21}$ calculated, respectively, in the framework of Q1D (closed symbols) and Q2D (semifilled symbols) spin transport. The lines show the dependence calculated using Equation 2.73 with $\sigma_0 = 2.7 \times 10^{-8}$ S/K s m, $k_3 = 3.1 \times 10^{12}$ s/K$^{9.5}$, and $n = 8.5$ (upper dashed line), those calculated using Equation 2.83 with $\sigma_0 = 3.95 \times 10^{-4}$ S/m K and $E_{ph} = 0.12$ eV (upper dash-dotted line) and $\sigma_0 = 1.75 \times 10^{-4}$ S/m K and $E_{ph} = 0.11$ eV (upper dotted line), and those calculated using Equation 2.77 with $\sigma_0 = 8.2 \times 10^{-19}$ S/K s$^{0.8}$ m and $E_a = 0.033$ eV (low-temperature region), $\sigma_0 = 3.9 \times 10^{-10}$ S/K s$^{0.8}$ m and $E_a = 0.41$ eV (high-temperature region) (lower dashed line), $\sigma_0 = 4.5 \times 10^{-12}$ S/K s$^{0.8}$ m and $E_a = 0.102$ eV (lower dash-dotted line), and $\sigma_0 = 3.1 \times 10^{-11}$ S/K s$^{0.8}$ m and $E_a = 0.103$ eV (lower dotted line). (From Krinichnyi, V.I., *Appl. Phys. Rev.*, 1(2), 021305/01, 2014. With permission.)

polaron pinning in heavily doped PANI–ES [477]. The strong temperature dependence σ_{3D} of the initial sample can probably be interpreted in the framework of the model for the activation charge transfer between the polymer chains described by Equation 2.77 with E_a equal to 0.033 and 0.41 eV for low- and high-temperature regions, respectively (Figure 4.30). The activation energy of interchain charge transfer in slightly doped samples is $E_a = 0.102$ eV for PANI:SA$_{0.01}$ and $E_a = 0.103$ eV for PANI:SA$_{0.03}$ (Figure 4.29).

Note that the spin diffusion coefficients and consequently the conductivities, calculated from the spin relaxation of PANI–ES with $y = 0.21$, in frameworks of one- and two-dimensional spin diffusion, are near to one another (Figures 4.29 and 4.30). This fact can possibly be interpreted as the result of the increase of system dimensionality above the percolation threshold lying around $y \approx 0.1$. However, this can be also due to the decrease in accuracy of the saturation method at high doping levels. In this case, the dynamic parameters of both spin and spinless charge carriers can be evaluated from the Dyson-like EPR spectrum of PANI with $y \geq 0.21$ using the method described earlier.

The g-factor of PC $P_{mob}^{\cdot+}$ in PANI:SA with $y \geq 0.21$ becomes isotropic and decreases from $g_{mob} = 2.00418$ down to $g_{iso} = 2.00314$. This is accompanied by a narrowing of the $P_{mob}^{\cdot+}$ line (Figure 4.31). Such effects can be explained by a further depinning of Q1D spin diffusion along the polymer chain and therefore spin delocalization and by the formation of areas with high spin density in which a strong exchange of spins on neighboring chains occurs. This is in agreement with the supposition [166,250,251,433,441,442] of formation in amorphous PANI–EB of high-conductive massive domains with 3D delocalized electrons.

Figure 4.31 depicts the dependence of linewidth of the PANI:SA on the temperature and doping level. The predominance of extremal $\Delta B_{pp}(T)$ curves evidences for the exchange of PC in the system described by Equation 2.24. The reason of such exchange can be an interaction of PC localized on neighboring polymer chains modulated by macromolecular librations. Assuming activation character of this chain motion, the dependencies presented in Figure 4.31 were fitted by Equation 2.25 with the parameters summarized in Table 4.6. The activation energy of spin–spin interaction E_a decreases at the increase of polarizing magnetic field and lies near activation energy of the macromolecular librations (0.015 eV). This is an evidence of the dependence of the spin–spin interaction on an external magnetic field and its correlation with macromolecular dynamics in the system. The $\Delta B_{pp}(T)$ dependencies presented evidence also of different charge transport mechanisms in this polymer with different doping levels. The mechanism affecting the linewidth, however, depends also on the electron precession frequency, so then the linewidth does not directly reflects the relaxation and dynamic parameters of PC in this polymer.

The doping of the PANI with sulfuric acid leads to an inverted Λ-like temperature dependence of an effective paramagnetic susceptibility (see Figure 4.32), as occurs in the case of polyaniline perchlorate [180]. However, this does not lead to a strong narrowing of the PC line (Figure 4.31). As in the case of, for example, PANI treated with ammonia, this should indicate a strong antiferromagnetic spin interaction due to a singlet–triplet equilibrium in the PANI:SA in which total paramagnetic susceptibility can be described by Equation 2.6. Indeed, Figure 4.32 shows that the paramagnetic susceptibility experimentally determined for the PANI:SA samples is well reproduced by Equation 2.6 with the parameters also presented in Table 4.6. The J_{af} value is close to that (0.078 eV) obtained for the ammonia treated PANI [180]. Note that $n(\varepsilon_F)$ determined for PANI:SA, consistent with those determined earlier for PANI heavily doped with other counterions [422,429,478].

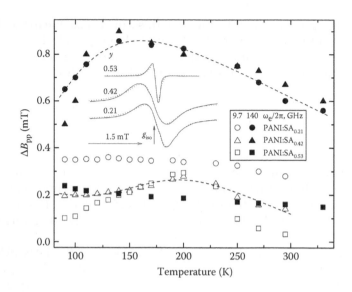

FIGURE 4.31
Temperature dependence of linewidth of paramagnetic centers (PCs) in PANI:SA with different doping levels registered at X-band and D-band electron paramagnetic resonance. Exemplary dashed lines show the dependencies calculated using Equation 2.25 with $E_r = 0.018$ eV and $J_{ex} = 0.19$ eV (top) and $E_r = 0.051$ eV and $J_{ex} = 0.64$ eV (down). Inset: Room temperature D-band absorption spectra of PCs in PANI:SA with different doping levels y. Top-down dotted lines represent the spectra calculated using Equations 2.27 through 2.29 with respective $D/A = 0.895$, $\Delta B_{pp} = 0.148$ mT, $D/A = 0.16$, $\Delta B_{pp} = 0.710$ mT, and $D/A = 0.53$, $\Delta B_{pp} = 0.586$ mT. The position of isotropic g-factor, g_{iso}, is shown. (From Krinichnyi, V.I., *Appl. Phys. Rev.*, 1(2), 021305/01, 2014. With permission.)

With the assumption of a metallic behavior, one can estimate that the energy of N_p Pauli spins in, for example, PANI–ES, with $0.21 \leq y \leq 0.53$ at the Fermi level $\varepsilon_F = 3N_p/2n(\varepsilon_F)$ [62] is to be 0.1–0.51 eV [185]. This value lies near to that (0.4 eV) obtained, for example, for PANI:CSA [428]. From this value, the number of charge carriers with mass $m_c = m_e$ in heavily doped PANI:SA [62], $N_c = (2m_c\varepsilon_F/\hbar^2)^{3/2}/3\pi^2 \approx 1.7 \times 10^{21}$ cm^{-3}, is evaluated. The N_c value is close to a total spin concentration in PANI:SA. This fact leads to the conclusion that all PCs take part in the polymer conductivity. For heavily doped PANI:SA samples, the concentration of spin charge carriers is less than that of spinless ones, due to the possible collapse of pairs of polarons into diamagnetic bipolarons. The velocity of the charge carrier near the Fermi level can be calculated as [62] $v_F = 2c_{1D}/\pi\hbar n(\varepsilon_F) = (3.3–7.2) \times 10^7$ cm/s that is typical for other conjugated polymers [118,354].

Both the *dc* and *ac* conductivities of the PANI:SA and PANI:HCA samples are presented in Figure 4.33 as function of temperature. The analysis shown that the *dc* conductivity in PANI:HCA and PANI:SA samples can be described in the framework of the models of Q1D VRH of charge carriers between

TABLE 4.6

ΔB_{pp}^0, ω_{hop}^0, E_r, J_{ex} Parameters Calculated from Equation 2.25, and the χ_P, $n(\varepsilon_F)$, C, c_{af}, J_{af} Values Determined from Equation 2.6 for Different Polyaniline Samples

Sample	ΔB_{pp}^0 (mT)	ω_{hop}^0 (rad/s)	Ea (eV)	J_{ex} (eV)	χ_P (emu/mol 1Ph)	$n(\varepsilon_F)$ (States/eV 1Ph)	C (emu K/mol 1Ph)	c_{af} (emu K/mol 1Ph)	J_{af} (eV)
PANI:SA$_{0.21}$[a]	0.45	7.5×10^{15}	0.021	0.72	—	—	—	—	—
PANI:SA$_{0.21}$[b]	0.46	9.6×10^{15}	0.018	0.19	3.1×10^{-5}	0.65	1.2×10^{-2}	4.2	5.1×10^{-2}
PANI:SA$_{0.42}$[a]	0.31	1.7×10^{17}	0.051	0.64	—	—	—	—	—
PANI:SA$_{0.42}$[b]	0.27	5.1×10^{16}	0.021	0.59	—	0.90	—	—	—
PANI:SA$_{0.53}$[a]	0.25	1.7×10^{17}	0.052	0.49	—	—	—	—	—
PANI:SA$_{0.53}$[b]	0.24	9.3×10^{16}	0.024	0.66	1.4×10^{-3}	1.4	1.6×10^{-2}	48.6	0.057
PANI:HCA$_{0.50}$[b]	—	—	—	—	—	1.9	—	—	—
PANI:AMPSA$_{0.4}$[a,c]	—	—	—	—	9.8×10^{-7}	—	4.5×10^{-4}	1.5×10^{-2}	0.001
PANI:AMPSA$_{0.4}$[a,d]	—	—	—	—	1.1×10^{-4}	0.42	7.2×10^{-3}	1.1×10^{-2}	0.005
PANI:AMPSA$_{0.6}$[a,c]	—	—	—	—	2.2×10^{-5}	—	1.7×10^{-2}	1.7×10^{-2}	0.004
PANI:AMPSA$_{0.6}$[a,d]	—	—	—	—	5.3×10^{-3}	3.5	6.5×10^{-1}	1.2×10^{-2}	0.006
PANI:CSA$_{0.5}$[a,c]	—	—	—	—	5.2×10^{-7}	—	2.3×10^{-4}	9.1×10^{-3}	0.005
PANI:CSA$_{0.5}$[a,d]	—	—	—	—	2.7×10^{-4}	1.2	2.7×10^{-2}	1.58	0.004
PANI:CSA$_{0.6}$[a,c]	—	—	—	—	8.5×10^{-7}	—	4.2×10^{-4}	6.6×10^{-3}	0.014
PANI:CSA$_{0.6}$[a,d]	—	—	—	—	7.1×10^{-5}	1.8	2.2×10^{-2}	2.13	0.004
PANI:TSA$_{0.5}$[a,e]	—	—	—	—	7.9×10^{-6}	0.6	1.3×10^{-3}	1.44	0.099
PANI:TSA$_{0.5}$[b,e]	—	—	—	—	3.3×10^{-6}	0.12	1.1×10^{-3}	0.29	0.041
PANI:TSA$_{0.5}$[a,f]	1.22	1.3×10^{19}	0.102	0.36	—	9.1	—	—	—
PANI:TSA$_{0.5}$[b,f]	0.17	9.6×10^{17}	0.058	0.28	5.6×10^{-4}	27	3.9×10^{-2}	—	—

[a] Determined at X-band EPR.

[b] Determined at D-band EPR.

[c] Determined for PC P$_{loc}^{+\cdot}$.

[d] Determined for PC P$_{mob}^{+\cdot}$.

[e] In nitrogen atmosphere.

[f] In air atmosphere.

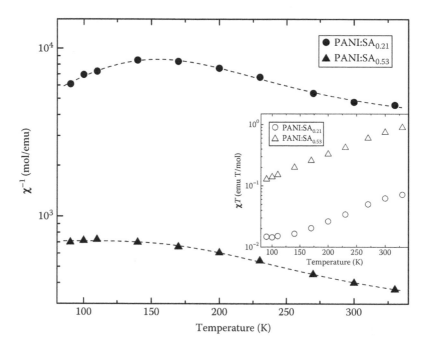

FIGURE 4.32
Temperature dependence of inversed paramagnetic susceptibility and χT product (inset) of PANI:SA samples with different doping levels. Upper and lower dashed lines show the dependencies calculated using Equation 2.6 with J_{af} equal to 0.051 and 0.057 eV, respectively, and other parameters presented in Table 4.6. (From Krinichnyi, V.I., *Appl. Phys. Rev.*, 1(2), 021305/01, 2014. With permission.)

crystalline high-conjugated regions through amorphous bridges and their scattering on the lattice optical phonons in metal-like clusters. Combining Equations 2.80 and 2.83, one gets for effective *dc* conductivity

$$\sigma_{dc}^{-1}(T) = k_1^{-1} T^{\pm 0.5} \exp\left[\left(\frac{T_0}{T}\right)^{\frac{1}{d+1}}\right] + k_2^{-1} T^{-1} \left[\sinh\left(\frac{E_{ph}}{k_B T}\right) - 1\right]^{-1}. \quad (4.7)$$

The parameters of Equation 4.7 determined from the fitting of experimental data are summarized in Table 4.7. It is seen from the table that both the percolation constant and lattice phonon energy of PANI:ES decrease as the polymer doping level increases. A transition from localization to delocalization of charge carriers occurs when $\pi^2 t_\perp/32^{1/2} T_0$ is equal to a unit [165]. This value was calculated for PANI:SA$_{0.21}$, PANI:SA$_{0.42}$, PANI:SA$_{0.53}$, and PANI:HCA$_{0.50}$ using [185] $t_\perp = 0.29$ eV and T_0 determined to be 0.31, 0.53, 2.1, and 3.1 eV, respectively. An increase in this value with y means increasing the charge carrier delocalization as a result of an increase in the interchain coherence.

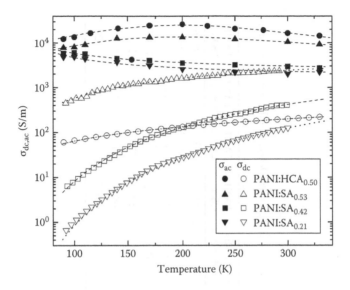

FIGURE 4.33
Temperature dependency of alternating current (filled symbols) and direct current (open symbols) conductivity calculated using Dyson-like electron paramagnetic resonance spectra of the PANI:HCA and PANI:SA samples with different doping levels y. Top and down dashed lines represent the dependencies calculated using Equations 2.92 and 2.93 with appropriate parameters summarized in Table 4.7. (From Krinichnyi, V.I., *Appl. Phys. Rev.*, 1(2), 021305/01, 2014. With permission.)

This value is higher than unity for heavily doped PANI:HCl [479] and decreases down to 0.1–0.4 for heavily doped derivatives of PANI, namely, poly(*o*-toluidine) and poly(*o*-ethylaniline) [480]. One can conclude that the insulator-to-metal transition in PANI:SA is of localization-to-delocalization type, driven by the increased structural order between the chains and through an increased interchain coherence. PANI:SA$_{0.53}$ and PANI:HCA$_{0.50}$ possess a more metallic behavior, and the properties of PANI:SA $y \leq 0.42$ demonstrate it to be near an insulator/metal boundary. The inherent disorder present in slightly doped PANI keeps the electron states localized on individual chains. At low y, the structural disorder in PANI localizes the charge to single chains (Curie-like carriers), and the higher doping leads to the appearance of delocalized electron states (Pauli-like carriers). This holds typically for the formation in PANI–ES with $y \geq 0.21$ metal-like domains, according to the island model proposed by Wang et al. [166,433]. This is consistent with that drawn earlier on the basis of data obtained with TEM and x-ray diffraction methods [469].

The energy of phonons interacting with a polaron determined from Equation 4.7 is near to that of superslow anisotropic librations obtained earlier from ST-EPR spectra. This means that macromolecular dynamics plays an important role in interacting processes taking place in spin reservoir.

TABLE 4.7

k_1, T_0, k_2, E_{ph} Values Determined from the Fitting of σ_{dc} by Equation 4.7, as well as σ_{01}, σ_{02}, and E_{ph}^l Ones Determined from the Fitting of σ_{ac} by Equation 4.8 for Polyaniline Samples with Different Counterions

Parameter	PANI:SA$_{0.21}$	PANI:SA$_{0.42}$	PANI:SA$_{0.53}$	PANI: HCA$_{0.50}$	PANI: AMPSA$_{0.4}$	PANI: AMPSA$_{0.6}$	PANI: CSA$_{0.5}$	PANI: CSA$_{0.6}$	PANI: TSA$_{0.5}$[a]	PANI: TSA$_{0.5}$[b]
k_1 (10^5 S K$^{\pm0.5}$ m)	730	35	9.2	0.84	1.2	1.1	5.6	5.5	—	0.026
T_0 (10^3 K)	19	11	2.8	3.6	0.209	0.277	0.521	0.753	—	46
k_2 (S K m)	0.92	1.3	53	51	31	18	170	29	—	19
E_{ph} (eV)	0.063	0.062	0.042	0.048	0.022	0.02	0.028	0.024	—	0.027
σ_{01} (S/K m)	41	48	86	150	168	162	199	227	1330	2490
σ_{02} (S/K m)	0.57	0.85	4.3	2.1	157	76	68	64	25	1080
E_{ph}^l (eV)	0.052	0.049	0.087	0.12	0.037	0.039	0.039	0.036	0.027	0.022

[a] Determined at X-band EPR.
[b] Determined at D-band EPR.

The lattice librations modulate the interacting spin exchange and consequently the charge transfer integral. Assuming that polaron is covered by both electron and excited phonon clouds, we can propose that both spin relaxation and charge transfer should be accompanied with the phonon dispersion. Such cooperating charge-phonon processes seem to be more important for the doped polymers the high-coupled chains of which constitute 3D metal-like clusters.

Aasmundtveit et al. [463] have shown that X-band EPR linewidth and consequently the spin–spin relaxation rate of PC in PANI depend directly on its *dc* conductivity. The comparison of $\Delta B_{pp}(T)$ and $\sigma_{ac}(T)$ functions presented in Figures 4.31 and 4.33 demonstrate the additivity of these values at least for the higher doped polymers. Using $T_2 = 1.7 \times 10^{-7}$ s, $\Sigma_{ij} = 1.2 \times 10^{57}$ m^{-6}, and $\omega_0 = 6.1 \times 10^{13}$ s^{-1} determined from experiment, a simple relation from Equation 2.26, $n_c \approx 55 \exp(n_s)$ is obtained. This means that on each chain at $L_{\parallel} = 7.0$ nm exist at least seven exchange spins which interact as spin-packet with $n_c = 20$ chains, i e. interchain hopping of spin charge carriers does not exceed distance more than $3d_{3D} < L_{\perp}$.

Figure 4.33 exhibits also the temperature dependence of the *ac* conductivity of highly doped PANI:SA and PANI:HCA samples determined from their Dysonian D-band EPR spectra using Equations 2.27 through 2.29 as well. The shape of the temperature dependencies presented demonstrates nonmonotonous temperature dependence with a characteristic point $T_c \approx$ 200 K. Such a temperature dependence can be attributed to the interacting charge carriers mentioned earlier with lattice phonons at high temperatures (the metallic regime) and by their Mott VRH at low temperatures (the semiconjugated regime). In this case, the charge transfer should consist of two successive processes, so then the *ac* conductivity should be expressed as a combination of Equations 2.81 and 2.83:

$$\sigma_{ac}^{\kappa}(T) = \left(\sigma_{0_1} T\right)^{\kappa} + \left\{\sigma_{0_2} T \left[\sinh\left(\frac{E_{ph}^{|}}{k_B T}\right) - 1\right]\right\}^{\kappa}. \qquad (4.8)$$

Figure 4.33 shows that the experimental σ_{ac} values obtained for PANI:SA and PANI:HCA are fitted well by Equation 4.8 with the appropriate parameters listed in Table 4.7. The energy determined for phonons in PANI sample lies near to that (0.066 eV) evaluated from the data obtained by Wang et al. [166,433].

RT σ_{ac} of heavily doped PANI:SA and PANI:HCA, estimated from the contribution of spin charge carriers, does not exceed 1.4×10^4 S/m. This value is much smaller than $\sigma_{ac}(\omega_e \rightarrow \infty) \cong 10^9$ S/m calculated theoretically [481]; however, it lies near that obtained for metal-like domains in PANI at 6.5 GHz [445]. The σ_{ac}/σ_{dc} ratio for these domains can be evaluated to be 80 for PANI:HSA$_{0.50}$ and 18, 7, and 4 for PANI:SA with $y = 0.21$, 0.42, and

0.53, respectively. Taking into account that the $\sigma_{ac} = \sigma_{dc}$ condition should be fulfilled for classic metals [62], one can conclude a better structural ordering of these domains in PANI:SA. The intrachain diffusion coefficient D_{1D} was determined for these polymers from relation $s_{ac} = e^2 n(\varepsilon_F) D_{1D} c_{1D}^2$ to vary within $(5-11) \times 10^{13}$ rad/s at RT that exceeds at least by an order of magnitude D_{1D}, such parameter obtained above for slightly doped samples. The RT mean free path $l_i = \sigma_{ac} m_c v_F/(Ne^2)$ [62] calculated for the highly doped PANI:SA and PANI:HCA lies near to 0.5 and 6.0 nm, respectively. These values are smaller than that estimated for oriented *trans*-PA [246] but also hold for extended electron states in these polymers. The energy of lattice phonons E_{ph}^l obtained from *ac* data lies near the energy J_{af} of the interaction between spins (see Table 4.6), which shows the modulation of spin–spin interaction by macromolecular dynamics in the system.

Indeed, the shape of EPR spectrum of PANI–ES depends on the nature of counterion. Figure 4.34 shows the EPR spectra of PANI films highly doped with camphorsulfonic and 2-acrylamido-2-methyl-1-propanesulfonic acids registered at different spin precession frequencies. For comparison, NMR spectra of PANI:CSA$_{0.5}$ registered at different temperatures are presented as well. In order to determine correctly the main magnetic resonance parameters (*g*-factor,linewidth, paramagnetic susceptibility) of PC with Dysonian contribution, all the spectra presented should be calculated using the method mentioned earlier. Konkin et al. have shown [186] that both the Dysonian line asymmetry factor A/B and the ratio D/A obtained for PC stabilized in the PANI:CSA and PANI:AMPSA samples are characterized by more complex dependence on the $2d/d$ ratio in Equations 2.30 and 2.31. The $D/A(2d/\delta)$ dependencies calculated for these polymer are presented in Figure 4.35. It was shown the applicability of such procedure for analysis of Dysonian EPR spectra of PC in PANI films while increasing their thickness in two, three and four times.

EPR spectra of PANI:CSA and PANI:AMPSA were then analyzed as sum of two different PCs coexisting in these polymers with Dysonian shape, namely, narrow EPR spectrum of PC $P_{loc}^{+\cdot}$ with $g = 2.0028$ localized in amorphous polymer matrix and broader EPR spectrum of PC $P_{mob}^{+\cdot}$ with $g = 2.0020$ and higher mobility in crystalline phase of the polymers. The cooling of the samples leads to the decrease in the relative concentration of PC $P_{mob}^{+\cdot}$ and to the monotonous increase in its linewidth, as it is seen in Figures 4.34 and 4.36. At the same time, the linewidth of PC $P_{loc}^{+\cdot}$ decreases monotonously and the sum spin concentration increases as the temperature decreases. Besides, NMR linewidth decreases at such a sample cooling (Figure 4.34) due possibly to the decrease of interaction of electron and proton spins. The RT ΔB_{pp} value of PC $P_{mob}^{+\cdot}$ in, for example, PANI:AMPSA$_{0.6}$ decreases from 5.4 down to 2.0 and then to 0.53 mT at the increase of registration frequency $\omega_e/2\pi$ from 9.7 up to 36.7 and then up to 140 GHz (Figure 4.34), so one can express this value as $\Delta B_{pp}(\omega_e) = 0.15 + 2.2 \times 10^8 \omega_e^{-0.84}$ mT.

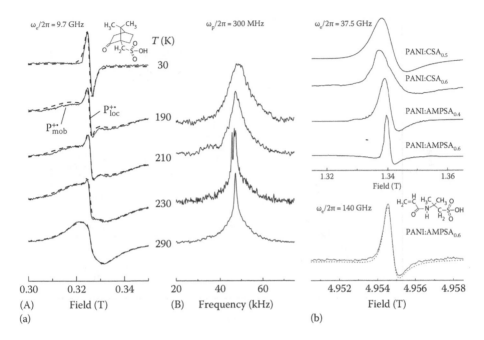

FIGURE 4.34

(a) X-band electron paramagnetic resonance (EPR) (A) and 300 MHz ^{1}H NMR (B) spectra of PANI:CSA$_{0.5}$ sample registered at different temperatures. Top-down dashed lines show the sum spectra of two lines calculated using Equations 2.27, 2.30, and 2.31 with the following respective values: $D_1/A_1 = 0.041$, $\Delta B_{pp1} = 1.99$ mT, $D_2/A_2 = 0.034$, $\Delta B_{pp2} = 49.2$ mT; $D_1/A_1 = 0.12$, $\Delta B_{pp1} = 1.81$ mT, $D_2/A_2 = 0.042$, $\Delta B_{pp2} = 17.3$ mT; $D_1/A_1 = 0.31$, $\Delta B_{pp1} = 2.17$ mT, $D_2/A_2 = 0.04$, $\Delta B_{pp2} = 17.2$ mT; $D_1/A_1 = 0.34$, $\Delta B_{pp1} = 2.41$ mT, $D_2/A_2 = 0.03$, $\Delta B_{pp2} = 15.2$ mT; and $D_1/A_1 = 0.26$, $\Delta B_{pp1} = 2.84$ mT, $D_2/A_2 = 0.02$, $\Delta B_{pp2} = 11.1$ mT. (b) Room temperature K-band and D-band EPR spectra of PANI:CSA and PANI:AMPSA samples with different doping levels y. The spectrum calculated with $D/A = 1.30$ and $\Delta B_{pp} = 0.533$ mT is shown by the dashed line as well (it is slightly shifted down only for the visibility). Counterions camphorsulfonic acid (CSA) and 2-acrylamido-2-methylpropane sulfonic acid (AMPSA) are also shown schematically. (From Krinichnyi, V.I., *Appl. Phys. Rev.*, 1(2), 021305/01, 2014. With permission.)

Such extrapolation reveals the dependence of spin–spin relaxation time on the registration frequency and allows estimating correct linewidth at $\omega_e \to$ 0 limit to be 0.15 mT.

The narrowing of the line on raising the PANI temperature can be explained by averaging of the local magnetic field caused by HFI between the localized spins whose energy levels lie near the Fermi level. The EPR line of PANI:CSA and PANI:AMPSA may also be broadened to some extent by relaxation due to the spin–orbital interaction responsible for linear dependence of T_1^{-1} on temperature [450]; however, this interaction seems to be rather weak in our case. It is significant that for both types of PCs, the linewidths are appreciably larger than those obtained previously for the fully oxidized powder-like and filmlike PANI:CSA (0.035 and 0.08 mT, respectively) [450], which

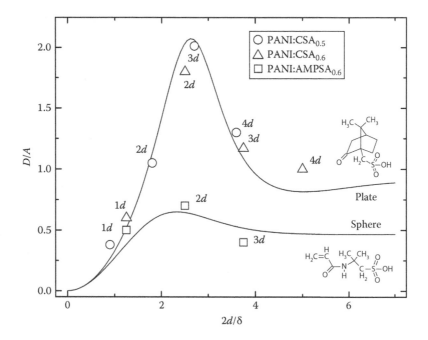

FIGURE 4.35

The theoretical $D/A(2d/\delta)$ dependencies calculated [186] and experimentally determined for the PANI:CSA and PANI:AMPSA films with different plate thickness (nd) at X-band electron paramagnetic resonance and $T = 300$ K. The structures of camphorsulfonic acid (CSA) and 2-acrylamido-2-methylpropane sulfonic acid (AMPSA) are shown schematically. (From Krinichnyi, V.I., *Appl. Phys. Rev.*, 1(2), 021305/01, 2014. With permission.)

indicates a higher conductivity of the samples under study. Comparison of the ΔB_{pp} values obtained for different PANI:CSA samples and presented in Figure 4.36 suggested that a crystalline phase is formed in the amorphous phase of the polymer, beginning with the oxidation level $y = 0.3$, and that the PCs of this newly formed phase exhibit a broader EPR spectrum. In the amorphous phase of the polymer, the PC $R1$ is characterized by less temperature-dependent linewidth and is likely not involved in the charge transfer being, however, as probes for whole conductivity of the sample. At the same time, the magnetic resonance parameters of radicals of the $P_{mob}^{+\cdot}$ type should reflect the charge transport in the crystalline domains of PANI:CSA and PANI:AMPSA. The linewidth of PANI appreciably decreases on replacement of the CSA anion by the AMPSA anion (see Figure 4.36), which is likely due to the shortening of spin–spin relaxation time of both PCs.

Figure 4.37 depicts the effective paramagnetic susceptibility of both the $P_{loc}^{+\cdot}$ and $P_{mob}^{+\cdot}$ PCs as a function of temperature. The J_{af} values obtained are much lower of the corresponding energy (0.078 eV) obtained for ammonia-doped PANI [180]. It is seen that at low temperatures when $T \leq T_c \approx 100$ K, the Pauli and Curie terms prevail in the total paramagnetic susceptibility

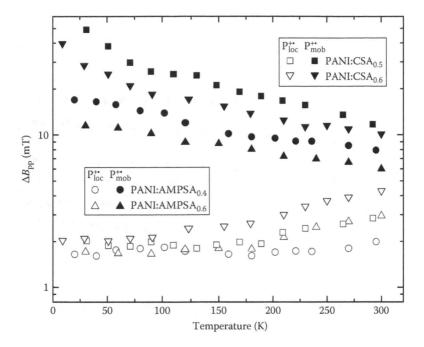

FIGURE 4.36

Temperature dependence of linewidth of the $P_{loc}^{\cdot+}$ (open points) and $P_{mob}^{\cdot+}$ (filled points) paramagnetic centers stabilized in the PANI:CSA and PANI:AMPSA samples with different doping levels y determined from their X-band electron paramagnetic resonance spectra with the Dysonian contribution. (From Krinichnyi, V.I., *Appl. Phys. Rev.*, 1(2), 021305/01, 2014. With permission.)

χ of both types of PCs in all PANI samples. At $T \geq T_c$, when the energy of phonons becomes comparable with the value $k_BT_c \approx 0.01$ eV, the spins start to interact that causes the appearance of the last term of Equation 2.6 in sum susceptibility as a result of the equilibrium between the spins with triplet and singlet states in the system. It is evident that the $P_{loc}^{\cdot+}$ signal susceptibility obeys mainly the Curie law typical for localized isolated PC, whereas the $P_{mob}^{\cdot+}$ susceptibility consists of the Curie-like and Pauli-like contributions. The data obtained from the fitting of experiment are summarized in Table 4.6.

The values of density of states $n(\varepsilon_F)$ at the Fermi level ε_F obtained for the charge carriers in PANI:CSA and PANI:AMPSA are also presented in Table 4.6. This value increases in the series PANI:AMPSA$_{0.4}$ → PANI:CSA$_{0.5}$ → PANI:CSA$_{0.6}$ → PANI:AMPSA$_{0.6}$. This parameter is in agreement with that obtained previously in the optical [279] (0.06–6 eV) and EPR [450] studies of polarons in PANI:CSA. The Fermi energy of the Pauli spins was calculated to be $\varepsilon_F \approx 0.2$ eV. This value is lower than the Fermi energy obtained for highly CSA- (0.4 eV [428]) and sulfur- (0.5 eV [184, 185]) doped PANI. Assuming again that the charge carrier mass in heavily doped polymer is

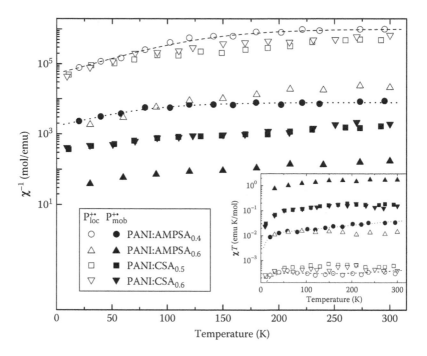

FIGURE 4.37

Temperature dependence of inversed effective paramagnetic susceptibility χ and χT product (inset) of the $P_{loc}^{\cdot+}$ (open points) and $P_{mob}^{\cdot+}$ (filled points) paramagnetic centers stabilized in the PANI:CSA and PANI:AMPSA samples with different doping levels y. Dashed and dotted exemplary lines show dependencies calculated using Equation 2.6 with $\chi_P = 9.8 \times 10^{-7}$ emu/mol, $n(\varepsilon_F) = 0.00$ states/eV, $C = 4.5 \times 10^{-4}$ emu K/mol, $c_{af} = 1.5 \times 10^{-2}$ emu K/mol, and $J_{af} = 1$ meV and $\chi_P = 1.1 \times 10^{-4}$ emu/mol, $n(\varepsilon_F) = 0.42$ states/eV, $C = 7.2 \times 10^{-3}$ emu K/mol, and $c_{af} = 1.1 \times 10^{-2}$ emu K/mol, and $J_{af} = 5$ meV, respectively. Respective parameters obtained for localized and mobile polarons in other doped polymers are summarized in Table 4.6. (From Krinichnyi, V.I., *Appl. Phys. Rev.*, 1(2), 021305/01, 2014. With permission.)

equal to the mass of free electron, that is, $m_c = m_e$, the number of charge carriers $N_c = (2m_c\varepsilon_F/\hbar^2)^{3/2}/3\pi^2$ [62] in such a quasi-metal was determined to be $N_c \approx 4.1 \times 10^{20}$ cm^{-3}. This is close to the spin concentration in this polymer; therefore, one can conclude that all delocalized PCs are involved in the charge transfer in PANI:CSA$_{0.6}$. The velocity of charge carriers near the Fermi v_F level in PANI:CSA was also calculated to be 3.8×10^7 cm/s that is close to those evaluated for this polymer from the EPR magnetic susceptibility data, $(2.8–4.0) \times 10^7$ cm/s [482,483] and 6.2×10^7 cm/s for PANI:AMPSA.

Thus, main PCs in the highly doped PANI:CSA and PANI:AMPSA samples are localized at $T \leq T_c$. This is the reason for the Curie type of susceptibility of the sample and should lead to the VRH charge transfer between the polymer chains. The spin–spin exchange is stimulated at $T \geq T_c$ due likely to the activation librations of the polymer chains. The activation energies of these librations lie within the energy range characteristic of PANI:CSA [484],

PANI:HCA [182,183], and PTTF [354,409]. The E_a value depends on the effective rigidity and planarity of the polymer chains that are eventually responsible for the electrodynamic properties of the polymer.

DC and *AC* conductivities of the highly doped PANI:CSA and PANI:AMPSA determined, respectively, by the *dc* conductometric method [186] and from the Dysonian spectra of the $P_{mob}^{\cdot+}$ PC are displayed in Figure 4.38 as a function of temperature. Charge carrier hops through amorphous part of the sample and then diffuses through its crystalline domain, so then the *dc* term of the total conductivity of the samples should be determined by 1D VRH between metal-like domains accompanied by the charge carriers scattering on the lattice phonons in these domains described by Equations 2.80 and 2.83. These processes occur parallel, so the effective conductivity can be expressed by Equation 4.7. Indeed, Figure 4.38 shows that *dc* conductivity of the polymers experimentally obtained is fitted well by Equation 4.7 whose fitting parameters are also summarized in Table 4.7. The v_F value was calculated for the PANI:CSA and PANI:AMPSA systems to be 2.4×10^7 and 6.0×10^7 cm/s, respectively [186].

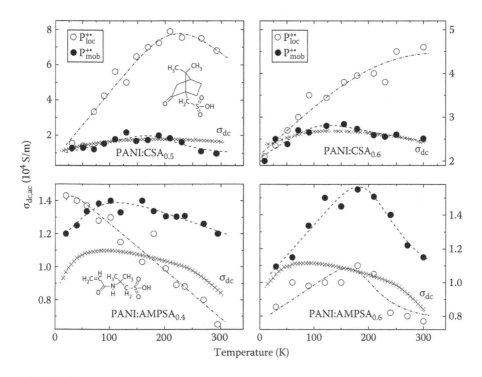

FIGURE 4.38

Temperature dependence of direct current (marked by the × symbol) and alternating current conductivity determined from Dysonian spectra of the $P_{loc}^{\cdot+}$ (open circles) and $P_{mob}^{\cdot+}$ (filled circles) paramagnetic centers stabilized in the PANI:CSA and PANI:AMPSA films. The lines show the dependencies calculated using Equations 4.1 and 4.2 with appropriate parameters presented in Table 4.7. (From Krinichnyi, V.I., *Appl. Phys. Rev.*, 1(2), 021305/01, 2014. With permission.)

In contrast with other conjugated polymers, lower T_0 parameter is characteristic for these samples. It should be noted that the 3D VRH model gives abnormal low T_0 value for all the samples. This is evidence of the longer averaged length of charge wave localization function in the samples. Indeed, $\langle L \rangle$ value was determined for PANI:CSA$_{0.6}$, PANI:CSA$_{0.5}$, PANI:AMPSA$_{0.6}$, and PANI:AMPSA$_{0.4}$ to be 17, 37, 264, and 239 nm, respectively. However, the $n(\varepsilon_F)$ value obtained for these polymers changes otherwise increasing in the series PANI:AMPSA$_{0.4}$ → PANI:CSA$_{0.5}$ → PANI:CSA$_{0.6}$ → PANI:AMPSA$_{0.6}$.

AC conductivity of the polymers also reflects the successive mechanisms mentioned earlier, so the experimental data can be circumscribed by Equation 4.7. Indeed, it is seen from Figure 4.38 that the $\sigma_{ac}(T)$ dependencies obtained experimentally for mobile PC $P_{mob}^{+\cdot}$ are fitted well by Equation 4.8 with the parameters summarized in Table 4.7. The energy determined for phonons in these PANI samples lies near to that obtained for other polymers [118,184] and evaluated (0.066 eV) from the data determined by Wang et al. for HCl-doped PANI [166,433]. It is evident that E_{ph}^{l} and E_a obtained earlier for the PANI:CSA$_{0.5}$ sample lie close. This means that protons situated in crystalline domains indeed sense electron spin dynamics. The data obtained can be evidence of the contribution of the $P_{loc}^{+\cdot}$ and $P_{mob}^{+\cdot}$ PCs in the charge transfer through, respectively, amorphous and crystalline parts of the polymers. RT σ_{ac} values determined from Dysonian spectra of $P_{mob}^{+\cdot}$ PC lie near respective σ_{dc} values that is characteristic for classic metals. The $\langle L \rangle$ and E_{ph} values determined earlier for mediatory doped samples correlate. This means that the higher the $\langle L \rangle$ value, the stronger the interaction of PC with phonons in metal-like crystallites. There is some tendency in the increase of RT σ_{dc} and σ_{ac} conductivities in the series PANI:AMPSA$_{0.6}$ → PANI:AMPSA$_{0.4}$ → PANI:CSA$_{0.5}$ → PANI:CSA$_{0.6}$ *feeling* by the $P_{mob}^{+\cdot}$ PC.

Thus, main PCs in the highly doped PANI:CSA and PANI:AMPSA samples are localized at low temperatures. This originates the Curie type of susceptibility of the samples and the VRH charge transfer between their polymer chains. The spin–spin exchange is stimulated at high-temperature region due likely to the activation librations of the polymer chains [485,486]. Both pinned and delocalized PCs are formed simultaneously in the regions with different crystallinity. An antiferromagnetic interaction in crystalline domains is stronger than that in amorphous regions of PANI:CSA and PANI:AMPSA. Charge transport between crystalline metal-like domains occurs through the disordered amorphous regions where the charge/spin carriers are more localized. The assumption that higher purity PANI coupled with homogeneous doping would give rise to no EPR signal, characteristic of a purely bipolaronic matrix, is in contradiction with the increase of *ac* conductivity with spin concentration in polymer systems. Both PANI:CSA and PANI:AMPSA reveal better electronic properties over PANI:SA and PANI:HCA, as shown by their electrical conductivity, which is both greater in magnitude and follows metallic temperature dependence. The change of conductivity with temperature is consistent with a disordered metal close to the critical regime of the metal–insulator transition with the Fermi energy close to the mobility edge [450,451].

PANI highly doped with *para*-toluenesulfonic acid, PANI:TSA$_{0.50}$, in nitrogen atmosphere at X-band EPR demonstrates Lorentzian exchange-narrowed lines in which an asymmetry factor A/B is 1.03 (Figure 4.39). The exposure of the samples to air was observed to lead to a reversible line broadening and an increase in the asymmetry factor up to 1.27. The asymmetry of the EPR line may be due to either unresolved anisotropy of the g-factor or the presence in the spectrum of the Dysonian term as in the case of other highly doped conjugated polymers. To verify these assumptions, the D-band EPR spectra of the sample were recorded. It is seen from Figure 4.39 that the polymer in this waveband EPR also exhibits a single asymmetric line, in which asymmetry factor varies at the exposition to air from 1.68 up to 1.95. This fact indicates substantial interaction of PC even in high fields; the line asymmetry of these PC indeed results from the interaction of an MW field with charge carriers in the skin layer.

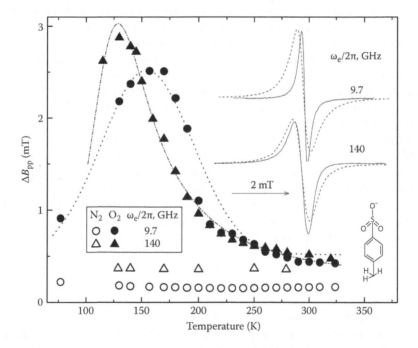

FIGURE 4.39
Inset: X-band and D-band electron paramagnetic resonance (EPR) spectra of the PANI:TSA$_{0.50}$ sample in nitrogen (solid lines) and air (dashed lines) atmospheres. Structure of the TSA counterion is also shown schematically. Temperature dependence of linewidth of paramagnetic centers in the PANI:TSA$_{0.50}$ sample with the presence of nitrogen and oxygen molecules registered at both the EPR wave bands. The lines show the dependencies calculated using Equation 2.25 with respective parameters presented in Table 4.6. (From Krinichnyi, V.I., *Appl. Phys. Rev.*, 1(2), 021305/01, 2014. With permission.)

Dysonian EPR spectra of the samples were calculated from Equations 2.27 through 2.29, and the main magnetic parameters were obtained.

As the spin precession frequency increases from 9.7 up to 140 GHz, the ΔB_{pp} value of PC in the PANI sample increases not more than by a factor of 2 (Figure 4.39). Such insignificant line broadening with the operating frequency increase was not observed in studies on other conjugated polymers, including PANI. This may be evidence for stronger exchange interaction between PCs in the polymer, which is not completely relieved in strong magnetic field. The temperature dependence of the effective absorption linewidth of the sample determined at both the X- and D-band EPR is presented in Figure 4.39. It is seen that ΔB_{pp} of PC in the PANI:TSA$_{0.50}$ sample containing nitrogen slightly depends on temperature. Air diffusion into the samples leads to the reversible extremal broadening of its EPR line. As shown in Figure 4.39, the T_c value described earlier characteristic for the $\Delta B_{pp}(T)$ dependencies presented shifts nearly from 160 to 130 K at the increase of polarizing frequency from 9.7 GHz up to 140 GHz. Such effect was interpreted as a result of exchange interaction of polarons with oxygen molecules possessing sum spin $S = 1$. The $\Delta B_{pp}(T)$ dependencies were fitted by Equation 2.25 with appropriate parameters listed in Table 4.6 (Figure 4.39). It is seen that the experimental data obtained can be rationalized well in terms of this theory. This result reveals that the oxygen biradical acts as a nanoscopic probe of the polaron dynamics. The obtained value of J_{ex} sufficiently exceeds the corresponding spin exchange constant for nitroxide radicals with paramagnetic ions, $J_{ex} \leq 0.01$ eV [152].

Figure 4.40 shows the temperature dependence for the paramagnetic susceptibility of PC in the PANI:TSA$_{0.50}$ sample in the absence and in the presence of oxygen in the polymer. An analysis of the paramagnetic susceptibility of the nitrogen containing sample determined at X- and D-band EPR showed that it can be described by Equation 2.6 with the parameters presented in Table 4.6. These data show that the transition of registration frequency from 9.7 to 140 GHz leads to a decrease in J_{af} determined for nitrogen filled PANI:TSA$_{0.5}$ sample from 0.099 down to 0.041 eV due possibly to the field effect. Figure 4.40 reveals that the effective susceptibility of the PANI:TSA$_{0.50}$ without oxygen, as determined from X-band EPR spectra, slightly varies with temperature. However, this value determined from the D-band EPR spectra noticeably decreases with a decrease in temperature. The exposure of the polymer to air increases proportionally all components of its magnetic susceptibility. This value of the sample exposed to air increases substantially and exhibits nonmonotonic temperature dependence in the X- and D-band EPR with a characteristic temperature maximum at 160 and 150 K, respectively. The nature of this effect is discussed earlier. The Pauli susceptibility of the sample is close to that determined at $n(\varepsilon_F) = 22.8$ eV^{-1} [281]. Note that the χ parameter measured for PANI:TSA$_{0.5}$ by more direct method [487] exhibits smaller temperature dependence. The velocity of charge carriers and the Fermi energy were calculated as $v_F = 2c_{1D}/\pi\hbar n(\varepsilon_F)$ and $\varepsilon_F = 3N_e/2n(\varepsilon_F)$ [62] to

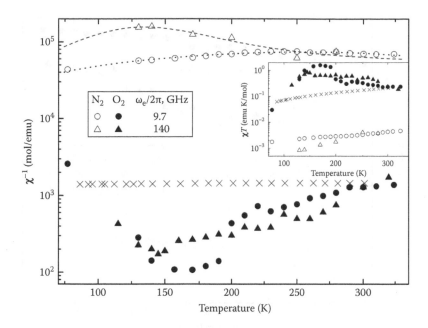

FIGURE 4.40

Temperature dependence of inversed effective paramagnetic susceptibility and χT product (inset) of paramagnetic centers stabilized in PANI:TSA$_{0.5}$ sample exposed to nitrogen (open symbols) and oxygen (filled symbols) determined at different spin precession frequencies. By the × symbol is shown the appropriate data [487] obtained for PANI:TSA$_{0.5}$ by using a *force* magnetometer in a direct current external magnetic field of 0.5 T. Dashed and dotted lines show the dependencies calculated using Equation 2.6 with respective $\chi_P = 3.3 \times 10^{-6}$ emu/mol, $C = 1.1 \times 10^{-3}$ emu K/mol, $c_{af} = 0.29$ emu K/mol, and $J_{af} = 0.041$ eV and $\chi_P = 7.9 \times 10^{-6}$ emu/mol, $C = 1.3 \times 10^{-3}$ emu K/mol, $c_{af} = 1.44$ emu K/mol, and $J_{af} = 0.099$ eV. (From Krinichnyi, V.I., *Appl. Phys. Rev.*, 1(2), 021305/01, 2014. With permission.)

be 3.1×10^6 cm/s and 0.16 eV, respectively. The latter value is less of Fermi energy obtained earlier for PANI:SA, PANI:CSA, and PANI:AMPSA.

Figure 4.41 depicts the temperature dependence of σ_{dc} determined for the PANI:TSA$_{0.50}$ sample by the *dc* conductometric method. An analysis of these dependencies leads to the conclusion that this polymer exhibits 1D VRH at low-temperature region, typical of a granular metal. As in the case of other PANI, the σ_{dc} value is governed by strong spin–spin interaction at high temperatures. Experimental data are shown from Figure 4.41 to be well described by Equation 4.7 with the parameters presented in Table 4.7.

The averaged length of charge wave localization $\langle L \rangle = 11$ nm exceeds the effective radius of a quasi-metallic domain equal to 4 nm [488]. It can be due to closely electronic properties of the metal-like domains in the polymer. It allows to evaluate charge transfer integral $t_\perp$ in such domains from relation connecting T_0 values at 1D VRH $T_0^{(3D)} = 256 T_0^{(1D)} \ln \left(2 T_0^{(1D)} / \pi t_\perp \right)$ [479] to be 0.10 eV. The most probable carrier hopping range $R = \left(T_0 / T \right)^{1/2} \langle L \rangle / 4$ was

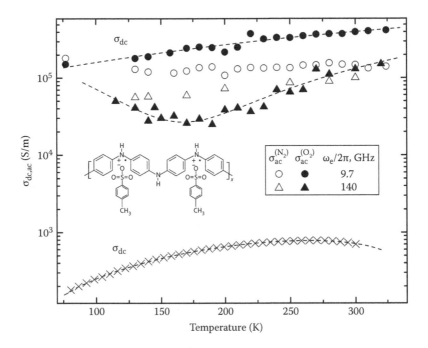

FIGURE 4.41
Temperature dependence of direct current and alternating current conductivities determined from X-band and D-band electron paramagnetic resonance Dysonian spectra of the PANI:TSA$_{0.5}$ sample in nitrogen (open symbols) and air (filled symbols) atmospheres. Dashed lines show the dependencies calculated using Equations 4.1 and 4.2 with respective parameters presented in Table 4.7. (From Krinichnyi, V.I., *Appl. Phys. Rev.*, 1(2), 021305/01, 2014. With permission.)

determined at RT to be 34 nm. The hopping energy W of a charge carrier in the polymer was determined in terms of the VRH theory, $W = k_B(T_0T^3)^{1/4}/2$, to be 0.034 eV, that is of the order of k_BT. The data make it possible to calculate the velocity of charge carriers $v_F = 4.0 \times 10^6$ cm/s moving near the Fermi energy $\varepsilon_F = 0.34$ eV. The latter value lies near to that determined for PANI:CSA (0.4 eV [428]) and PANI:SA (0.5 eV [181,185]). This is in agreement with the supposition earlier made by Pelster et al. [488] that the charge transport in PANI:TSA takes place *via* two contributions: metallic conduction through a crystalline core of 8 nm and thermally activated tunneling (hopping) through an amorphous barrier of 1–2 nm diameter.

Spin–lattice and spin–spin relaxation times measured by the saturation method at X-band EPR for the PANI:TSA$_{0.50}$ by using Equations 2.32 and 2.33 are, respectively, 1.2×10^{-7} and 3.1×10^{-8} s (in the nitrogen atmosphere) and 1.1×10^{-7} and 1.6×10^{-8} s (in the air) [473,474]. If one supposes that the polarons in this polymer possess mobility and diffuse along and between polymer chains with the diffusion coefficients D_{1D} and D_{3D}, respectively, $D_{1D} = 3.5 \times 10^8$ and $D_{3D} = 1.1 \times 10^9$ rad/s (in the nitrogen atmosphere) and $D_{1D} = 8.1 \times 10^{11}$ and $D_{3D} = 2.3 \times 10^8$ rad/s (in the air) are evaluated from Equations 2.46, 2.51, and 2.52. Corresponding conductivities due to so possible polaron mobility calculated from Equation (2.71) are, respectively, $\sigma_{1D} = 2.5 \times 10^{-2}$ S/m, $\sigma_{3D} = 2.3 \times 10^{-2}$ S/m and $\sigma_{1D} = 2.9 \times 10^3$ S/m, $\sigma_{3D} = 0.24$ S/m. This means that $D_{1D} < D_{3D}$ in the sample without oxygen; however, the conductivity appears to be practically isotropic in character. The D_{1D}/D_{3D} ratio for the sample exposed to air increases to ~10^4, which substantially exceeds the value $D_{1D}/D_{3D} \sim 50$ obtained for highly doped PANI:HCA [452]. The conductivity of this sample also becomes anisotropic, σ_{1D}/σ_{3D} ~10^4, and is also determined mainly by the diffusion of paramagnetic center along the polymer chain. The data obtained can be compared with those evaluated from the Dysonian EPR spectra.

The σ_{ac} values determined for the PANI:TSA$_{0.50}$ sample from their Dysonian X- and D-band EPR spectra by using Equations 2.27 through 2.29 are shown in Figure 4.41 versus temperature. An intrinsic conductivity of the sample visibly increases at its exposition to air (Figure 4.41). RT conductivity obtained for the sample at X-band EPR is two orders of magnitude higher than σ_{1D} and σ_{3D} calculated earlier in terms of Q1D polaron diffusion along the *single conjugated chain* [452]. Hence, it may be concluded that the conductivity in this nanomodified polymer, as in PANI with other counterions, is mainly governed by the mobility of 3D delocalized electrons in metal-like domains in which paramagnetic polarons are localized on parallel chains due to their strong exchange interaction. The temperature dependence of intrinsic conductivity can be interpreted in terms of the VRH mechanism of charge carriers and their scattering on the polymer lattice phonons, respectively, in amorphous and crystalline phases of the samples. Analogous to other PANI, the charge carrier crosses these phases one after another, so the resulting conductivity should be described by Equation 4.8. Figure 4.41 evidences that the $\sigma_{ac}(T)$ dependence evaluated for the PANI sample exposed to air follows well Equation 4.8 with the parameters listed in Table 4.6. The $E_{ph}^{\mid}$ value obtained for the sample correlate with E_{ph} determined from the fitting of its $\sigma_{dc}(T)$ dependence in terms of the same charge transport mechanism (Table 4.7). This fact confirms additionally supposition mentioned earlier made on the existence of strong spin dipole–dipole interaction in crystalline domains.

A decrease in σ_{ac} with an increase in the registration frequency may be the result of, for example, the influence of external magnetic field on the spin exchange process in the polymer or deeper penetration of MW field into the polymer bulk at D-band EPR. Indeed, the intrinsic conductivity should be higher if the skin layer is formed on metal-like domains with a smaller radius in PANI particles.

In general, as in the case of main conjugated polymers, two types of PCs are formed in PANI, polarons localized on chains in amorphous polymer regions and polarons moving along and between polymer chains. During the polymer doping, the number of the mobile polarons increases and the conjugated chains become crystallization centers for the formation of the massive metal-like domains of strongly coupled chains with 3D delocalized charge carriers. This process is accompanied by the increase of the electron–phonon interaction, crystalline order, and interchain coupling. The latter factor plays an important role in the stabilization of the metallic state, when both 1D electron localization and *Peierls instability* are avoided. Above the percolation threshold, the interaction between spin charge carriers becomes stronger and their mobility increases, so part of the mobile polarons collapses into diamagnetic bipolarons. Besides, the doping changes the interaction of the charge carriers with the lattice phonons and therefore the mechanism of charge transfer. It also results in an increase of the number and size of highly conjugated domains containing charge carriers of different types and mobilities, which lead to an increase in the conductivity and Pauli susceptibility. Such process is modulated by macromolecular dynamics and is accompanied by an increase in the crystalline order (or dimensionality) and planarity of the system.

In the initial PANI, the charges are transferred isoenergetically between solitary chains in the framework of the Kivelson formalism. The growth of the system dimensionality leads to the scattering of charge carriers on the lattice phonons. The charge 3D and 1D hops between these domains in the medium and heavily doped PANI, respectively. In heavily doped PANI, the charge carriers are transferred according to the Mott VRH mechanism that is accompanied by their scattering on the lattice phonons. This is in agreement with the concept of the presence of 3D metal-like domains in PANI–ES rather than the supposition that 1D solitary conjugated chains exist even in heavily doped PANI.

In contrast with PANI:SA and PANI:HCA characterized as a Fermi glass with electronic states localized at the Fermi energy due to disorder, PANI:CSA, PANI:AMPSA, and PANI:TSA are disordered metals on the metal–insulator boundary. The metallic quality of ES form of PANI grows in the series PANI:HCA → PANI:SA → PANI:TSA → PANI:AMPSA → PANI:CSA.

4.7 Polythiophene

Polythiophene (PT) shown in Figure 1.2 and its composites are also considered as perspective systems for designing the various molecular devices [12,50,290,489]. Pristine PT demonstrates a single symmetric line with $g = 2.0026$ and $\Delta B_{pp} = 0.8$ mT at X-band, showing that the spins do not belong to a sulfur-containing moiety and are localized on the polymer chains [104,105]. The low concentration of PC ($n \cong 66$ ppm or 6.6×10^{-5} spin per monomer unit) is consistent with a relatively high purity of material, containing few chain defects. The doping of polymer leads to the formation on its chains of polarons in which spins are delocalized along eight thiophene units [490]. This process also provokes the appearance in the gap of asymmetric states with $\Delta E_1 = 0.32$ eV and $\Delta E_2 = 0.47$ eV [369] (see Figure 1.4). Stafström and Brédas [369] found that these energies increase up to 0.57 and 0.73 eV, respectively, as pairs of polarons collapse into diamagnetic bipolarons. Kaneto et al. [491] obtained a maximum number of spins in PT doped with BF_4^- up to $y \approx 0.03$ by EPR study, suggesting a crossover from polaron to bipolaron at this doping level. On the other hand, a fairly small number of spins in PT doped by iodine were reported [492, 493]. Chen et al. [494] found only a vanishingly small EPR signal in PT electrochemically doped with ClO_4^- up to $y = 0.14$, suggesting a bipolaron ground state.

Spin dynamics in PT electrochemically doped with ClO_4^- was studied at RT by EPR [495]. It was shown that highly doped sample reveals temperature dependence of linewidth due to the Elliott mechanism [248], characteristic of metals. RT coefficients of spin diffusion along and between polymer chains as well as the conductivity terms due to such spin diffusion were obtained to be, respectively, $D_{1D} = 1.9 \times 10^{15}$ rad/s, $D_{3D} = 5.5 \times 10^9$ rad/s and $\sigma_{1D} = 1.2 \times 10^5$ S/m, $\sigma_{3D} = 3.7$ S/m at $n(\varepsilon_F) = 0.12$ states per eV per C atom.

The conductivity of $PT:ClO_4^-$ changes as T^{-2} at high temperatures, whereas a metal–insulator (or semiconductor) transition takes place in this sample at 30 K [496]. The temperature dependence of conductivity of $PT:BF_4^-$ supports the Mott's VRH mechanism [497]. Temperature dependence of activation energy indicated that the charge carrier hopping is the dominating mechanism of charge transport in this sample.

Powder-like nanocomposites of PT synthesized electrochemically from monothiophene (PT) and dithiophene (PdT) and counterions BF_4^-, ClO_4^-, and I_3^- as a dopant were studied at both the X- and D-band EPR [331].

At the X-band EPR, these samples demonstrate a symmetric single line with effective $g \cong g_e$ and the width, slightly changing in a wide temperature range (Table 4.8). However, the spectrum of the $PT:I_3^-$ sample is broadened significantly with the temperature increase. At this waveband, EPR spectrum of $PdT:ClO_4^-$ appears as a single symmetric line, in which the width decreases smoothly monotonically from 0.70 down to 0.25 mT as the temperature decreases from RT down to 77 K.

D-band EPR spectra of these conjugated polymers demonstrate a greater variety of line shape (Figure 4.42). The figure shows that PCs stabilized in $PT:BF_4^-$ and $PT:ClO_4^-$ samples are characterized by an axially symmetric spectrum typical for PC localized on a polymer backbone. The analogous situation seems to be realized also for $PT:I_3^-$, for which the broadening and overlapping of canonic components of EPR spectrum can take place due to a stronger spin–orbit interaction of PCs with counterions. $PdT:ClO_4^-$ sample also demonstrates a single EPR line at this waveband EPR in a wide temperature range, thus indicating the domination of delocalized PCs in this polymer.

TABLE 4.8

g Tensor Components, $g_{\parallel}$ and $g_{\perp}$, the Energy of Electron Excitation State $\Delta E\sigma_{\pi*}$, Spin Susceptibility χ, Linewidth ΔB_{pp}, Charge Carrier Q1D Diffusion Rate D_{1D}, and Intrinsic Conductivity σ_{ac} of Polythiophene Samples at Room Temperature

Parameter	PT:J$_3^-$	PT:BF$_4^-$	PT:ClO$_4^-$	PdT:ClO$_4^{-a}$
$g_{\parallel}$	2.00679[b]	2.00412	2.00230	2.00232[b]
$g_\perp$	2.00232[b]	2.00266	2.00239	2.00364[b]
$\Delta E\sigma_{\pi*}$ (eV)	1.6	4.0	7.1	4.5
χ (emu/mol)	2.7×10^{-5}	6.8×10^{-5}	4.4×10^{-5}	8.7×10^{-5}
ΔB_{pp}^{c} (mT)	0.75	0.23	0.46	0.70
$\Delta B_{pp}^{c,d}$ (mT)	0.80	0.34	0.52	0.25
ΔB_{pp} (mT)	6.52	1.52	2.63	0.51
D_{1D} (rad/s)	1.2×10^{13}	1.1×10^{13}	1.0×10^{14}	4.1×10^{13}
σ_{ac} (S/m)	3.2×10^2	1.2×10^3	6.8×10^3	5.5×10^3

[a] Synthesized from bithiophene.
[b] The values are calculated from the equation $g_{iso} = 1/3(g_\perp + 2g_{ge})$.
[c] Measured at X-band EPR.
[d] Measured at $T = 77$ K.

Magnetic resonance parameters, calculated for these samples from their D-band EPR spectra, are also presented in Table 4.8. It derives from the analysis of the data that the energy of an excited configuration $\Delta E_{\sigma\pi} \propto \Delta g^{-1}$ determined from Equation (2.3) increases more than four times at the transition from I$_3^-$ to BF$_4^-$ and to ClO$_4^-$ anions. Such a transition also leads to the growth of film conductivity and a sufficient PC concentration change. This may be the evidence for the realization of charge transfer in PT both by polarons and bipolarons, whose concentrations depend on the origin of anion, introduced into a polymer. The width of EPR spectral components of PT is susceptible to the registration frequency change (Table 4.8), indicating a strong spin–spin exchange in PT.

As the temperature decreases, a Dyson-like line is displayed in the region of a perpendicular component of the PT:BF$_4^-$ EPR spectrum without a noticeable change of signal intensity (Figure 4.42). Further temperature decrease results in the increase of the line asymmetry factor A/B without reaching the extreme in the 100–300 K temperature range, thus being the evidence for the growth of ac conductivity as it occurs in case of other semiconductors of lower dimensionality. Therefore, an intrinsic conductivity of a sample can be determined from Equations 2.27 through 2.29 using the characteristic size of sample particles. Then applying relationship (2.71), one can determine the mobility μ_{1D} and the rate of Q1D diffusion D_{1D} of charge carriers in PT. RT σ_{ac}, μ_{1D} and D_{1D} values calculated for PT:BF$_4^-$ according to this procedure amount to 3.6×10^2 S/m, 0.11 cm^2 V^{-1} s^{-1} and 3.2×10^{12} s^{-1}, respectively.

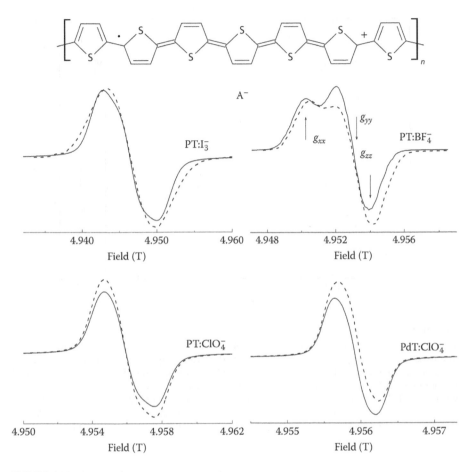

FIGURE 4.42
X-band absorption electron paramagnetic resonance spectra of electrochemically synthesized polythiophene (PT) and polydithiophene and registered at $T = 300$ (solid line) and 200 K (dashed line). The components of **g** tensor are shown. At the top, the PT-based nanocomposite with polaron charge carrier is shown schematically. (Modified from Krinichnyi, V.I., *Synth. Metals*, 108(3), 173, 2000. With permission.)

D-band EPR spectrum of the PdT:ClO$_4^-$ sample also demonstrates Dyson-like line, being the reflection of the change of its intrinsic conductivity. Using the procedure mentioned earlier in the suggestion that the total concentration of charge carriers of both types is constant, one can establish the main magnetic resonance parameters (Table 4.8). With the temperature growth, a linewidth of the PdT:ClO$_4^-$ sample first increases and starts to decrease as T becomes lower than $T_c \approx 170$ K (Figure 4.43). This is accompanied by the resembling change in an inverted paramagnetic susceptibility χ^{-1} (Figure 4.43). Such a fact is evidence that the $\Delta B_{pp}(T)$ and $\chi(T)$ dependencies obtained for this sample are not symbasis. An extremal change in $\Delta B_{pp}(T)$ can be

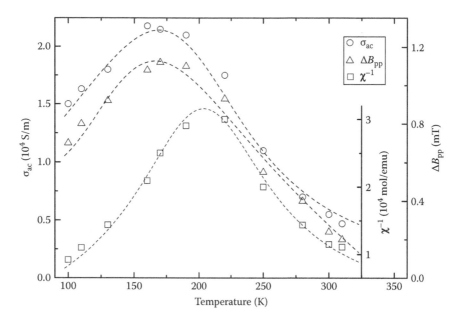

FIGURE 4.43

Temperature dependence of the intrinsic conductivity σ_{ac}, linewidth ΔB_{pp}, and inverted magnetic susceptibility of PdT–ClO$_4^-$ determined from its Dysonian D-band electron paramagnetic resonance spectra. Top-down dashed lines represent the dependencies calculated using Equation 4.8 with $\sigma_{0_1} = 1.4 \times 10^2$ S/K m, $\sigma_{0_2} = 0.31$ S/K m, and $E_{ph}^l = 0.13$ eV; Equation 2.25 with $\Delta B_{pp}^0 = 0.81$ mT, $\omega_{hop}^0 = 1.4 \times 10^{17}$ s^{-1}, $E_r = 0.040$ eV, and $J_{ex} = 0.57$ eV; and Equation 2.6 with $\chi_P = 8.2 \times 10^{-5}$ emu/mol, $C = 2.1 \times 10^{-2}$ emu K/mol, $c_{af} = 10.2$ emu K/mol, and $J_{af} = 0.071$ eV. (Modified from Krinichnyi, V.I., *Synth. Metals*, 108(3), 173, 2000. With permission.)

interpreted in terms of the Houzé–Nechtschein model [153] of the exchange interaction of spins localized on neighboring polymer chains in Q1D polymer system. Figure 4.43 shows the adaptability of this model for the PdT:ClO$_4^-$ sample and evidences for the strong and weak interaction between spins below and above T_c, respectively. The energy of activation of such interaction obtained from Equation 2.25, $E_a = 0.040$ eV, lies near that determined for macromolecular librations in other conjugated polymers [118,122,124]. This leads to a reversible collapse of polaron pairs into bipolarons at $T \leq T_c$ and to bipolaron decay to polarons at higher temperatures. Figure 4.43 also shows that the temperature dependence of an effective magnetic susceptibility of the sample follows Equation 2.6 with $\chi_P = 8.2 \times 10^{-5}$ emu/mol, $C = 2.1 \times 10^{-2}$ emu K/mol, and $J_{af} = 0.071$ eV. Assuming a linear dependence for the bipolaron decay rate and frequency of polymer chain librations, the activation energy of the latter process can be evaluated from $\chi(T)$ dependence to be equal to $E_a = 0.025$ eV at $T \geq 200$ K. Such a complex characteristic of polaron–bipolaron transformation and spin–spin interaction seems to explain the narrowing of X-band EPR spectrum mentioned earlier of this sample with temperature.

The intrinsic conductivity of the samples was also evaluated from their Dysonian D-band EPR spectra and then the rate of 1D diffusion of charge carrier was determined (Table 4.8). RT $D_{1D} = 4.1 \times 10^{13}$ rad/s obtained for the PdT:ClO$_4^-$ sample is less considerably than that obtained for PT:ClO$_4^-$, 1.0×10^{14} rad/s [331] and 1.9×10^{15} rad/s [495] but close on the order of value to that determined for polarons in doped PANI, in which interchain charge transfer dominates. The conductivity of PdT:ClO$_4^-$ sample reveals an extremal character with bending point $T_c \approx 170$ K characteristic for the $\Delta B_{pp}(T)$ dependence (Figure 4.43). As in the case of other conjugated polymers, the charge transfer in the PdT:ClO$_4^-$ sample was analyzed to be explained in terms of 1D Mott's hopping and electron scattering on the lattice phonons. As Figure 4.43 evidences, $\sigma_{ac}(T)$ is approximated well by Equation 4.8 with $E_{ph}^| = 0.13$ eV. The additivity of the ΔB_{pp} and σ_{ac} values supports the Houzé–Nechtschein model [153] predicted linear dependence of these values.

4.8 Poly(3-Alkylthiophenes)

Poly(3-alkylthiophenes) (Figure 1.2) were shown [498,499] to be suitable model system for understanding the electronic and optical properties of sulfur-based Q1D conjugated polymers with nondegenerate ground states [498–500]. Later, these polymers appeared to be the most effective active matrix for organic electronic and photonic devices [35,50–54].

P3AT in undoped state are semiconductors, whose energy bandgap is determined by the presence of the π-orbital conjugation along the main polymer axis. For P3AT, the gap amounts ca. 2 eV at ambient temperatures giving rise to its characteristic red color. This parameter, however, demonstrates temperature dependence, so that this color transforms to yellow at the temperature change (*thermochromism*). McCullough et al. [501] have reported the maximum RT σ_{dc} of I$_2$-doped P3AT to be 6×10^3 S/m for poly(3-hexylthiophene) (P3HT, $m = 6$ in Figure 1.2), 2×10^4 S/m for poly(3-octylthiophene) (P3OT, $m = 8$ in Figure 1.2), and 1×10^5 S/m for poly(3-dodecylthiophene) (P3DDT, $m = 12$ in Figure 1.2). This shows the correlation of the P3AT *dc* conductivity with the alkyl group length and the morphology of the sample. It is well known that the transport properties of this class of materials are mainly governed by the presence of positively charged mobile polarons originating from the synthesis and adsorption of oxygen from ambient atmosphere (spontaneous *p*-type doping) [20,502]. These charge carriers are subsequently partly trapped by the impurities, as opposed to inorganic semiconductors where, in the case of *p*-type doping, the hole is transferred from the impurity to the valence band. The presence of polaron in PT and its derivatives was revealed by optical absorption measurements and EPR [105].

When the concentration y of dopant (oxygen, iodine, etc.) increases, the number of polarons increases and, starting from some doping level, the polarons combine to form diamagnetic bipolarons. The energy levels associated with the bipolarons are empty and are located closer to the bandgap than those associated with the polarons [503]. Kunugi et al. [504] have shown by the electrochemical study of charge transport that the RT carrier mobility in a regioregular P3OT film is 5×10^{-7} m^2/V s at $y = 1.4 \times 10^{-4}$. This value decreases down to 5×10^{-8} m^2/V s at $y = 1.0 \times 10^{-2}$ due to the scattering of polarons by ionized dopants and the formation of immobile π-dimers. Then it increases up to 0.5 cm^2/V^{-1}s^{-1} at $y = 0.23$ due to the formation of bipolarons, followed by the evolution of the metal-like conduction.

Polarons stabilized in P3AT possess spin $S = \frac{1}{2}$ as well that also stipulates their wide investigation by different magnetic resonance methods. ^{1}H NMR proton spin-lattice relaxation time study of an initial and ClO$_4$-doped P3OT samples have shown [505] that the molecular motion of the chains and side octyl groups occurs at different temperatures. At X-band EPR, the spectrum of polaron in P3AT is characterized by a single line with the linewidth $\Delta B_{pp} \approx$ 0.6–0.8 mT and the g-factor close to the g-factor of a free electron [105]. EPR signal of a slightly BF$_4$-doped poly(3-methylthiophene) (P3MT, $n = 1$ in Figure 1.2) was found to be a superposition of Gaussian line with $g_1 = 2.0035$ and $\Delta B_{pp} \cong 0.7$ mT attributed to the presence of localized PC and a Lorentzian one with $g_2 = 2.0029$ and $\Delta B_{pp} = 0.15$ mT due to delocalized PC [506]. In this sample, the total concentration of PCs amounts to about 3×10^{19} cm^{-3}, that is, about one spin per 300 thiophene rings. After doping, only one symmetric Lorentzian component of the former spectrum is observed. This line is symmetric at $y \leq 0.25$ and demonstrates a Dyson-like line at higher y [507]. This process is accompanied by a sufficient decrease of electron both spin–lattice and spin–spin relaxation times [506], which may indicate the growth of system dimensionality upon doping process. The analysis of $\chi(y)$ dependence shows that polarons are formed predominantly at low doping level and then start to combine into bipolarons at higher y.

P3OT is also expected to be a suitable and perspective material for molecular electronics [374,508–512], for example, polymer sensors [513] and polymer–fullerene solar cells [514–518]. The piezoelectric effect has also been registered in P3OT [519]. DC conductivity of a synthesized P3OT is about 10^{-4} S/m [520]. A constriction of the P3OT bandgap was observed [521] due to the decrease of the torsion angle between its adjacent thiophene rings and the enhancement of interchain interactions between parallel polymer planes. More detail information on the magnetic and electronic properties of the initial P3OT sample and treated by an annealing (P3OT-A) and by both recrystallization and annealing (P3OT-R) was obtained at X- and D-band EPR [522,523]. The transition temperature of P3OT is close to 450 K, so respective treatment was made at this temperature.

Figure 4.44 shows EPR spectra of the initial and treated P3OT samples obtained at both wavebands EPR. At the X-band, the samples show a single

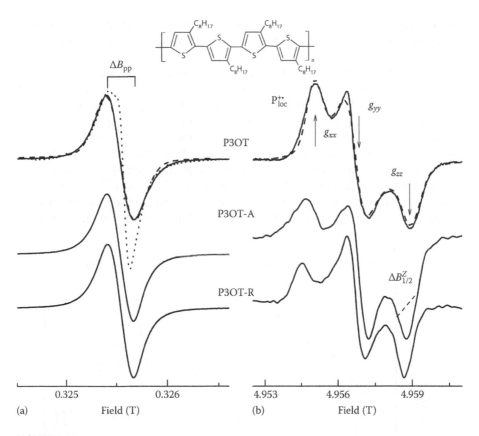

FIGURE 4.44

Room temperature X-band (a) and D-band (b) absorption electron paramagnetic resonance (EPR) spectra of the poly(3-octylthophene) (P3OT), P3OT-A, and P3OT-R samples. The X-band EPR spectrum of the P3OT stored for 2 years is shown on the left by the dotted line. The magnetic resonance parameters measured are shown. The spectra calculated using $g_{xx} = 2.004089$, $g_{yy} = 2.003322$, $g_{zz} = 2.002322$, $\Delta B_{pp}^{x} = \Delta B_{pp}^{y} = \Delta B_{pp}^{z} = 0.25$ mT, and Lorentzian/Gaussian = 0.9 line shape ratio (a) and $\Delta B_{pp}^{x} = 0.82$ mT, $\Delta B_{pp}^{y} = 0.78$ mT, $\Delta B_{pp}^{z} = 0.88$ mT, and Lorentzian/Gaussian = 0.4 line shape ratio (b) are shown by the dashed lines. The sum spectrum of the lines calculated with $g_{\parallel} = 2.00387$, $g_{\perp} = 2.00275$, and $\Delta B_{pp} = 0.11$ mT and $g_{iso} = 2.00312$ and $\Delta B_{pp} = 0.35$ mT with amplitude ratio of 2.5:1 is shown by the dotted line on the right as well. (From Krinichnyi, V.I. and Roth, H.K., *Appl. Magn. Reson.*, 26, 395, 2004. With permission.)

nearly Lorentzian EPR line with $g_{eff} = 2.0019$. The asymmetry factor of the lines is $A/B = 1.1$, and their peak-to-peak linewidth was determined to be $\Delta B_{pp} = 0.27$ (P3OT), 0.27 (P3OT-A), and 0.26 mT (P3OT-R). The ΔB_{pp} value is smaller than $\Delta B_{pp} \approx 0.6$–0.8 mT obtained for PT, and poly(3-methylthiophene) [105], however, is close to 0.32 mT registered for polarons in regioregular P3OT [524]. The X-band EPR spectrum of the P3OT sample stored for 2 years is also shown in Figure 4.44a by the dotted line. Its computer modeling has shown that it consists of anisotropic and isotropic spectra.

The effective *g*-factors of these samples are close; therefore, one can conclude that localized PCs with more anisotropic magnetic parameters appear during the polymer storage.

The RT total spin concentration of P3OT increases during the polymer treating from 3.9×10^{19} cm^{-3} in P3OT up to 3.1×10^{20} cm^{-3} in P3OT-A and 3.5×10^{20} cm^{-3} in P3OT-R or from 0.013 to 0.11 spin per a monomer unit, respectively. Their inverse effective paramagnetic susceptibility χ^{-1} and the χT value are presented in Figure 4.45 as function of temperature. In the high-temperature region, the spin susceptibility of the P3OT and P3OT-R samples seems to include a contribution due to a strong spin–spin interaction as it was revealed in PANI [185,363], PATAC [351], and P3DDT [364–367]. This contribution disappears at low temperatures due to the phase transition opening an energy gap at the Fermi level [105], so then the susceptibility

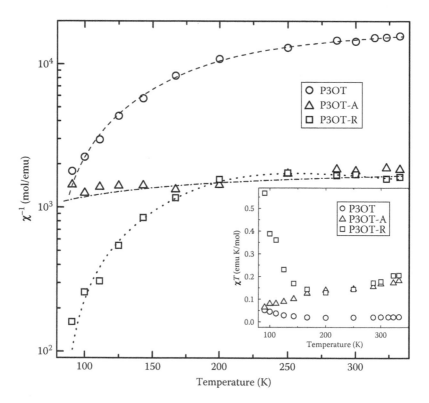

FIGURE 4.45
The $1/\chi(T)$ and $\chi T(T)$ (inset) dependencies of paramagnetic susceptibility of the poly(3-octylthophene) (P3OT), P3OT-A, and P3OT-R samples. The dependencies calculated using Equation 2.6 with $\chi_P = 0$, $C = 0.72$ emu K/mol, $c_{af} = 10.6$ emu K/mol, and $J_{af} = 0.0062$ eV (dashed line); $\chi_P = 4.5 \times 10^{-4}$ emu/mol, $C = 0.044$ emu K/mol, and $c_{af} = 0$ (dotted line); $\chi_P = 0$, $C = 4.1$ emu K/mol, $c_{af} = 52.1$ emu K/mol, and $J_{af} = 0.011$ eV (dash-dotted line), respectively, are shown as well. (From Krinichnyi, V.I. and Roth, H.K., *Appl. Magn. Reson.*, 26, 395, 2004. With permission.)

demonstrates the Curie behavior. Figure 4.45 shows that the total spin susceptibility of all P3OT samples follows Equation 2.6. As in the case of sulfonated PANI [525], these results evidence that the annealing of the P3OT polymer affects an effective exchange coupling between spins, which results in the change of an effective number of the Bohr magnetons per a monomer unit; however, the recrystallization neglects this effect.

At the D-band EPR, the samples demonstrate the superposition of more broadened convoluted Gaussian and Lorentzian lines (with the Lorentzian/ Gaussian line shape ratio ca. 0.4) with the anisotropic g-factor (Figure 4.44b) as it is typical for PC in some other conjugated polymers with heteroatoms (e.g., PATAC, PP, PTTF, PANI, and PT described earlier). The linewidth of the spectra increases by a factor of ca. 3 at the increase of the registration frequency (Figure 4.46). It is seen that the linewidths of the PCs in the samples depend on the polymer treatment and temperature. The treatment leads to the decrease in the averaged linewidth $\langle \Delta B_{pp} \rangle$ from 0.82 mT in P3OT down to 0.78 mT in P3OT-A and then to 0.67 mT in P3OT-R confirming the supposition mentioned earlier of the growth of the system crystallinity. The linewidth of the P3OT increases as the temperature increases from 90 K up to the phase transition characteristic temperature $T_c = 200$ K and then decreases at the further temperature growth. The analogous tendency demonstrates P3OT-R; however, its T_c value decreases down to ca. 170 K (Figure 4.46). This effect of the linewidth decrease below T_c was detected also in the study of doped PANI [182, 185] and was not registered in other conjugated polymers; it can be interpreted, for example, as the manifestation of defrosting of molecular motion and/or acceleration of relaxation processes at low temperatures. On the other hand, the ΔB_{pp}^x value of P3OT-A decreases as the temperature increases up to $T_c = 200$–250 K and starts to increase at $T \geq T_c$, whereas the other spectral components are broadened linearly with the temperature growth from 100 K (Figure 4.46). If one supposes that molecular dynamics and/or electron relaxation should stimulate activated broadening of the ith line, $\Delta B_{pp}^i = \Delta B_{pp}^{i(0)} \exp(E_a / k_B T)$, from the slopes of these dependencies, it is possible to determine separately the parameters of electron relaxation and molecular dynamics near the principal macromolecular axes. The preexponential factor and the energy for activation of molecular motion near the principal x-, y- and z-axes in the samples are presented in Table 4.9.

From the RT spectra of P3OT, the main components of its **g** tensor have been determined to be $g_{xx} = 2.00409$, $g_{yy} = 2.00332$, and $g_{zz} = 2.00235$. The treatment of the samples leads to the change of these parameters to $g_{xx} = 2.00404$, $g_{yy} = 2.00315$, and $g_{zz} = 2.00231$ for P3OT-A and to $g_{xx} = 2.00402$, $g_{yy} = 2.00313$, and $g_{zz} = 2.00234$ for P3OT-R. The structure of a polymer should affect the distribution of an unpaired electron in polaron changing the principal values of its PC **g** tensor and hyperfine structure. The shift of the g_{xx} and g_{yy} values from the g_e-factor for a free electron can be compared with that calculated from Equation 2.3 with the constant of spin–orbit interaction of the electron spin with the sulfur nuclear λ_s equal to 0.047 eV.

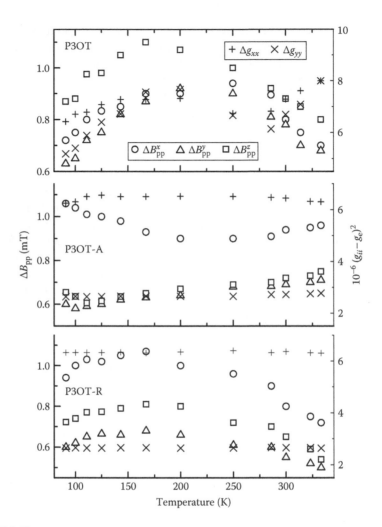

FIGURE 4.46

Temperature dependencies of the linewidth of the main spectral components of the poly(3-octylthophene) (P3OT), P3OT-A, and P3OT-R D-band electron paramagnetic resonance spectra. Temperature dependencies of the squared shifts of the g_{xx} and g_{yy} values from $g_e = 2.00232$ are shown as well. (From Krinichnyi, V.I. and Roth, H.K., *Appl. Magn. Reson.*, 26, 395, 2004. With permission.)

The effective g-factor of the P3AT is higher than that of the most hydrocarbonic conjugated polymers; therefore, one can conclude that in P3AT the unpaired electron interacts with sulfur atoms. This is typical for other sulfur-containing compounds, for example, PTTF [118], PATAC [351], and benzotrithioles [359], in which sulfur atoms are involved in the conjugation. In such organic solids, an unpaired electron is localized mainly on the sulfur atom, so its effective g-factor is 2.014 < g_{iso} < 2.020 [1,357–359,361]. The g-factor

TABLE 4.9

Preexponential Factor ΔB_{pp}^0 and Activation Energy E_a of Molecular Motion near the Principal Axes as well as Parameters σ_{01}, α, E_a Calculated from Equation 2.77, σ_{02}, E_{ph} Calculated from Equation 2.83, and τ_{c0}^x, E_a, k_1, α, β Calculated from Equation 4.9 for the Initial and Treated Poly(3-Octylthophene) Samples

Parameter	P3OT	P3OT-A	P3OT-R
ΔB_{pp0}^x (mT)	1.14	1.20	1.28
ΔB_{pp0}^y (mT)	1.18	0.72	0.78
ΔB_{pp0}^z (mT)	1.48	0.75	0.91
E_a^x (meV)	3.6	7.2	2.4
E_a^y (meV)	4.9	2.0	1.9
E_a^z (meV)	4.3	2.1	1.8
σ_{01} (S s/m K)	9.1×10^{-9}	3.1×10^{-5}	1.3×10^{-7}
α	2.1	3.6	2.2
E_a (meV)	180	190	190
σ_{02} (S/m K)	1.1×10^{-3}	5.5×10^{-6}	7.1×10^{-4}
E_{ph} (meV)	180	200	130
τ_{c0}^x (s)	1.3×10^{-8}	1.1×10^{-8}	6.3×10^{-8}
E_a (meV)	69	54	73
k_1 (s/K$^\beta$)	3.1×10^{-10}	7.2×10^{-12}	3.1×10^{-13}
α	1	-1	-1
β	1.8	2.5	3.5

of PC in P3OT is much smaller so one can expect a higher spin delocalization in the monomer units. Indeed, assuming $\Delta E_{n\pi^*} \approx 2.6$ eV as typical for sulfuric solids, then the g-factor components of the initial P3OT yield the decrease in $\rho_s(0)$ by a factor of 1.8 as compared with PATAC [351] and by a factor of 2.6–3.9 as compared with PTTF [118]. One can evaluate the lower limit of RT probability (in frequency units) for the duration of the stay of spin at the sulfur site in the P3OT samples by the total shift of the spectral components registered at the positions g_{ii}^{P3OT} mentioned earlier relative to the g_{ii}^S value typical for the sulfuric radical from Equation 4.1 to be $D_{1D}^0 \geq 3.4 \times 10^9$ rad/s.

As in the case of other organic radicals, the g_{zz} values of the P3OT samples are close to the g-factor for a free electron so they feel the change in the system properties weakly. In contrast to X-band EPR, high spectral resolution achieved at D-band EPR allows one to register separately the structure and/or dynamic changes in all spectral components. Figure 4.46 evidences that the g_{xx} and g_{yy} values reflect more efficiently the properties of the radical microenvironment. These values of the initial P3OT decrease as the temperature decreases from 333 to 280 K possibly due to the transition to the more planar conformation of the polymer chains. Below 280 K, these values increase at the sample freezing down to 160–220 K and then also decrease at the further temperature decrease. The decrease of g_{xx} and g_{yy} values at low temperatures can be explained by a harmonic vibration of macromolecules that evokes the crystal field modulation and is characterized by the $g(T) \propto T$ dependence [526]. The temperature polymer treatment weakly affects these parameters. This fact can be interpreted by the growth of the system crystallinity due to the higher chain packing in the treated polymers.

The analysis of the data presented in Figure 4.46 shows that the linewidth can also correlate with the spin–orbit coupling in the framework of the Elliot mechanism playing an important role in the charge transfer in organic ion-radical salts [1] and conductive polymers with pentamerous unit rings [105]. Indeed, $\Delta B_{pp}^{x} \propto \Delta g_{xx}^{2}$ and $\Delta B_{pp}^{y} \propto \Delta g_{yy}^{2}$ dependencies are valid for, for example, P3OT at least at $T \leq T_c$. This means that different mechanisms can affect the individual components of the P3OT spectrum and that the scattering of charge carriers (see Figure 4.48c) should be governed by the potential of the polymer backbone.

The inset in Figure 4.47 exhibits in-phase and $\pi/2$-out-of-phase terms of D-band EPR dispersion spectra of P3OT registered at different temperatures. It is seen that the bell-like contribution with Gaussian spin packet distributions is registered in both these dispersion terms. The appearance of such a component is attributed to the adiabatically fast passage of saturated spin packets by a modulating magnetic field, as discussed earlier. The intensities of the $\pi/2$-out-of-phase spectral components change with the temperature. This effect evidences the appearance of the saturation transfer over the quadrature spectrum due to superslow macromolecular dynamics. Figure 4.47 exhibits temperature dependencies of correlation time τ_c^x of libration motion of the chain macromolecular segments near the principal molecular x axis determined from Equation 2.65 for the initial and treated P3OT samples. This value of P3OT and P3OT-A decreases with the temperature increase of up to $T_c \approx 150$ K and increases above this critical temperature. The opposite temperature dependence is characteristic for τ_c^x of P3OT-R with the close T_c value. Note that Masubuchi et al. [505] have observed ^{1}H NMR T_1 temperature dependence with the same critical temperature that was attributed to the defrosting of molecular motion of alkyl chain end groups. The dependencies obtained can be interpreted in the frames of the superslow activation 1D libration of the polymer chains together with polarons at low temperatures when $T \leq T_c$, whereas their high-temperature part can be explained by the defrosting of the collective 2D motion at $T \geq T_c$. In this case, the effective correlation time is determined as

$$\tau_c^x = \left\{ \left[\tau_{c0}^x \exp\left(\frac{E_a}{k_B T} \right) \right]^\alpha + \left(k_1 T^\beta \right)^\alpha \right\}^{\frac{1}{\alpha}}. \tag{4.9}$$

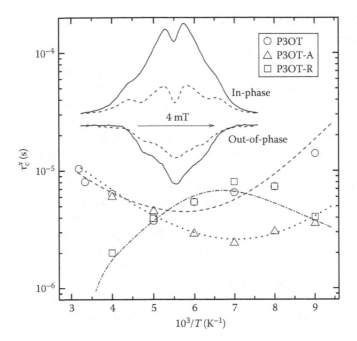

FIGURE 4.47
Inset: In-phase and $\pi/2$-out-of-phase terms of the D-band dispersion electron paramagnetic resonance spectra of the initial poly(3-octylthophene) (P3OT) sample at $T = 100$ K (solid lines) and 145 K (dashed lines). Temperature dependence of correlation time of the super slow librations of macromolecules near main x-axis in the P3OT, P3OT-A, and P3OT-R samples. The dependencies calculated in framework of activation motion from Equation 4.9 with respective parameters presented in Table 4.9 are also shown.

Figure 4.47 shows that the dependencies obtained experimentally for P3OT, P3OT-A, and P3OT-R are well fitted by Equation 4.9 with respective parameters summarized in Table 4.9. The activation energies obtained are close to those of macromolecular librations in other organic conjugated polymers [118]. The preexponential factors are the lowest limit for the respective correlation times in these samples. The linear compressibility of an initial P3OT with planar chains is strongly anisotropic, being 2.5 times higher for the direction along molecular a-axis than along the b-axis [527]. It was proved that the low- and high-frequency modes exist in PTs [528]. These modes differently superposed in P3OT above and below T_c should lead to the change of the α exponent in Equation 4.9 from 1 for *successive* macromolecular dynamics in P3OT and P3OT-A to −1 for *parallel* molecular librations in P3OT-R. Osterbacka et al. [529] have found that the interchain coupling existing in self-assembled lamellae in P3AT drastically changes the properties of the polaron excitations and that the traditional self-localized polaron in one dimension is delocalized in two dimensions, resulting in a much reduced relaxation energy and

multiple absorption bands. The upper limit for the correlation time of anisotropic molecular motion in the P3OT registered by the ST-EPR method was
evaluated from Equation 2.61 to be $\tau_c^x \leq 4.4 \times 10^{-4}$ s at 66 K.

It is evident that the polymer structure and the polymer treatment influence spin charge relaxation and dynamics. At $T \geq 200$ K, the inequality $\omega_m T_1 <$
1 holds for polarons in all P3OT samples. The opposite inequality is fulfilled
at lower temperatures when the dispersion spectrum is determined by the
last two terms of Equation 2.35. The semilogarithmic temperature dependence of the relaxation times determined from the saturated EPR dispersion
spectra using Equations 2.36 through 2.39 are shown in Figure 4.48a. One can
conclude from these data that the polymer treatment leads to the acceleration
of the PC effective relaxation possibly due to the increase of the interaction
of polaron charge carriers with the lattice phonons. The relaxation times of
the samples increase simultaneously as the temperature decreases from 333
to ca. 250 K (P3OT, P3OT-R) and 150 K (P3OT-A) (Figure 4.48). The spin–spin
relaxation is accelerated below this point leading to the appropriate change
of the spectral component linewidth (Figure 4.46).

The spin–spin relaxation time of the P3OT samples determined from their
X-band EPR absorption spectra as $T_2 = 2/\sqrt{3}\,\gamma_e \Delta B_{pp}$ and from their saturated
D-band EPR dispersion spectra increases approximately by a factor of six.
This means that the experimental data can rather be explained in terms of
a modulation of spin relaxation by the polaron intra- and interchain motion
in the polymers. In the framework of such approach, the relaxation time of
the electron or proton spins in the sample should vary as $T_{1,2} \propto \omega_e^{1/2}$ [162,205]
that should lead to the increase in relaxation times approximately by a factor
of four. Note that Q2D spin motion should lead to $T_{1,2} \propto \ln(\omega_e)$ dependence
[162,205] or to the increase of the relaxation times by a factor of about two at
the transition from X-band to D-band EPR.

Figure 4.48b shows the temperature dependencies of the effective dynamic
parameters D_{1D} and D_{3D} calculated for PC in the initial and treated P3OT
samples from the data presented in Figure 4.48a with Equations 2.46, 2.51,
and 2.52 at $L \approx 5$ [75]. The RT D_{3D} value obtained is about $D \approx 2.1 \times 10^{10}$ s^{-1}
evaluated from the charge carrier mobility in slightly doped P3OT [504] and
the D_{1D} value exceeds by one to two orders of magnitude the lower limit of
the spin motion D_{1D}^0. The RT anisotropy of spin dynamics D_{1D}/D_{3D} increases
from 6 in P3OT up to 18 in P3OT-A and decreases down to 2 in P3OT-R. As
the temperature decreases down to 200 K, this value increases up to $2.5 \times$
10^3, 1.4×10^2, and 3.9×10^2, respectively, and then up to 3.8×10^8, 6.5×10^7, and
3.4×10^{10}, respectively, as the temperature decreases down to 100 K (Figure
4.48b).

The $D_{1D}(T)$ dependence calculated from the ^{1}H 50 MHz spin–lattice relaxation data obtained by Masubuchi et al. [505] for the initial P3OT with
Equation 2.60 is also presented in Figure 4.48b. It is seen from the figure
that the D_{1D} value calculated from the NMR data is changed weaker with

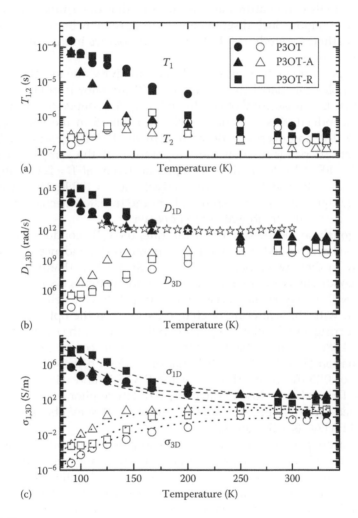

FIGURE 4.48

(a) Temperature dependencies of the relaxation times T_1 (filled symbols) and T_2 (open symbols) of the poly(3-octylthophene) (P3OT), P3OT-A, and P3OT-R samples. (b) Temperature dependencies of the spin/charge diffusion coefficients along (D_{1D}, filled symbols) and between (D_{3D}, open symbols) polymer chains in the P3OT, P3OT-A, and P3OT-R samples determined using Equations 2.46, 2.51, and 2.52. Open stars show the $D_{1D}(T)$ dependence calculated using Equation 2.59 with ^{1}H NMR data for the P3OT. (From Masubuchi, S. et al., *Synth. Metals*, 101(1–3), 594, 1999.) (c) Temperature dependencies of the conductivity due to the spin/charge diffusion along (σ_{1D}, filled symbols) and between (σ_{3D}, open symbols) polymer chains in the P3OT, P3OT-A, and P3OT-R samples. The lines show the dependencies calculated using Equation 2.77 with respective parameters σ_{01}, α, and E_a (dotted lines) and using Equation 2.83 with respective parameters σ_{02} and E_{ph} (dashed lines) presented in Table 4.9.

temperature. Besides, this value considerably exceeds the D_{1D} value obtained by EPR at high temperatures and is close to that determined for the low-temperature region. Such discrepancy occurs probably because NMR is not a direct method for studying electron spin dynamics in this and other conjugated polymers.

The temperature dependencies of the conductivity due to polaron motion in the samples calculated from Equation 2.71 are shown in Figure 4.48c. It is seen that the anisotropy of conductivity σ_{1D}/σ_{3D} depends on the system treatment. Indeed, this RT value increases from 15 in the P3OT up to 48 in the P3OT-A and decreases down to 4 in the P3OT-R (Figure 4.48c). The σ_{1D}/σ_{3D} ratio depends also on the temperature of the samples increasing up to 6.8×10^3, 3.7×10^2, and 1.2×10^3, respectively, at $T = 200$ K and up to 1.0×10^9, 1.2×10^8, and 1.1×10^{10}, respectively, at $T = 100$ K. This means that the charge dynamics and the spin and charge transfer mechanism depend on the structure and treatment of the polymer.

Analogously to the 1D spin diffusion rate, the Q1D conductivity of these samples is characterized by the strong temperature dependence, $\sigma_{1D}(T) \propto T^{-(7-9)}$ especially in the low-temperature region, when $T \leq T_c$ (Figure 4.48c). Note that the conductivity of organic salts frequently exhibits $\sigma \propto T^{-2}$ relationship [1] and $\sigma \propto T^{-1}$ one typical for classical metals [62]. Such a behavior is usually associated with the scattering of charge carriers on the optical lattice phonons occurring also in other conjugated polymers (see the earlier text).

Figure 4.48c evidences that the intramolecular *ac* conductivity calculated with Equation (2.71) follows well Equation 2.83 with the energy of optical phonons of 0.13 eV for P3OT, 0.20 eV for P3OT-A, and 0.18 eV for P3OT-R samples. This value is close to the energy of lattice phonons of 0.01–0.15 eV determined for PANI–ES (see Table 4.7) and 0.04–0.06 eV obtained for laser-modified PATAC (see Section 4.2).

The polaron–phonon interaction seems to play an important role also in the interchain charge transfer at $T \geq T_c$. However, another mechanism should prevail at lower temperatures. Figure 4.48c shows that the interchain conductivity σ_{3D} increases as the temperature increases at the low-temperature region and then slightly decreases at $T \geq T_c$. The analogous dependence with the characteristic temperature $T_c \approx 150$ K was obtained for *dc* conductivity of P3HT [530] typical for the systems with a strong coupling of the charge with the lattice phonons. In this case, the strong temperature dependence of the hopping conductivity is more evidently displayed in the *ac* conductivity. This interaction should lead to the narrow energy gap E_a much higher than the RT thermal energy ($k_B T \approx 0.026$ eV). The strong temperature dependence for σ_{ac} at low temperatures can be due to the thermal activation of the charge carriers from widely separated localized states in the gap to closely localized states in the tails of the valence and conjugated bands. In this case, *ac* conductivity of P3OT can be described by Equation 2.77. Typical activation energies for the semiconjugated, for example, organic ion-radical salts are of the order of 0.1 eV [1]. The increase of the dimensionality of the polymer system should lead to the decrease in E_a. The relation $\kappa = 1 - \alpha k_B T / E_a$ with $\alpha = 1$ and $E_a = 0.18$ eV is typical for P3MT [226], so we can use this approach for the explanation of the 3D conductivity in the P3OT samples as well. The temperature dependencies calculated from Equation 2.77 for the P3OT samples are also shown in Figure 4.48c for comparison. The activation energies obtained are close to the energy of lattice phonons of these samples determined earlier and also to that determined for the charge transfer in the PATAC, PTTF-C_2H_5-C_6H_4:HCA$_{0.50}$, PANI:HCA$_{0.50}$, P3HT:PC$_{61}$BM, and PCDTBT:PC$_{61}$BM polymer matrices (see Sections 4.2, 4.5, 4.6, 5.4 and Tables 4.5, 4.7, 5.2)..

DC conductivity of P3OT is close to 10^{-4} S/m [520] and is mainly determined by the activation hopping of a charge carrier between high-conductive crystalline domains. Its intrinsic conductivity is defined by the σ_{1D} and σ_{3D} values, so the $\sigma_{1,3D} \gg \sigma_{dc}$ relation should hold for all the samples.

Thus, the interaction of the spin charge carrier with the heteroatom in sulfurous organic polymer semiconductors P3OT leads to the appearance of the *g*-factor anisotropy in their D-band EPR spectra. Spin relaxation and dynamics are determined by the interaction of the mobile polaron with optical phonons of the polymer lattice. Polymer modification leads to a distinct change in magnetic, relaxation, and dynamic properties of the polaron and its microenvironment. The near values of the energies of spin interchain transport, dipole–dipole interaction, and optical lattice phonons indicate the correlation of charge transport and molecular dynamics in P3OT. These energies increase at the polymer treatment indicating the increase of its effective crystallinity. It can be concluded that the two types of charge transport mechanism can be associated with change in the lattice geometry at the characteristic (the polymer-glass transition) temperature T_c, for example, with its thermochromic effect.

Let us consider the advantages of the method in the respective study of the other P3AT conjugated polymers and their nanocomposites.

5

Charge Transfer by Spin Carriers in Polymer:Fullerene Nanocomposites

P3AT being modified with a fullerene (see Section 1.2) or other suitable nanoadducts becomes an active matrix of organic composite capable of converting visible light into electricity [35,51–57]. As a complex exciton is created in such system, there are different pathways its decay into the charge transfer state. The desired pathway would be exciton separation into extractable carriers, which should be displaced from the polymer:fullerene interface. This can happen by diffusion and relaxation of the polarons and electrons into higher- and lower-lying energy levels, respectively, which usually exist, because the energies of the HOMO and LUMO levels of conjugated polymers exhibit a broadened density of states. Relaxation of the carriers into the tails of the density of states can provide sufficient energy to overcome the Coulombic binding energy. With increasing distance from the material interface, the Coulombic attraction becomes less, and finally, the charge carriers become independent of each other. However, there are concurring processes, among them geminate recombination, which means the recombination of spin charge carriers originating from the same photoinitiated exciton. It was found [531] that the mobility and stability of charge carriers become higher considerably in the formation of bulk heterojunctions (BHJ), for example, by the chains of regioregular poly(3-hexylthiophene) with the globes of methanofullerene $PC_{61}BM$ as compared with other polymer:fullerene composites. A much longer charge carrier lifetime achieved in the P3HT:$PC_{61}BM$ BHJ should, therefore, lead to higher concentration of charge carriers and their reduced recombination rate. Specific nanomorphology of such composites could result in a screened Coulombic potential between the radical pairs photoexcited in their BHJ and facilitate their splitting into noninteracting charge carriers with a reduced probability of their further annihilation. It was explained by better structural order in the presence of interface dipoles provoking the creation of a potential barrier for carrier recombination in this composite. This implies that a longer charge carrier lifetime can be achieved at the same concentrations, which finally results in higher photocurrent and larger power conversion efficiency of such solar cells. This is why $PC_{61}BM$ has appeared to be the most suitable electron acceptor to be used for a long time in prototypes of plastic solar cells and other molecular device.

Quantum efficiency of light conversion (the number of electrons per each photon absorbed by the system) has already attained about 3% for the P3HT:PC$_{61}$BM BHJ [532] and around 6%–8% for other organic solar cells [533]. This parameter is governed by different factors. The first limitation originated from the high binding energy of polarons photoinduced in conjugated polymers upon light excitation, so by blending in an electron acceptor, it becomes energetically favorable for the electron to escape from a polymer macromolecule and to transfer to an acceptor. This requires the LUMO$_D$ to be 0.3–0.5 eV higher than the LUMO$_A$ [534,535]. However, such energy difference can be much higher for some polymer matrices, which decreases optimal open-circuit voltage, since the latter is ultimately limited by the difference between the HOMO$_D$ and LUMO$_A$ [536,537]. Raising, for example, the LUMO$_A$, it becomes real to increase the efficient factor of plastic solar cells without affecting their light absorption. This approach is theoretically more beneficial for a single-layer solar cell and results in an estimated efficiency of 8.4% when the lowest unoccupied molecular orbital level (LUMO) offset is reduced to 0.5 eV [538].

The structures of donor and acceptor as well as the conformation of the respective BHJ can also affect charge transport and recombination [539]. Lenes et al. [540] have suggested to use as electron acceptor *bis*-PC$_{62}$BM methanofullerene, *bm*F$_{62}^-$, in which the fullerene cage is functionalized by two methano-bridged PBM side groups, with a higher (by ~0.1 eV) LUMO$_A$ than that of PC$_{61}$BM. Indeed, quantum efficiency of plastic solar cells appeared [541,542] to improve as PC$_{61}$BM is replaced by *bis*-PC$_{61}$BM. However, it was shown [543] that photoluminescence dynamics becomes slower at such replacement due to the reorganization of BHJ. Besides, quantum efficiency can be reduced due to possible formation of triplets from intersystem crossing excitons before the charge separation and further recombination [544]. Another way to improve this important parameter can be seen by decreasing the bandgap of the active polymer matrix (approximately 1.9 eV for P3HT), which limits absorbance of light photons with higher energy. So to harvest more solar photons, thereby increasing the power conversion efficiency, one should use polymers with lower bandgap energy E_g in such device.

Poly[N-9″-hepta-decanyl-2,7-carbazole-alt-5,5-(4′,7′-di-2-thienyl-2′,1′,3′-benzothiadiazole)] (PCDTBT) (see Figure 1.2) with $E_g < 1.9$ eV [545,546] has been discovered [547] to be one of the most efficient low-bandgap semiconjugated polymers to be used in organic thin-film transistors and solar cells [548–550]. The light conversion efficient of the PCDTBT:PC$_{71}$BM composite layer has reached 7.2% [291] due to a relatively low HOMO$_D$ level and an internal quantum efficiency approaching to 100% [551]. Such outstanding results were explained [81] mainly by the ultrafast charge separation in the PCDTBT:PC$_{71}$BM composite before localization of the primary excitation to form a bound exciton in contrast with, for example, a P3HT-based one, where photoinduced charge separation happens after diffusion of the polymer exciton to a fullerene interface. The morphology of the PCDTBT:PC$_{71}$BM

BHJ, another important property, has demonstrated [291,552] to be laterally oriented with a *column-like* bilayer-ordered polymer matrix with methanofullerene embedded between its chains that improves the mobility of charge carriers [553]. The dimensionality of the PCDTBT backbone with such morphology should be higher than that of P3AT matrices. Gutzler and Perepichka [554] stated that higher π-overlapping in 2D thiophene-based polymers hinders their torsional twisting and, therefore, lowers their bandgap. This allows holes to hop through such well-ordered bilayer of PCDTBT surfaces to the anode and electrons to move to the cathode inside methanofullerene pools located between these bilayers. This is evidence that charge dynamics is another important parameter affecting device light conversion efficiency. Higher charge carrier mobility of the polymer increases the diffusion length of electrons and photoinitiated holes and decreases the probability of their recombination in the active layer.

A real polymer:fullerene system consists of domains with different bandgaps (i.e., different $LUMO_A$—$HOMO_D$) determining its energetic disordering with Gaussian-distributed density of states [555]. Another constraint comes due to the finite number and mobility of charge carriers in organic solar cells, which are lower as compared with conventional semiconductors. These main parameters depend on the structure and properties of a polymer matrix and fullerene derivative embedded [57,540,556–559]. This is the reason why their power conversion efficiency appears to be governed also by an ultrafast electron transfer from a photoexcited polymer to fullerene [80], a large interfacial area for charge separation due to intimate blending of the materials [77], and an efficient carrier transport across a thin film. Unambiguously, to increase power conversion efficiency, it is necessary to photoinitiate a higher density of charge carriers. However, an increased carrier density causes a reduced lifetime due to bimolecular recombination, and the efficiency of solar cells might be reduced [559].

Planarity and regioregularity of polymer matrix, governed by the structure of the polymer and methanofullerene side substitutes, play an important role in charge separation and transition of the polymer:fullerene composites. In addition, the presence of substituents can result in torsional and energetic disorder of polymer chains, thereby changing effective mobility of charge carriers. Side chain groups accelerate torsional and librational chain dynamics modulating intrachain and interchain charge transfer, respectively. It was shown [560] that the torsional reordering of the backbone rings of conjugated polymers determines their electronic structure and charge-transfer mechanism. The average torsion angle between adjacent thiophene rings θ is sensitive to steric repulsion and electron delocalization that is detected in the optical spectra of these materials. The less a torsion angle θ, the higher intrachain transfer integral and effective crystallinity can be reached. The increase in planarity reduces bandgap and increases charge mobility, stability, and interactions between parallel polymer planes. Modification of the $P3HT:PC_{61}BM$ composite with N- or B-doped carbon nanotubes [561] and self-assembled dipole molecule deposition in plastic light-emitting diodes [562] can, in principle, enhance their power conversion efficiency. Thermal annealing can also modify morphological structure of BHJ and increase its light conversion efficiency [563]. Such a treatment leads to the formation of crystalline regions in an amorphous polymer matrix that is accompanied by the shift of its light absorption maximum, for example, the P3HT:PC$_{61}$BM composite from $E_{ph} \approx 2.5$–2.8 eV to the lower photon energies, 1.9–2.3 eV. The observed shift indicates also the increase in the conjugation length in crystallites, because the polymer molecules within such crystallites are perfectly oriented and there are no defects like chain kinks, which limit the conjugation length. This process can be controlled, for example, by the UV/vis spectroscopy, grazing-incidence x-ray diffraction, and atomic force microscopy [564–566].

Charge recombination is considered to be predominantly a nongeminate process governing efficiency of polymer:fullerene solar cells [567–569]. Normally, the delay of charge carriers consists of prompt and persistent contributions [76,570]. The excitation light intensity of a prompt process is dependent on the activation bimolecular type and implies mutual annihilation within the initially created radical pair [571]. The persistent contribution is independent of the excitation intensity and originates from deep traps due to disorder [570]. Bimolecular and quadrimolecular recombination were shown [572] to be dominant in the P3AT:fullerene composites at lower and higher intensity of the excited light, respectively. The lifetime of charge carriers is usually estimated from photocurrent transients after the excitation by a short light pulse. However, this method seems to be inaccurate in the case of organic materials, because the photocurrent transients depend not only on the decay of charge carrier concentration but also on the mobility relaxation within the broad density of states [555]. The estimation of lifetimes from transient absorption techniques is difficult because of the very large dispersion observed leading to power law decays [567]. So the photoexcitation of charge carriers and their recombination are the most interesting points.

The F$_{61}^-$ charge carriers like P$^{+\cdot}$ possess uncompensated spin $S = \frac{1}{2}$. This accounts for the wide use of light-induced electron paramagnetic resonance (LEPR) spectroscopy as a direct method for the investigation of charge photoexcitation, separation, transfer, and recombination in fullerene-modified conjugated polymers [55,57,571–573]. Generally, polaron intramolecular Q1D diffusion and its intermolecular Q3D hopping between the chains and/or fullerene domains as well as a rotational librative motion of fullerene globes are realized in polymer:fullerene system. All spin-assisted molecular and electronic processes are expected to correlate. LEPR measurements revealed the existence of two radicals with different line shapes, magnetic resonance parameters, and saturation behaviors. The photoinduced spins should coact with its own charged microenvironment through exchange or dipole–dipole interaction. Such interactions are not registered in LEPR spectra of plastic solar cells that can be interpreted as the recession of mobile polarons on a conjugated polymer backbone and the fullerene anions with the rate faster than 10^{-9} s. That is the reason why both the charge carriers excited in main polymer:fullerene composites are characterized by a considerable long lifetime and can be registered separately. Understanding the basic physics underlying the electron relaxation and dynamic behavior of fullerene-modified organic polymers is essential for the optimization of devices based on these materials.

Hence, the optimization of structure and nanomorphology of BHJ as well as the understanding of photoexcitation, dynamics, and recombination of charge carriers in such systems is of fundamental interest for controlled fabrication of optimal molecular photovoltaic devices. Understanding the charge separation and charge transport in such materials at a molecular level is crucial for improving the efficiency of the solar cells. However, they are not yet understood in detail, and there is no generally applicable model describing molecular, electronic, and relaxation processes in different polymer:fullerene composites, and there are no generally applicable models available.

Described in the present part are the results of a multifrequency (9.7–140 GHz) LEPR study of magnetic, relaxation, and charge transport properties of spin charge carriers stabilized and photoinduced in different organic polymer:fullerene composites.

5.1 Line Shape and G-Factor

No electron paramagnetic resonance (EPR) signal has been found in pure fullerene derivatives at X-band in an entire temperature range. Some initial regioregular P3AT samples, for example, P3HT, demonstrate at X-band EPR Lorentzian exchange–narrowed nearly symmetrical line with effective $g_{eff} = 2.0029$ (Figure 5.1a). As in case of other regioregular P3AT with longer side chains, this fact was interpreted as stabilization in the polymer of mobile polarons due to its treating by oxygen of air [556,557]. As the sample is modified by methanofullerenes mF_1–mF_5, in its spectrum appear two additional lateral lines (Figure 5.1a). This demonstrates the localization of a part of polarons probably at cross-bonds and/or at the ends of polymer chains during such modification. The intensity of these components decreases in the series P3HT:mF_1 → P3HT:mF_6 → P3HT:mF_2 → P3HT:mF_3 → P3HT:mF_4 → P3HT:mF_5. High-field/frequency EPR study shows [522,523,574] that the interaction of an unpaired electron of $P^{+\cdot}$ with sulfur heteroatoms involving the P3AT backbone leads to anisotropy of its magnetic resonance parameters.

In order to determine and analyze the main magnetic resonance parameters of all paramagnetic center (PC) stabilized in the samples, the spectra

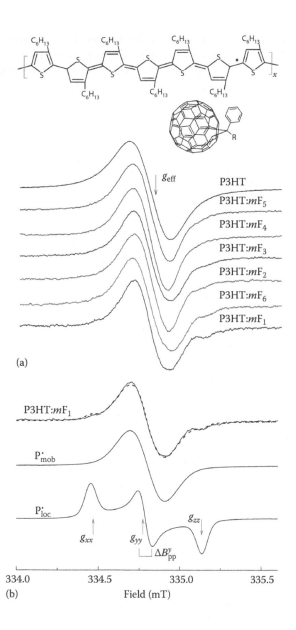

FIGURE 5.1

(a) Room temperature electron paramagnetic resonance (EPR) spectra of the initial poly(3-hexylthophene) (P3HT) and P3HT:mF_1 P3HT:mF_6 composites. (b) Experimental EPR spectrum of the sample P3HT:mF_1 (solid line) compared with sum theoretical spectrum (dashed line) of localized P'_{loc} and mobile P'_{mob} polarons with relative concentration ratio $[P'_{loc}]/[P'_{mob}] = 0.089$ and I_L^0/I_G^0 ratios of 0.3 and 0.2, respectively. The formation of polaron in P3HT is shown schematically. The magnetic resonance parameters measured are shown as well. (Reprinted from *Spectroscopy of Polymer Nanocomposites*, Ponnamma, D., Rouxel, D., Thomas, S., and Krinichnyi, V.I., EPR spectroscopy of polymer:fullerene nanocomposites, pp. 202–275, Chapter 9, Copyright 2016, with permission from Elsevier.)

presented were deconvoluted as it was made in the case of other organic systems [264,558,575–579]. This gives the conclusion that there is stabilization of two types of PC in the P3HT:mF_i samples, namely, polarons localized at cross-bonds and/or on the short π-conjugated polymer chains P'_{loc} with $g_{xx} = 2.0049$, $g_{yy} = 2.0030$, $g_{zz} = 2.0010$, and linewidth $\Delta B_{pp} = 0.066$ mT and a polaron moving along the main π-conjugated polymer chain P'_{mob} with $g_{iso} = 1/3(g_{xx} + g_{yy} + g_{zz}) = 2.0029$, and $\Delta B_{pp} = 0.215$ mT. The best fit of the P'_{loc} signal was achieved using a nearly Gaussian line shape. This means that the transitions are inhomogeneously broadened mainly due to unresolved hyperfine interaction of unpaired spin with protons. Simulated spectra of P'_{loc} and P'_{mob} are also shown also in Figure 5.1b. The isotropic g-factor of polarons P'_{loc} lies near to that of the P'_{mob} ones. This fact supports the supposition made earlier about the nature of PC. Spin concentration ratio $[P'_{loc}]/[P'_{mob}]$ is approximately 0.089 for P3HT:mF_1 and decreases for other compounds (see Figure 5.1a). Note that the existence of such polarons with different relaxation and dynamics was also determined in other conjugated polymers [118] and polymer composites [55].

As a polymer with embedded fullerene derivative is irradiated at $T \le 200$ K by IR-vis photons directly in the cavity of the EPR spectrometer, two overlapping contributions appear in LEPR spectra, whose shape, relative intensity, and position depend on spin precession frequencies ω_e. Figure 5.2 shows exemplary LEPR spectra of the P3HT:PC_{61}BM and PCDTBT:PC_{61}BM composites background irradiated by monochromic light at different ω_e values [57,558,559,580–582]. Subsequent LEPR measurement cycles of heating up to room temperature, cooling down to $T \le 200$ K, illumination, switching the light off, and heating up again yield identical data. The spectra registered in low and high fields can be attributed to positively charged polarons P^+ and negatively charged methanofullerene mF_{61}^- photoinduced in the bulk heterojunction (BHJ) of composites, respectively. Both PCs are characterized by anisotropic magnetic resonance parameters manifesting higher ω_e as it follows from Equation 2.2. These parameters measured at different spin precession frequencies ω_e are summarized in Table 5.1. G-factor of the fullerene ion radicals lies near that of other fullerene anion radicals [583]. As in the case of the initial C_{60} molecule [584], the deviation of the mF_{61}^- g-factor from that of the free electron $g_e = 2.00232$ is due to the fact that the orbital angular moment is not completely quenched. Due to the dynamical Jahn–Teller effect accompanying the structural molecular deformation, the isotropic nature of the icosahedral C_{60} molecule is distorted after formation of the mF_{61}^- anion radical, resulting in an axial or even lower symmetry [585]. This is also realized in the case of the mF_{61}^- anion radical [586], where the high symmetry of the molecule is already decreased by the bond to the phenyl side chain prior to electron accepting. Asymmetrical distribution of spin density in polaron and fullerene anion radical leads also to tensor character of their linewidths [574,586]. This should be taken into account in order to calculate more precisely an effective LEPR spectrum of the P3AT:PC_{61}BM system.

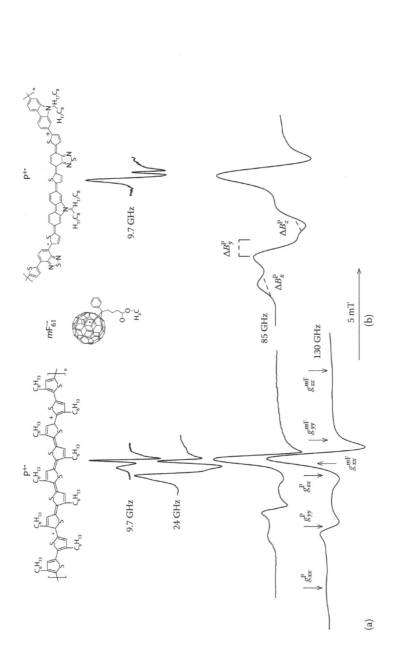

FIGURE 5.2

Light-induced electron paramagnetic resonance spectra of the P3HT:PC$_{61}$BM (a) and PCDTBT:PC$_{61}$BM (b) composites irradiated by laser registered at different spin precession frequencies ω_e shown at appropriated spectra and low temperature. The appearance of a polaron P$^{+•}$ on polymer chain and methanofullerene ion radical mF$_{61}^{-•}$ embedded between polymer chains are shown. The main values of **g** tensors of these paramagnetic centers are shown as well. (Reprinted from *Spectroscopy of Polymer Nanocomposites*, Ponnamma, D., Rouxel, D., Thomas, S., and Krinichnyi, V.I., EPR spectroscopy of polymer:fullerene nanocomposites, pp. 202–275, Chapter 9, Copyright 2016, with permission from Elsevier.)

TABLE 5.1

The Main Magnetic Resonance Parameters Obtained from Polymer Composites by the Light-Induced EPR Method at Different Wave Bands and Low (20–80 K) Temperatures When All Spin Motions Are Considered to Be Frozen

Sample	g_{xx}	g_{yy}	g_{zz}	g_{iso}	ΔB_{PP}^{x} (mT)	ΔB_{PP}^{y} (mT)	ΔB_{PP}^{z} (mT)	ΔB_{PP}^{iso} (mT)	WB[a]	References
P3HT	2.0049	2.0030	2.0010	2.0030	0.66	0.66	0.66	0.66	X	[556]
P3HT	2.0030	2.0021	2.0011	2.0021	0.16	0.15	0.16	0.16	K	[587]
P3HT	2.0028	2.0019	2.0009	2.0019	1.07	0.53	0.64	0.75	W	[580]
P3HT	2.00380	2.00230	2.00110	2.00240	—	—	—	—	D	[577]
P3OT	2.00409	2.00332	2.00232	2.00324	0.82	0.78	0.88	0.83	D	[523]
P3DDT	2.0026	2.0017	2.0006	2.0016	0.25	0.14	0.15	0.18	X	[558]
PCDTBT	2.0031	2.0026	2.0010	2.0022	—	—	—	1.4	X	[582]
PCDTBT	2.00320	2.00240	2.00180	2.00247	—	—	—	—	D	[581]
M3EH-PPV	2.0034	2.0025	2.0024	2.0028	0.22	0.27	0.28	0.25	K	[587]
M3EH-PPV	2.00377	2.00275	2.00220	2.00291	0.40	0.40	0.40	0.40	W	[559]
MDMO-PPV	2.00341	2.00341	2.00241	2.00308	0.80	0.50	0.50	0.60	W	[586]
MDMO-PPV	2.0033	2.0022	2.0022	2.0026	0.96	0.82	0.82	0.87	W	[580]

(Continued)

TABLE 5.1 (Continued)

The Main Magnetic Resonance Parameters Obtained from Polymer Composites by the Light-Induced EPR Method at Different Wave Bands and Low (20–80 K) Temperatures When All Spin Motions Are Considered to Be Frozen

Sample	g_{xx}	g_{yy}	g_{zz}	g_{iso}	ΔB_{pp}^x (mT)	ΔB_{pp}^y (mT)	ΔB_{pp}^z (mT)	ΔB_{pp}^{iso} (mT)	WB[a]	References
bis-PC$_{61}$BM	—	—	—	2.0007	—	—	—	1.3	X	[588]
PC$_{61}$BM	2.0005	2.0004	1.9988	1.9999	0.12	0.11	0.29	0.17	K	[587]
PC$_{61}$BM	2.00031	2.00011	1.99821	1.99954	0.23	0.13	0.88	0.41	W	[586]
PC$_{61}$BM	2.00021	2.00000	1.99860	1.99960	0.50	0.40	1.7	0.87	W	[559]
PC$_{61}$BM	2.00058	2.00045	1.99845	1.99983	—	—	—	—	D	[577]
PC$_{71}$BM	2.0062	2.0031	2.0027	2.0040	—	—	—	0.14	X	[579]
PC$_{71}$BM	2.0056	2.0023	2.0022	2.0034	—	—	—	—	X	[589]
PC$_{71}$BM	2.00592	2.00277	2.00211	2.00360	—	—	—	—	D	[577]
PC$_{61}$BM–O–C$_{60}$	2.0004	2.0002	1.9984	1.9997	0.36	0.36	0.44	0.39	K	[587]
PC$_{61}$BM–O–C$_{60}$	2.00045	2.00004	1.99860	1.99970	0.57	0.58	1.6	0.92	W	[559]
MDHE-C$_{61}$	2.00016	2.00000	1.99940	1.99985	0.21	0.19	1.0	0.47	W	[559]
MDHE-C$_{61}$–O–C$_{60}$	2.00050	2.00023	1.99910	1.99994	0.21	2.3	0.90	1.14	W	[559]

Source: Reprinted from *Spectroscopy of Polymer Nanocomposites*, Ponnamma, D., Rouxel, D., Thomas, S., and Krinichnyi, V.I., EPR spectroscopy of polymer:fullerene nanocomposites, .pp. 202–275, Chapter 9, Copyright 2016, with permission from Elsevier.

a The wave bands (WB) correspond to spin precession frequency $\omega_e/2\pi$ and resonant magnetic field B_0 of 9.7 GHz and 0.34 T (X), 24 GHz and 0.86 T (K), 85 GHz and 3.3 T (W), and 140 GHz and 4.9 T (D), respectively.

If one includes Coulombic interactions, this should affect the activation energy for either defrosting or thermally assisted tunneling by an amount $U_c = e^2/4\pi\varepsilon\varepsilon_0 r$, where r is a charge pair separation. Assuming, for example, $\varepsilon = 3.4$ for P3HT [590], the minimum separation of charge carriers is equal to the radius a of π-electrons on the C atoms, which is two times longer than the Bohr radius, that is, 0.106 nm, r equal to interchain separation, 0.38 nm [520], one obtains the decrease in U_c from ~0.4 down to 0.02 eV during dissociation of an initial radical pair. Therefore, both the photoinduced polaron and the anion radical indeed should be considered as noninteracting that prolongs their life.

Electronic properties of a plastic solar cell can also be improved, for example, by increasing its light absorption coefficient. Photoluminescence and atomic force microscopy studies showed [591–593] that a wider and stronger light absorption is reached in polymer:PC$_{71}$BM composites. Since optical absorption is closely related to crystallinity of such systems, it was inferred that, for example, P3HT:PC$_{71}$BM composite is more crystalline than P3HT:PC$_{61}$BM one and, therefore, demonstrates higher (by ~33%) current density and power conversion efficiency. So the understanding of the elementary processes of exciton initiation, charge separation, stabilization, and recombination should be a prerequisite for improving the efficiency of such photovoltaic systems. Indeed, the formation of C$_{70}$ anion radicals initiates a subgap photoinduced absorption band at 0.92 eV [594], hidden in the spectra of polymer:PC$_{71}$BM composites, which allows more exact studies of charge separated states in such systems. On the other hand, comparative multifrequency EPR investigation of various polymer:fullerene composites has demonstrated [577,581] significant difference in deconvoluted LEPR spectra of both charge carriers. Indeed, the isotropic g-factor of the $m\mathrm{F}_{61}^{\cdot-}$ and $m\mathrm{F}_{71}^{\cdot-}$ methanofullerene anion radicals was obtained to be equal to 1.99983 and 2.00360, respectively. Taking into account that the isotropic (effective) g-factor g_{iso} of polarons is approximately 2.003, this should mean the decrease in spectral resolution at replacing the PC$_{61}$BM by PC$_{71}$BM counterions. So the effective g-factor of different $m\mathrm{F}_{61}^{\cdot-}$ anion radicals is normally less than g-factor of the free electron [558,577,578,586,588]; g_{iso} value of the $m\mathrm{F}_{71}^{\cdot-}$ anion radical exceeds g_e [577,589]. This is in agreement with the study of respective anion radicals in crystalline ($g_{iso} = 2.0047$) [595,596] and dissolved [596–599] fullerene C$_{70}$. Such effect has been supposed [600] to appear due to different Jahn–Teller dynamics of C$_{60}$ and C$_{70}$ molecules, which might contribute to different signs of the g-value shifts. According to the classical Stone theory of g-factors [601], negative deviation of the g-factor from g_e is due to spin–orbit coupling with empty p- or d-orbitals, while spin–orbit coupling with occupied orbitals leads to positive g-factor deviation. The latter case is typical for most organic radicals. Thus, a difference in g-values of $m\mathrm{F}_{61}^{\cdot-}$ and $m\mathrm{F}_{71}^{\cdot-}$ anion radicals indicates the different electronic structure of their molecular orbitals. Positive shift of the g-factors in solution for $m\mathrm{F}_{61}^{\cdot-}$ relative $m\mathrm{F}_{71}^{\cdot-}$ can be explained in the framework of the static Jahn–Teller effect [600, 602]. Jahn–Teller dynamics in the solid phase seems to be quite different for C_{60} and C_{70} globes, which might contribute to different signs of their g-value shifts [600]. However, there is no jet unified theory that can explain **g** tensors of both the $mF_{61}^{\cdot-}$ and $mF_{71}^{\cdot-}$ anion radicals.

Nevertheless, the contribution of the $mF_{71}^{\cdot-}$ charge carriers can be obtained by using the *light on–light off* method accompanied by the deconvolution of the sum of LEPR spectra of both charge carriers into two individual spectra, which can then be compared with those obtained at millimeter wave band EPR. In Figure 5.3 are drawn X-band LEPR spectra of the P3HT:PC$_{71}$BM and PCDTBT:PC$_{71}$BM composites illuminated at $T = 77$ K [582,603,604]. Assuming that each optical photon absorbed by the nanocomposite initiates positively and negatively charged carriers, the $mF_{71}^{\cdot-}$ spectrum may simply be obtained by the extraction of the P$^+$ spectrum from the initial LEPR one shown in Figure 5.3.

In order to analyze in details all magnetic resonance parameters as a function of different effects, the sum of LEPR spectra were deconvoluted by using numerical simulation as it was made in the case of other polymer:fullerene systems [57,558,577–579]. Such algorithm in combination with the *light on–light off* method allowed us to determine appropriate parameters of both charge carriers photoinduced in the P3HT:PC$_{71}$BM composite (see Table 5.1). These values lie near those determined at higher spectral resolution [577,580]. This was used also for separate determination of all main magnetic resonance parameters of charge carriers stabilized and photoinitiated in other analogous BHJ at a wide range of the temperature, photon energy, and registration frequency. The best fit of such LEPR spectra was achieved using a convolution of Gaussian and Lorentzian line shapes, which means that electron excitation leads to inhomogeneous and homogeneous line broadening, respectively, due to unresolved hyperfine interaction of an unpaired spin with neighboring protons and also to its different mobility.

It was demonstrated [118,122,124] that the main parameters of spin charge carriers stabilized in conjugated polymers are governed by their structure as well as by different adducts embedded into polymer matrix [42,55,57,446]. A similar effect should be probably reached by varying the number of electrons trapped by each acceptor of a polymer:fullerene composite. In order to test this assumption, the charge carriers photoinitiated in some BHJ formed by the poly[2,5-dimethoxy-1,4-phenylene-1,2-ethenylene-2-methoxy-5-(2-ethylhexyloxy)-(1,4-phenylene-1,2-ethenylene)] (M3EH-PPV shown in Figure 1.2) macromolecules with different mono- and di-C_{60}-fullerene derivatives were studied at W-band EPR [559]. Figure 5.4 exhibits W-band LEPR spectra of the M3EH-PPV:PC$_{61}$BM, M3EH-PPV:PC$_{61}$BM–O–C$_{60}$, M3EH-PPV:MDHE–C$_{61}$, and M3EH-PPV:MDHE–C$_{61}$–O–C$_{60}$ composites. The main magnetic resonance parameters determined for both charge carriers in these systems are also summarized in Table 5.1. It was shown that

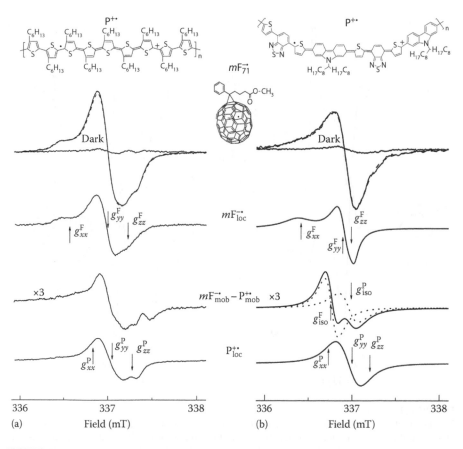

FIGURE 5.3

X-band light-induced electron paramagnetic resonance spectra of the P3HT:PC$_{71}$BM (a) and PCDTBT:PC$_{71}$BM (b) composites and their contributions due to mobile quasi-pairs $m\mathrm{F}^{\rightarrow}_{\mathrm{mob}} - \mathrm{P}^{+\bullet}_{\mathrm{mob}}$ and localized polarons $\mathrm{P}^{+\bullet}_{\mathrm{loc}}$ and methanofullerene anion radicals $m\mathrm{F}^{\rightarrow}_{\mathrm{loc}}$ background photoinduced by photons with $h\nu_{\mathrm{ph}} = 2.10$ eV at $T = 77$ K. Dashed and dotted lines show the spectra calculated using the appropriate terms of **g** tensor presented in Table 5.1. Photoinitiation of a polaron on a polymer chain accompanied by electron transfer to a methanofullerene globe is schematically shown. The positions of photoinduced radicals and terms of their g-factors are shown as well. (Reprinted from *Spectroscopy of Polymer Nanocomposites*, Ponnamma, D., Rouxel, D., Thomas, S., and Krinichnyi, V.I., EPR spectroscopy of polymer:fullerene nanocomposites, pp. 202–275, Chapter 9, Copyright 2016, with permission from Elsevier.)

the increase of a number of elemental negative charges on each fullerene molecule leads to the appearance of dispersion term in LEPR spectra of respective ion radicals (see Figure 5.4). This effect was interpreted as the formation of highly ordered fullerene domains (metal-like crystallites) in the amorphous phase of M3EH-PPV matrix. Interaction of conduction electrons with microwave (MW) field originates the appearance on the

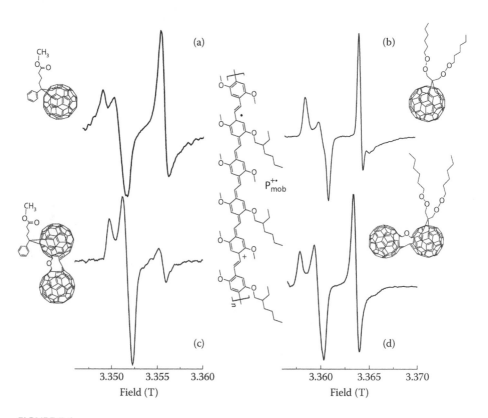

FIGURE 5.4

X-band light-induced electron paramagnetic resonance spectra of charge carriers photoinitiated in bulk heterojunctions formed by M3EH-PPV macromolecules with [6,6]-phenyl-C61-butanoic acid methyl ester ($PC_{61}BM$) (a), [6,6]-malonedihexyle-fullerene ester ($MDHE-C_{61}$) (b), $PC_{61}BM-O-C_{60}$ (c), and $MDHE-C_{61}-O-C_{60}$ (d) irradiated by laser photons with the energy of 2.3 eV (wavelength $\lambda_{ph} = 530$ nm) at low temperature when molecular motion is frozen. The polaron formed on the polymer after the transfer of a charge from its chain to a fullerene molecule is shown schematically. (Reprinted from *Spectroscopy of Polymer Nanocomposites*, Ponnamma, D., Rouxel, D., Thomas, S., and Krinichnyi, V.I., EPR spectroscopy of polymer:fullerene nanocomposites, pp. 202–275, Chapter 9, Copyright 2016, with permission from Elsevier.)

surface of the sample of skin depth δ and, therefore, Dysonian contribution in its LEPR spectrum described earlier. Figure 5.5 shows that the data obtained from M3EH-PPV modified with both di-fullerene derivatives are well fitted by the dependence calculated from Equations 2.27, 2.30, and 2.31. The conclusion made earlier is also confirmed by the scanning electron microscopy [559]. This allowed a more complete investigation of magnetic, relaxation, and dynamic parameters of spin charge carriers depending on different properties of polymer composites and their ingredients. The skin-layer depth δ for di-C_{62}-fullerene was estimated to be approximately 10 μm at $\omega_e/2\pi = 95$ GHz. This allows evaluating *ac* conductivity of the di-C_{60}-fullerene domains.

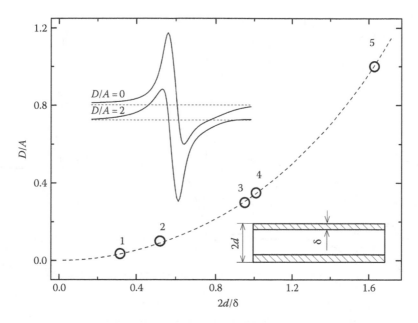

FIGURE 5.5
Dependence D/A(2d/δ) calculated using Equations 2.27, 2.30, and 2.31 (solid line) and obtained experimentally for the M3EH-PPV:PC$_{61}$BM-O-C$_{60}$ (open points 2, 3, and 5) and M3EH-PPV:MDHE-C$_{61}$-O-C$_{60}$ (open points 1 and 4) composites at their different plane thicknesses 2d and spin precession frequencies $\omega_e \sim \delta^{-2}$. In the insets, X-band light-induced electron paramagnetic resonance spectra and parameters of the samples are shown. (Reprinted from *Spectroscopy of Polymer Nanocomposites*, Ponnamma, D., Rouxel, D., Thomas, S., and Krinichnyi, V.I., EPR spectroscopy of polymer:fullerene nanocomposites, pp. 202–275, Chapter 9, Copyright 2016, with permission from Elsevier.)

Let us show how the energy of excitation photons $h\nu_{ph}$ governs the main resonance parameters of both charge carriers.

Figure 5.6 demonstrates the effect of this parameter on the LEPR spectrum of the P3DDT:PC$_{61}$BM nanocomposite at $T \le 200$ K. As in the case of other polymer:fullerene composites, sum spectra were attributed to radical quasi-pairs of polarons P$^+$ with $g_{iso} = 2.0023$ and negatively charged anion radicals $mF_{61}^{\cdot-}$ with $g_{iso} = 2.0001$ [558,573,605–611]. The sum spectra calculated with the fitting magnetic parameters presented in Table 5.1 are shown in Figure 5.6 as well.

In order to analyze the conjoint effect of the structure of fullerene derivative and the energy of initiating photons on electronic properties of a polymer:fullerene composite, P3HT:PC$_{61}$BM, and P3HT:*bis*-PC$_{61}$BM composites were studied at X-band LEPR [578,588,612]. Their detached ingredients were shown to characterize by the absence of both *dark* and photoinduced LEPR signals over the temperature range of 77–340 K. As they form polymer:fullerene BHJ and irradiate by visible light directly in a cavity of the

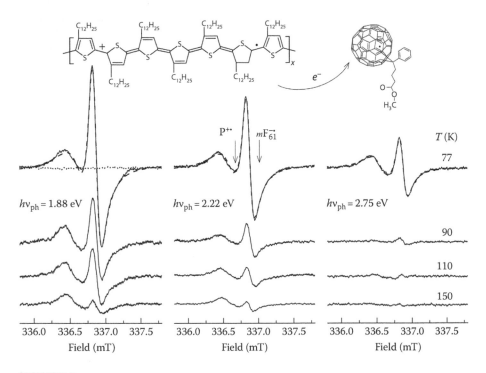

FIGURE 5.6

X-band LEPR spectra of the radical quasi-pairs photoinduced in the P3DDT:PC$_{61}$BM system shown in the top inset by steady-state laser irradiation with different photon energies $h\nu_{ph}$ as function of temperature. By the left dotted line, the *dark* spectrum obtained before laser irradiation is shown. Dashed lines show sum light-induced electron paramagnetic resonance spectra calculated using magnetic resonance parameters presented in Table 5.1.

EPR spectrometer, two overlapping LEPR lines appear at $T \leq 200$ K (Figure 5.7). As in the case of other polymer:fullerene systems, low- and high-field lines photoinduced in the P3HT:PC$_{61}$BM composite consists of two Lorentzian contributions of mobile polarons, $P_{mob}^{\cdot+}$, and methanofullerene anion radicals, $mF_{mob}^{\cdot-}$ (shown in Figure 5.7 as radical quasi-pairs 2, $P_{mob}^{\cdot+} - mF_{mob}^{\cdot-}$), as well as two Gaussian contributions of localized polarons, $P_{loc}^{\cdot+}$, and methanofullerene anion radicals, $mF_{loc}^{\cdot-}$, pinned in polymer traps. Analogous localized polarons $P_{loc}^{\cdot+}$ with $g_{iso}^{P} = 2.0023$ and quasi-pairs with $g_{iso}^{mF} = 2.0007$ also contribute to LEPR spectra of the P3HT:*bis*-PC$_{61}$BM composite; however, there is absent contribution of pinned *bis*-methanofullerene radicals, $bmF_{loc}^{\cdot+}$ (Figure 5.7). These values differ slightly from those obtained for P3HT:PC$_{61}$BM but, however, are close to appropriate parameters determined for polarons stabilized in other fullerene-modified conjugated polymers [55,57,606,607,613,614] and fullerene anion radicals [583]. The absence of localized anion radicals in the P3HT:*bis*-PC$_{61}$BM composite implies that the number of deep traps able to capture is sufficiently lower than that in the P3HT:PC$_{61}$BM one due to the

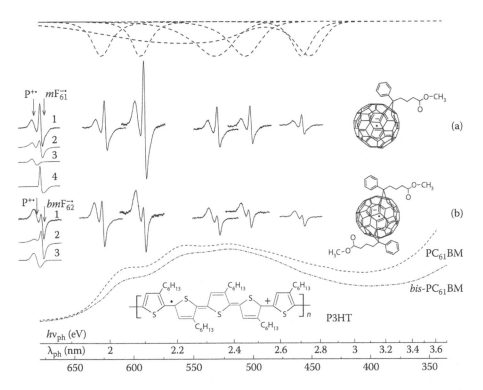

FIGURE 5.7
Normalized light-induced electron paramagnetic resonance (LEPR) spectra of charge carriers background photoinduced at 77 K in bulk heterojunctions formed by macromolecules of regioregular poly(3-hexylthophene) with globes of [6,6]-phenyl-C61-butanoic acid methyl ester ($PC_{61}BM$) (a) and bis-$PC_{61}BM$ (b) as function of the photon energy hv_{ph} (wavelength λ_{ph}). From left to right, the spectra obtained at irradiation of the samples by the white light and by the light with photon energies of 1.98, 2.10, 2.33, 2.45, and 2.72 eV are shown. Top dashed lines show irradiation spectra of the light sources. At the bottom, absorption spectra of the P3HT:$PC_{61}BM$ and P3HT:bis-$PC_{61}BM$ composites are shown by the dashed and dash-dotted lines, respectively. At the left, theoretical sum spectra (1) and their Lorentzian contribution of mobile radical quasi-pairs, $P^{+\cdot}_{mob} - bmF^{-\cdot}_{mob}$ (2), Gaussian contributions caused by localized polarons $P^{+\cdot}_{loc}$ (3), and methanofullerene $mF^{-\cdot}_{loc}$ (4) calculated using respective magnetic resonance parameters presented in Table 4.1, with concentration ratios $[P^{+\cdot}_{loc}] : [P^{+\cdot}_{mob} - bmF^{-\cdot}_{mob}] : [mF^{-\cdot}_{loc}]$ of 6:57:1 for P3HT:$PC_{61}BM$ and 15:87 for P3HT:bis-$PC_{61}BM$ are also shown. The polaron formed on the poly(3-hexylthophene) (P3HT) chain after transfer of its charge to the methanofullerene is shown schematically. The positions of LEPR spectra of polarons, $P^{+\cdot}$, and methanofullerene anion radicals, $bmF^{-\cdot}_{mob}$, are shown as well. (From Krinichnyi, V.I. and Yudanova, E.I., *Sol. Energy Mater. Sol. Cells*, 95(8), 2302, 2011. With permission.)

better ordering of the former; however, their depth depends on the energy of photons.

PCDTBT, in contrast with P3AT, is characterized by *dark* EPR spectrum typical for localized PCs, which changed slightly under its irradiation by visible light over wide temperature range. Figure 5.8 shows LEPR spectra of charge

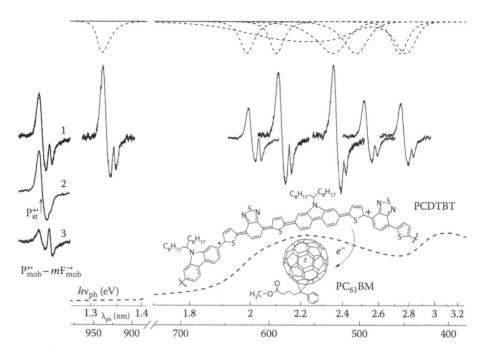

FIGURE 5.8

Light-induced electron paramagnetic resonance (LEPR) spectra of charge carriers background photoinduced at $T=77$ K in bulk heterojunctions formed by macromolecules of poly[N-9″-hepta-decanyl-2,7-carbazole-alt-5,5-(4′,7′-di-2-thienyl-2′,1′,3′-benzothiadiazole)] (PCDTBT) with globes of [6,6]-phenyl-C61-butanoic acid methyl ester (PC$_{61}$BM) as function of the photon energy $h\nu_{ph}$ (linewidth λ_{ph}) normalized to the intensity of light sources. From left to right, the spectra obtained at irradiation of the samples by the white light and by the light with photon energies $h\nu_{ph}$ equal to 1.32, 1.98, 2.10, 2.34, 2.46, and 2.73 eV are shown. At the left, theoretical sum spectra (1, dashed lines) and their Lorentzian contributions caused by stabilized polarons $P_{st}^{+\cdot}$ (2) and highly mobilized pair radicals, $P_{mob}^{+\cdot}$ and $mF_{mob}^{-\cdot}$, (3) numerically calculated using $\Delta B_{pp}^{P} = 0.137$ mT, $\Delta B_{pp}^{mF} = 0.113$ mT, and a concentration ratio $[P_{st}^{+\cdot}]:[P_{mob}^{+\cdot} - mF_{mob}^{-\cdot}] = 1:14$ are shown. The dashed lines at the top show irradiation spectra of the light sources and the dashed lines at the bottom show IR-Vis absorption spectrum of the PCDTBT:PC$_{61}$BM composite. The charge transfer from PCDTBT to the methanofullerene accompanied by the formation on the polymer chain having positively charged polaron $P^{+\cdot}$ and negatively charged ion radical $mF_{61}^{-\cdot}$ both with spin $S=\frac{1}{2}$ is shown schematically. The positions of LEPR spectra of all charge carriers are shown as well. (Reprinted with permission from Krinichnyi, V. I., Yudanova, E.I., and Denisov, N.N., The role of spin exchange in charge transfer in low-bandgap polymer:fullerene bulk heterojunctions, *J. Chem. Phys.*, 141(4), 044906/01. Copyright 2014, American Institute of Physics.)

carriers stabilized and reversibly photoinitiated in the PCDTBT:PC$_{61}$BM sample under its background irradiation in a microwave cavity of the EPR spectrometer by light with photon energy of 1.32–2.73 eV in comparison with its IR–vis absorption spectrum [582]. In the latter, the low energy, broad, and featureless absorption band with a peak at $h\nu_{ph} = 2.23$ eV ($\lambda_{ph} = 560$ nm) corresponds to the intramolecular charge-transfer transition, whereas meanwhile the pronounced absorption lines in the higher energy region $h\nu_{ph} = 3.09, 3.72$, and 4.70 eV ($\lambda_{ph} = 400, 333$, and 264 nm) (the two latter lines are not shown) are attributed to the fullerene units. This leads to an increase in the intensity of the initial EPR spectrum and the appearance of the second line at higher magnetic fields (see Figure 5.8). Following previous studies, these low- and high-field signals can be also assigned to $P_{mob}^{\cdot+} - mF_{mob}^{\cdot-}$ quasi-pairs.

There were some recorded unusual effects that should be emphasized. The most important of them is that the LEPR signal of the sample appears under its irradiation even in the near-infrared region, $h\nu_{ph} = 1.32$ eV ($\lambda_{ph} = 940$ nm), where the absorption band is nearly nulled (Figure 5.8). The intensity of this signal is comparable to that obtained at higher photon frequencies. This fact can be derived, for example, by a nonlinear optical effect in the bulk of the PCDTBT:PC$_{61}$BM composite converting the photon frequency/energy into a higher value; however, such a supposition was not supported under second harmonic illumination of the composite. So it can be assumed that the formation of spin quasi-pairs indeed occurs in the sample BHJ under their excitation by infrared quanta. Besides, the comparison of an intensity of all LEPR and optical spectra presented shows that the intensity of the former does definitely not correlate neither with the number of optical quanta reaching the sample surface nor with those absorbed by the sample. This does not confirm the conclusion made by Tong et al. [615] that the efficiency of carrier generation in the PCDTBT:fullerene BHJ should be essentially independent of the excitation wavelength. Various hypotheses can be supposed for explanation of these effects. One of them could be the interaction of charge carriers with microwave field. Indeed, only for those relatively isolated excitons, there is a reasonably high probability that metastable PCs will result from the optical production of an electron–hole pair by means of trapping of one carrier and the hopping away of the other [616]. Thus, the separation and, therefore, lifetime of photoinitiated radical quasi-pairs $P_{mob}^{\cdot+} - mF_{mob}^{\cdot-}$ will generally increase with electron spin precession frequency ω_e. One, therefore, can take into account a combination of different processes affecting electronic transport through PCDTBT:PC$_{61}$BM BHJ.

The best fit was achieved supposing stabilization in a polymer backbone of polarons $P_{loc}^{\cdot+}$, characterized by the anisotropic Lorentzian spectrum 2 shown in the left of Figure 5.8 with the main magnetic resonance parameters presented in Table 5.1. The anisotropic nature of the $P_{loc}^{\cdot+}$ spectrum is *prima facie* evidence for its perceptible spin–orbit interaction with the nucleus of nitrogen and sulfur heteroatoms in the polymer network as well as slow mobility. Under light irradiation of the composite, the number of such polarons somewhat increases, and additionally, highly mobile radical quasi-pairs $P_{mob}^{\cdot+} - mF_{mob}^{\cdot-}$ with $g_{iso}^{P} = 2.0022$ and $g_{iso}^{mF} = 2.0006$ (shown as contribution 3 in the left of Figure 5.8) appear. The latter value lies close to that obtained from other fullerene anion radicals [577,581,583,617,618]; however, it slightly exceeds that obtained from PC$_{61}$BM embedded into P3AT matrices

[55,558,609,613,614]. The absence of a Gaussian contribution in LEPR spectra of these charge carriers is argued, as in the case of the P3HT:*bis*-PC$_{61}$BM composite, in favor of a smaller number of spin traps, as well as their faster dynamics in this sample. The two main differences of this composite from the known systems should be emphasized. Hyperfine polaron interaction with neighboring hydrogen and heteroatoms should broaden its spectrum and cause its Gaussian shape. Such a line shape is also a characteristic of spins captured by energetically deep traps. This leads to the appearance of appropriate Gaussian contributions in LEPR spectra of P3HT:PC$_{61}$BM and analogous composites. However, both charge carriers stabilized and photo-initiated in the PCDTBT:PC$_{61}$BM composite are characterized by Lorentzian line shapes. This fact was not previously detected in LEPR study of various polymer:fullerene composites and should likely indicate a lower number of spin traps and higher spin dynamics in the PCDTBT:PC$_{61}$BM system.

Methanofullerene anion radicals photoinduced, for example, in the P3HT:*bis*-PC$_{61}$BM, P3HT:PC$_{61}$BM, and PCDTBT:PC$_{61}$BM systems, demonstrate nearly temperature-independent g_{iso}^{F}. On the other hand, this parameter of polaronic charge carriers, g_{iso}^{P}, happened to be a function of temperature and, to a lesser extent, of photon energy $h\nu_{ph}$ (see Figures 5.7 and 5.8). Figure 5.9 demonstrates that the temperature increase leads to the decrease in g_{iso}^{P} especially in the two latter composites. It can be noted that as PC$_{61}$BM counterions are replaced by *bis*-PC$_{61}$BM ones, the scattering in $g_{iso}^{P}(T)$ dependencies decrease possibly due to the ordering of the polymer:fullerene composite. In other words, this parameter is strongly governed by the structure and conformation of conjugated π-electron system. Indeed, the highest occupied molecular orbital level (HOMO) energy level depends on the overlap of adjacent thiophene molecular orbits and, therefore, is expected to shift with ring angle [619] similar to the valence band involved in the π–π* transition. The bandgap, LUMO–HOMO, slightly depends on both temperature [159] and torsion angle θ [620], being approximately 30° in regioregular P3HT [287]. A decrease in g_{iso}^{P} occurs at electron excitation from the unoccupied shell to the antibonding orbit, π → σ* [135]. Comparing the data obtained, one may conclude that the energy of antibonding orbits decreases as *bis*-PC$_{61}$BM is embedded into the P3HT matrix instead of PC$_{61}$BM. This increases g_{iso}^{P} of the P3HT:*bis*-PC$_{61}$BM composite and decreases the slope of its temperature dependency characteristic of a more ordered system. Indeed, the changes in total energy with the angle θ appear as an effective steric potential energy. The angular dependence of this energy is nonharmonic, with larger angles becoming more probable with the temperature increase. In this case, the decrease of molecular regioregularity or a greater distortion of the thiophene rings out of coplanarity reduces charge mobility along the polymer chains [158]. This is usually attributed to a decrease in the effective conjugation lengths of the chain segments. The intrachain transfer integral t_{1D} is primarily governed by the degree of overlap between the p$_z$ atomic

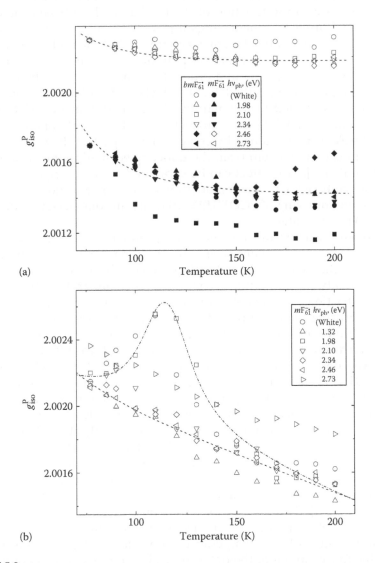

FIGURE 5.9
The value of g_{iso}^P for polarons photoinduced in the P3HT:*bis*-PC$_{61}$BM, P3HT:PC$_{61}$BM (a), and PCDTBT:PC$_{61}$BM (b) bulk heterojunctions as a function of photon energy $h\nu_{ph}$ and temperature. Dashed line shows the dependences calculated using Equation 2.4 with E_1 equal to 9.5 meV (*a*, top line), 8 meV (*a*, bottom line), and 8 meV (*b*). The dash-dotted line (b) is drawn arbitrarily only for illustration to guide the eye. (Reprinted from *Spectroscopy of Polymer Nanocomposites*, Ponnamma, D., Rouxel, D., Thomas, S., and Krinichnyi, V.I., EPR spectroscopy of polymer:fullerene nanocomposites, pp. 202–275, Chapter 9, Copyright 2016, with permission from Elsevier.)

orbitals of the carbon atoms forming polymer units and, therefore, should evolve a square-cosine function of the torsion angle between the planes of the neighboring thiophene rings θ [157]. This allows one to evaluate the decrease in the θ value by approximately 12° at the replacement of the $PC_{61}BM$ by *bis*-$PC_{61}BM$ in appropriate polymer:fullerene system. Therefore, this indicates a more planar and ordered polymer matrix in the P3HT:*bis*-$PC_{61}BM$ composite than in the P3HT:$PC_{61}BM$ one. Peculiar extremal $g_{iso}^P(T)$ dependences obtained from polarons photoinitiated in PCDTBT:$PC_{61}BM$ BHJ by wideband white and monochromic light with photon energy $h\nu_{ph} = 1.98$ eV were explained by competitive impact of the change in the energy transition of a spin from the unoccupied shell to the antibonding orbit $\pi \rightarrow \sigma^*$ [135]. This effect can be explained by the change of effective dimensionality of the polymer matrix [554] and is an additional indication for the better ordering of PCDTBT:$PC_{61}BM$ BHJ in comparison with analogous P3AT composites.

Monotonic temperature dependences presented in Figure 5.9 can be explained *inter alia* by joint harmonic librations of polymer units and chains together with localized polarons, which change the backbone dimensionality [554], modulate charge-transfer integrals [140], lift symmetry restrictions, and enable intrachain and interchain spin relaxation to cause broader polaron linewidth due to the relation $\Delta B_{pp}^P \propto (\Delta g_{iso}^P)^2$ [1]. Figure 5.9 indicates that the dependences calculated from Equation 2.4 with their respective E_1 summarized in Table 5.2 are fitted well in the main experimental data. Such libration dynamics in the P3HT:*bis*-$PC_{61}BM$ and PCDTBT:$PC_{61}BM$ composites occur mainly with the lower activation energy than in the P3HT:$PC_{61}BM$ one (see Table 5.2). Temperature sensitivity of the polaron g-factor decreases at $T \geq 200$ K. Besides, the concentration of both types of charge carriers decreases dramatically at this temperature range that limits significantly the precision of determination of their main magnetic resonance parameters. Such effect can be attributed to fluctuations in local symmetry of side groups relative to the main polymer axis. These groups begin to move at the glass transition of the polymer matrix at $T_g \approx 200$ K [621]; their local relaxation contributes to the topological disorder in the polymer structure and leads to an increase in energy barriers for charge transport. The temperature dependence of g-factors is argued to be due to a coupling of the holes to local vibrations of the chains and/or side groups along a backbone of the polymer matrix. This means that photoinitiated spins act as a nanoscopic probe of molecular and polaron dynamics in polymer:fullerene composites. More detailed information can be obtained at higher spin precession frequencies ω_e, that is, at higher spectral resolution. So the absence in the LEPR spectra of Gaussian signals of localized charge carriers can indicate for higher ordering of PCDTBT layers and also for a low number of energetically deep spin traps that facilitates charge transfer from the polymer chain to the methanofullerene cage. This is similar to cation radical single crystals with alternating

TABLE 5.2

n_p and n_f Values Determined for Polarons and Fullerene Ion Radicals, Energies E_l Determined from Equation 2.4, ΔE_{ij} Determined from Equation 2.15, E_0 Determined from Equation 2.13, E_r Determined from Equation 2.20, E_{ph} Determined from Equation 2.83, E_a Determined from Equation 2.77, and E_0 and E_t Determined from Equation 2.85 for the P3HT:PC$_{61}$BM, P3HT:*bis*-PC$_{61}$BM and PCDTBT:PC$_{61}$BM Composites Irradiated by Polychromatic White and Monochromatic Light with Different Photon Energy/Wavelength $h\nu_{ph}/\lambda_{ph}$

		$h\nu_{ph}/\lambda_{ph}$ (eV/nm)					
Parameter	White	1.32/940	1.98/625	2.10/590	2.34/530	2.46/505	2.73/455
	P3HT:PC$_{61}$BM						
n_p	4.9×10^{-5}	—	3.9×10^{-5}	9.7×10^{-5}	4.2×10^{-5}	4.3×10^{-5}	2.4×10^{-5}
n_f	4.2×10^{-5}	—	2.5×10^{-5}	7.7×10^{-5}	3.1×10^{-5}	3.4×10^{-5}	2.1×10^{-5}
E_l (eV)	0.0098	—	0.0100	0.0097	0.0099	0.0056	0.0095
ΔE_{ij} (eV)[a]	0.0092	—	0.0054	0.0028	0.0017	0.0048	0.0066
ΔE_{ij} (eV)[b]	0.0607	—	0.0307	0.0476	0.0663	0.0793	0.0744
ΔE_{ij} (eV)[c]	0.0304	—	0.0258	0.0438	0.0540	0.0558	0.0569
E_0 (eV)[a]	0.0134	—	0.0519	0.0430	0.0088	0.0082	0.0318
E_0 (eV)[b]	0.0103	—	0.0181	0.0208	0.0113	0.0075	0.0163
E_0 (eV)[c]	0.0210	—	0.0236	0.0296	0.0307	0.0186	0.0169
E_r (eV)	0.0465	—	0.0460	0.0452	0.0351	0.0412	0.0473
E_{ph} (eV)	0.1132	—	0.1218	0.0822	0.1042	0.0709	0.0823
E_0 (eV)	0.0086	—	0.0144	0.0212	0.0109	0.0133	0.0183
E_t (eV)	0.1409	—	0.1441	0.1356	0.1335	0.1267	0.1404
E_a (eV)	0.0504	—	0.1024	0.0739	0.0551	0.0363	0.0454
	P3HT:*bis*-PC$_{61}$BM						
n_p	9.7×10^{-5}	—	7.6×10^{-5}	1.6×10^{-4}	7.0×10^{-5}	6.5×10^{-5}	4.9×10^{-5}
n_f	7.2×10^{-5}	—	5.2×10^{-5}	7.7×10^{-5}	3.3×10^{-5}	3.2×10^{-5}	2.2×10^{-5}
E_l (eV)	0.0053	—	0.0093	0.0066	0.0097	0.0088	0.0081
ΔE_{ij} (eV)[a]	0.0027	—	0.0014	0.0288	0.0140	0.0162	0.0177
ΔE_{ij} (eV)[c]	0.0171	—	0.0239	0.0404	0.0350	0.0375	0.0413
E_0 (eV)[a]	0.0309	—	0.0409	0.0272	0.0184	0.0144	0.0107
E_0 (eV)[c]	0.0080	—	0.0202	0.0136	0.0198	0.0222	0.0227
E_r (eV)	0.0453	—	0.0479	0.0495	0.0468	0.0509	0.0503
E_{ph} (eV)	0.0615	—	0.0765	0.0751	0.0909	0.0875	0.0650
E_0 (eV)	0.0193	—	0.0167	0.0219	0.0174	0.0160	0.0206
E_t (eV)	0.1228	—	0.1283	0.1294	0.1275	0.1189	0.1250
E_a (eV)	0.0627	—	0.0770	0.0423	0.0786	0.0565	0.0588

(Continued)

TABLE 5.2 (*Continued*)

n_p and n_f Values Determined for Polarons and Fullerene Ion Radicals, Energies E_l Determined from Equation 2.4, ΔE_{ij} Determined from Equation 2.15, E_0 Determined from Equation 2.13, E_r Determined from Equation 2.20, E_{ph} Determined from Equation 2.83, E_a Determined from Equation 2.77, and E_0 and E_t Determined from Equation 2.85 for the P3HT:PC$_{61}$BM, P3HT:*bis*-PC$_{61}$BM and PCDTBT:PC$_{61}$BM Composites Irradiated by Polychromatic White and Monochromatic Light with Different Photon Energy/Wavelength $h\nu_{ph}/\lambda_{ph}$

Parameter	White	\multicolumn{6}{c}{$h\nu_{ph}/\lambda_{ph}$ (eV/nm)}
		1.32/940	1.98/625	2.10/590	2.34/530	2.46/505	2.73/455
\multicolumn{8}{l}{PCDTB:PC$_{61}$BM}
E_l, eV	0.006[d]	0.004	0.008[d]	0.003	0.002	0.004	0.004
ΔE_{ij} (eV)[a]	0.018	0.005	0.056	0.050	0.032	0.072	0.059
ΔE_{ij} (eV)[c]	0.048	0.051	0.068	0.060	0.076	0.065	0.058
E_r (eV)[a]	0.028	0.117	0.001	0.001	0.017	0.003	0.015
E_r (eV)[c]	0.013	0.013	0.014	0.019	0.020	0.018	0.018
E_{ph} (eV)	0.121	0.098	0.049	0.074	0.080	0.043	0.120
E_a (eV)	0.079	0.093	0.067	0.081	0.110	0.141	0.178

Source: Reprinted from *Spectroscopy of Polymer Nanocomposites*, Ponnamma, D., Rouxel, D., Thomas, S., and Krinichnyi, V.I., EPR spectroscopy of polymer:fullerene nanocomposites, pp. 202–275, Chapter 9, Copyright 2016, with permission from Elsevier.

[a]   Determined for $P_{loc}^{\cdot+}$.
[b]   Determined for $mF_{loc}^{\cdot-}$.
[c]   Determined for $mF_{mob}^{\cdot-}$.
[d]   Determined for monotonic part of the curve.

organic and inorganic layers possessing Lorentzian EPR spectra of their charge carriers, characteristic of free electrons [1].

5.2 Photoinitiation and Recombination of Charge Carriers in Polymer:Fullerene Nanocomposites

Another important parameter of nanomodified polymer composite is spin susceptibility χ proportional to the sum number of charge carriers with non-compensated electrons.

In darkened P3HT:mF_1 BHJ, spin concentration ratio $[P_1^{\cdot+}]/[P_1^{\cdot+}]$ of coupled polarons is approximately 0.089 (see Figure 5.1a) [556,557] and is lower for other compounds. Under light illumination, the number of spin charge carriers increases sharply in composites, and, however, the charge carriers tend to recombine. This means that illuminating a polymer:fullerene bulk heterojunction, one can register only the net PCs as a difference of forward initiating (fast) and reversed recombination (slow) processes [569]. The number of such centers depends mainly on the maximum number of polarons able to be formed at the first photoinitiation stage and also on the recombination coefficient of charge carriers. To determine the limiting number of polarons able to be stabilized in a polymer matrix and to compare magnetic resonance parameters of polarons reversibly and irreversibly initiated in a polymer matrix, its modification by, for example, iodine molecules [580,586], can be used. However, this allows one to modify an initial polymer only. Such a procedure was used to obtain the limiting number of both polarons n_p and methanofullerene anion radicals n_f simultaneously formed per each polymer unit of the P3HT:PC$_{61}$BM and P3HT:*bis*-PC$_{61}$BM composites [578]. The former parameter was obtained from these systems at $T=77$ K to be 2.8×10^{-3} and 3.8×10^{-3}, respectively. Note that the n_p values obtained are considerably lower than that, $n_p \approx 0.05$, estimated for polarons excited in doped polyaniline [153]. Limiting paramagnetic susceptibility χ was determined for these systems to be 8.7×10^{-6} and 1.1×10^{-5} emu/mol, respectively, at $T=310$ K. As temperature decreases down to 77 K, these values increase to 8.7×10^{-5} and 9.9×10^{-5} emu/mol, respectively. Besides, the analysis showed that the cooling of the samples leads to the appearance of anisotropic Gaussian term in their sum of EPR spectra attributed to strongly frozen polarons. The ratio of a number of mobile to frozen polarons at 77 K is approximately 8:1 for P3HT:PC$_{61}$BM and 7:1 for P3HT:*bis*-PC$_{61}$BM. Mobile polarons initiated in these systems by the I$_2$-doping at 310 K demonstrate single Lorentzian EPR spectrum with linewidth ΔB_{pp} of 0.40 and 0.56 mT, respectively, which are much broader than that obtained from polarons in other conjugated polymers [118]. The broadening of the EPR transitions is most likely due to the interaction between neighboring charged polarons. The contribution to linewidth due to such interaction can be estimated as $\Delta B_{dd} = \mu_B R_0^{-3} = 4/3 \pi \mu_B n_p$, where R_0 is a distance between polarons proportional to their concentration n_p on the polymer chain. At the transition from PC$_{61}$BM to *bis*-PC$_{61}$BM counteranion, the ΔB_{pp} value of mobile and trapped polarons characterized by Lorentzian and Gaussian distribution of spin packets, respectively, changes at 77 K from 0.19 and 0.23 mT down to 0.18 and 0.19 mT. Assuming intrinsic linewidth of polarons $\Delta B_{pp}^0 = 0.15$ mT in regioregular P3HT [622], one can obtain from the line broadening as a result of dipole–dipole interaction $R_0 \approx$ 1.6 nm for P3HT:PC$_{61}$BM and 1.3 nm for P3HT:*bis*-PC$_{61}$BM. Intrinsic concentration of doping-initiated polarons determined only upon polymer fractionation of these composites was determined to be 1.6×10^{19} and 2.2×10^{19} cm^{-3}, respectively, at 77 K. This value is approximately 2×10^{19} cm^{-3} obtained from the acceptors of ZnO-treated P3HT [623] but, however, is less sufficient than 10^{21} cm^{-3} for maximum polaron concentration in regioregular P3HT [76]. Effective concentrations calculated for both polymer and fullerene phases in the P3HT:PC$_{61}$BM and P3HT:*bis*-PC$_{61}$BM composites are 1.8×10^{18} and 2.1×10^{18} cm^{-3}, respectively. This allowed us to evaluate an effective number of both types of charge carriers per each polymer unit initiated in these polymer:fullerene composites by I_2-doping described earlier and also by light irradiation. These values obtained from photoinitiated charge carriers are summarized in Table 5.2.

Electronic properties of polymer:fullerene nanocomposites depend mainly on the number, mobility, and lifetime of spin quasi-pairs photoinitiated in their BHJ. Spin charge carriers localized in polymer matrix take part in electron transfer only indirectly. Contributions of each charge carriers for effective spin susceptibility of such systems are governed not only by the energy of photons inducing spin traps in their matrix but also by their morphology and ordering. Indeed, a comparative study showed [612] that ultrasonic (US), microwave, and thermal treatment of a composite leads to a change in the main magnetic resonance parameters of all spin packets taking part in charge transport through its BHJ. Figure 5.10 shows that the concentrations of photoinduced polarons and radical anions of methanofullerene are characterized by a bell-like dependence with a maximum of approximately 2.1 eV. This value is close to the energy gap $E_g = 1.92$ eV obtained from P3HT [624]. It should be noted that the number of anion radicals of fullerenes $mF_{loc}^{\cdot-}$ trapped in the P3HT:PC_{61}BM BHJ is practically independent of both the energy of initiating photons and the method of composite treatment. The treatment of both composites leads to the increase in the number of charge carriers (Figure 5.10). In this case, the probability of immobilization of polarons $P_{loc}^{\cdot+}$ in polymer matrix increases upon the illumination of P3HT:PC_{61}BM composite by photons with the energy of $h\nu_{ph} = 2.1$–2.5 and P3HT:bis-PC_{61}BM BHJ by photons with the energy of $h\nu_{ph} \approx 2.1$ eV.

Because electronic properties of a polymer:fullerene composite are determined largely by mobile charge carriers, the contribution of such carriers seems to be more important. The concentration ratio $[mF_{mob}^{\cdot-}]/[P_{loc}^{\cdot+}]$ obtained from both initial and treated P3HT:PC_{61}BM and P3HT:bis-PC_{61}BM BHJ is shown in Figure 5.11 as function of the photon energy. This ratio was determined at photon energy $h\nu_{ph} = 1.7$–2.2 eV for the initial P3HT:PC_{61}BM composite to be 4:1. This value increases up to 10:1 and 14:1 at its MW and US treatment but, however, decreased down to 2:1 as the sample is annealed. On the other hand, the concentration ratio obtained from P3HT:bis-PC_{61}BM BHJ, $[mF_{mob}^{\cdot-}]/[P_{loc}^{\cdot+}] = 2:1$, appeared to be less sensitive to the thermal, MW, and US treatment of the sample. This can be explained by a more ordered P3HT:bis-PC_{61}BM BHJ as compared with the P3HT:bis-PC_{61}BM ones. One can conclude that the treatment of the sample does not change notably the number, distribution, and energetic depth of spin traps reversibly induced in its matrix. Such conclusion was confirmed by the data obtained by the scanning electron microscopy [612].

Figure 5.12 illustrates the changes in LEPR spectra of the initial P3HT:PC_{61}BM and P3HT:bis-PC_{61}BM composites upon heating and shows temperature dependences of all contributions into sum χ. Since concentration of main charge carriers decreases dramatically at $T > 200$ K, the precision of the determination of their spin susceptibility falls significantly. The fitting of

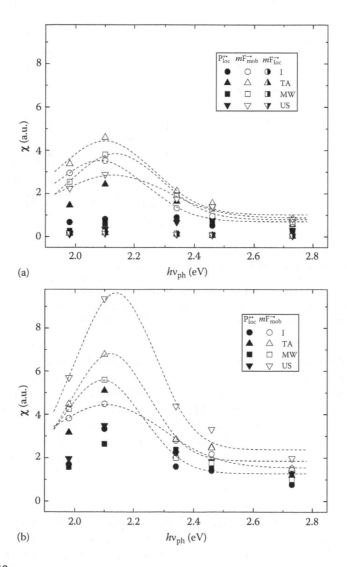

FIGURE 5.10
The contributions of localized polarons (filled symbols) and localized and mobile metha-nofullerene radical anions (semifilled and open symbols, respectively) to the effective para-magnetic susceptibility of the initial, thermally annealed, microwave, and ultrasonic treated P3HT:PC$_{61}$BM (a) and P3HT:*bis*-PC$_{61}$BM (b) nanocomposites as function of the energy of initiat-ing photons $h\nu_{ph}$. The registration temperature is $T = 77$ K. The dashed lines connecting experi-mental points obtained for mobile methanofullerene radical anions are drawn arbitrarily only for illustration to guide the eye.

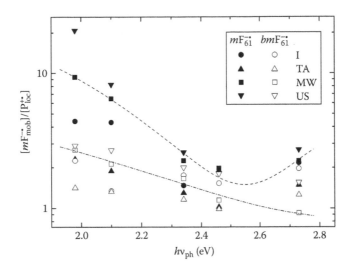

FIGURE 5.11

The ratio of the number of mobile methanofullerene radical anions to that of localized polarons, $[m\mathrm{F}_{mob}^{-\bullet}]/[\mathrm{P}_{loc}^{+\bullet}]$, photoinduced at 77 K in the initial, thermally annealed, microwave, and ultrasonic treated P3HT:PC$_{61}$BM and P3HT:*bis*-PC$_{61}$BM nanocomposites versus photon energy $h\nu_{ph}$. Exemplary dashed and dash-dotted lines connecting experimental point obtained for these systems, respectively, are drawn arbitrarily only for illustration to guide the eye.

their double integrated sum of LEPR spectra allowed us to obtain separately all terms of effective paramagnetic susceptibility χ. This value consists of the contributions of mobile and localized polarons χ_P and methanofullerene anion radicals χ_F. In χ of the P3HT:*bis*-PC$_{61}$BM composite, the contribution of localized fullerene anion radicals is absent within the entire temperature range used. Assuming the absence of a dipole–dipole interaction between fullerene anion radicals, one can evaluate energy ΔE_{ij} from Equation 2.15 for all charge carriers from temperature dependences of their paramagnetic susceptibility as function of the energy of photons $h\nu_{ph}$ (see Table 5.2). As it is seen from Figure 5.12, the net electronic processes in the composites can be described in terms of spin exchange with ΔE_{ij} presented in Table 5.2.

It is evident that the energy required for polaron trapping in the P3HT matrix is lower than that obtained from other charge carriers. ΔE_{ij} evaluated from $\chi(T)$ for mobile radicals increases considerably indicating higher energy required for their trapping in the system. This value becomes larger for methanofullerene after its pinning in bulk of P3HT:PC$_{61}$BM matrix. The data obtained indicate once more that all spin-assisted processes are governed mainly by the structure of anion radical as well as by the nature and dynamics of charge carriers photoinduced in BHJ of a composite. It is seen that the χ value of both charge carriers becomes distinctly higher at characteristic energy $h\nu_{ph} \approx 2.1$ eV lying close to the bandgap of P3AT [624]. Such a dependence of spin concentration on photon energy can be explained either

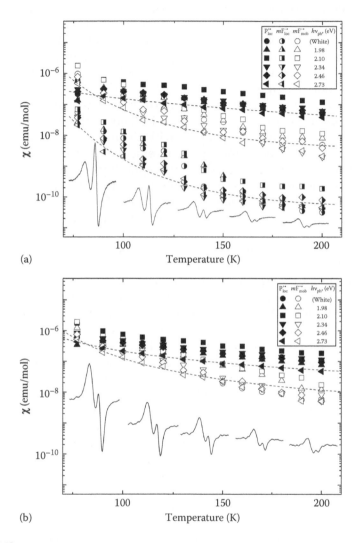

FIGURE 5.12
Temperature dependence of paramagnetic susceptibility of the $P_{loc}^{+\cdot}$, $mF_{loc}^{-\cdot}$, and $mF_{mob}^{-\cdot}$ charge carriers photoinduced in bulk heterojunctions formed by poly(3-hexylthophene) chains with the [6,6]-phenyl-C61-butanoic acid methyl ester (PC$_{61}$BM) (a) and *bis*-PC$_{61}$BM (b) methanofullerenes by photons with different energies $h\nu_{ph}$. Exemplary dashed lines show dependences calculated using Equation 2.15 for different charge carriers photoinitiated at $h\nu_{ph}=2.73$ eV. The ΔE_{ij} parameters obtained for these paramagnetic centers at other photon energies are summarized in Table 5.2. light-induced electron paramagnetic resonance spectra of these heterojunctions registered at respective temperatures are shown at the bottom. (From Krinichnyi, V.I. and Yudanova, E.I., *Sol. Energy Mater. Sol. Cells*, 95(8), 2302, 2011. With permission.)

by the formation of spin pairs with different properties in homogeneous (higher ordered) composite fragments or by the excitation of identical charge carriers in heterogeneous (lower ordered) domains of the system. Different spin pairs can be photoinduced as a result of the photon-assisted appearance of traps with different energy depths in a polymer matrix. However, the revealed difference in the parameters of radicals seems to be a result of their interaction with their microenvironment in domains inhomogeneously distributed in a polymer:fullerene composite. Different ordering of these domains can be the reason for variation in their bandgap energy leading, hence, to their sensitivity to photons with definite but different energies. This can give rise to the change in the interaction of charge carriers with a lattice and other spins. Effective spin susceptibility of the P3HT:bis-PC$_{61}$BM composite somewhat exceeds that obtained from the P3HT:PC$_{61}$BM one. This effect and the absence of trapped anion radicals of the former support additionally the existence in this system of a more ordered BHJ, which interfere in the formation of traps in its matrix.

Figure 5.13 depicts appropriate contributions of charge carriers into an effective spin susceptibility of the PCDTBT:PC$_{61}$BM composite as a function of temperature and photon energy. The data presented can also be interpreted in terms of the exchange interaction of both charge carriers with the spin flip-flop probability p_{ff} inserted in Equation 2.14 earlier. As it is seen from Figure 5.13, the spin–spin interaction processes in the composite can indeed be described by Equations 2.9 and 2.15, so the energies ΔE_{ij} evaluated for both charge carriers from their dependences can be determined (Table 5.2). It is evident that polaron diffusion in the PCDTBT matrix requires lower energy ΔE_{ij} as compared with mobile methanofullerene anion radicals. It is seen also that the χ value of anion radicals becomes more temperature dependent when $h\nu_{ph} \approx 1.98$ and 2.73 eV. The former value lies near the bandgap obtained from PCDTBT (1.88 eV) [546]. Such dependence of spin concentration on photon energy should indeed be a result of different interactions of charge carriers with their microenvironment in domains inhomogeneously distributed in the composite causing sensitivity to photon energy. Figure 5.13 illustrates that the illumination of the PCDTBT:PC$_{61}$BM composite by photons with $h\nu_{ph} = 1.98$ eV induces a minimum number of traps for polarons that leads to the formation of a larger number of mobile spin quasipairs at low temperatures. Such photon frequency selectivity is governed by polymer structure, effective dimensionality, and also the properties of an acceptor involved in BHJ. It can be used, for example, in plastic sensoric photovoltaics, whereas the composites with low selectivity seem to be more suitable for a higher efficient conversion of solar energy.

When initiating background illumination is switched off, photoinitiation of charge carriers in polymer:fullerene BHJ stops, and the concentration of spin charge carriers excited starts to decrease. This is demonstrated in Figure 5.14 where the decay of spin charge carriers in the P3HT:PC$_{61}$BM, P3HT:bis-PC$_{61}$BM, and P3DDT:PC$_{61}$BM systems is shown. A lifetime of charge carriers

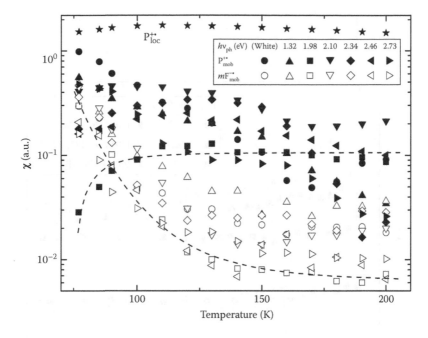

FIGURE 5.13
Temperature dependence of paramagnetic susceptibility of polarons $P_{loc}^{+\cdot}$ and $P_{mob}^{+\cdot}$ (filled points) and methanofullerene anion radicals $mF_{mob}^{-\cdot}$ (open points) photoinduced in PCDTBT:PC$_{61}$BM bulk heterojunctions by photons with different energies $h\nu_{ph}$ normalized to illumination intensity of the light sources. Dashed lines show dependences calculated as an example for $h\nu_{ph} = 1.98$ eV from Equation 2.9 with $\Delta E_{ij} = 0.056$ eV and Equation 2.15 with $\Delta E_{ij} = 0.068$ eV. (Reprinted with permission from Krinichnyi, V.I., Yudanova, E.I., and Denisov, N.N., The role of spin exchange in charge transfer in low-bandgap polymer:fullerene bulk heterojunctions, *J. Chem. Phys.*, 141(4), 044906/01. Copyright 2014, American Institute of Physics.)

seems to be much longer than $t \sim 0.1$ µs obtained by optical absorption spectroscopy for relevant recombination times of mobile photoexcitations in organic solar cells [625,626]. So the data obtained are mainly pertinent to carriers that are time dependent, separated, or trapped in polymer matrix [57,558,578,588,609,627,628].

The decay of spin susceptibility of the P3DDT:PC$_{61}$BM composite shown in Figure 5.14 was interpreted [558] in terms of the earlier described approach of recombination of charge carriers with different effective localization radii, $n_p = 1.2 \times 10^{-4}$, $n_f = 6.3 \times 10^{-5}$, and separated by a time-dependent distance R_0 [150]. Figure 5.14 shows that the dependences calculated from Equation 2.12 with respective $n_0 a^3$ products and τ_0 values fit well with the experimental data obtained. Therefore, the decay of long-living spin quasi-pairs photoinduced in the P3DDT:PC$_{61}$BM composite can indeed be described in terms of this model in which the low-temperature recombination rate is particularly strongly dependent on the spatial distance between photoinduced polarons and methanofullerene ion radicals. The long lifetimes are solely ascribed to

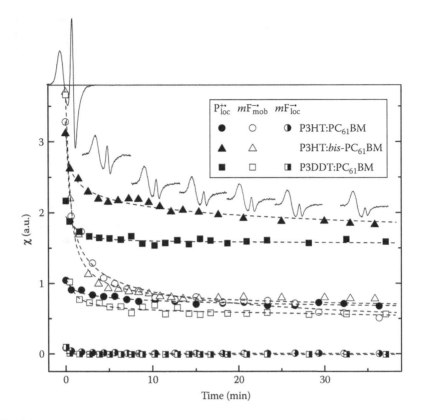

FIGURE 5.14

Decay of spin susceptibility of pinned polarons $P_{loc}^{+\cdot}$ (filled points) as well as pseudo-rotating ($mF_{mob}^{-\cdot}$, open points) and pinned ($mF_{loc}^{-\cdot}$, semifilled points) methanofullerenes photoinduced in the bulk heterojunctions of P3HT:PC$_{61}$BM (circles), P3HT:*bis*-PC$_{61}$BM (triangles), and P3DDT:PC$_{61}$BM (squares) systems at 77 K. Dashed lines show the dependences calculated for these carriers in the latter sample from Equation 2.12 with $n_0 a^3 = 2.1 \times 10^{-5}$ and $\tau_0 = 0.27$ min, $n_0 a^3 = 1.4 \times 10^{-7}$ and $\tau_0 = 7.2 \times 10^{-5}$ min, and $n_0 a^3 = 6.4 \times 10^{-6}$ and $\tau_0 = 2.2 \times 10^{-5}$ min, respectively, as well as from Equation 2.13 with E_0 summarized in Table 5.2. Typical changes in the light-induced electron paramagnetic resonance spectrum of a polymer:fullerene composite at appropriate time t are shown as well. (Reprinted from *Spectroscopy of Polymer Nanocomposites*, Ponnamma, D., Rouxel, D., Thomas, S., and Krinichnyi, V.I., EPR spectroscopy of polymer:fullerene nanocomposites, pp. 202–275, Chapter 9, Copyright 2016, with permission from Elsevier.)

the large spatial distances that build up among the remaining photoinduced charge carriers, which did not recombine at a shorter time. The product $n_0 a^3$ decreases considerably as the methanofullerene acceptor is replaced, for example, by the azagomofullerene one [558]. This can be due to the higher probability of excitation of mobile polarons in the P3DDT:PC$_{61}$BM system characterizing with prolonged radiative lifetime of spin quasi-pairs τ_0 that also corresponds to the lower constant of bimolecular recombination in this system.

The other data presented in Figure 5.14 can be described in the frameworks of Tachiya's concept [151] of bulk recombination of charge carriers during their repeated trapping into and detrapping from trap sites with different depths. The dependences calculated with E_0 presented in Table 5.2 for charge carriers photoinduced in the BHJ studied are also shown in Figure 5.14. It is shown that Equation 2.13 fits well with the experimental data presented in the figure. Therefore, the decay of long-lived charge carriers originated from initial spin pairs photoinduced in the polymer:fullerene composite can successfully be described in terms of the earlier model in which the low-temperature recombination rate is strongly governed by temperature and the width of energy distribution of trap sites. The analysis of the experimental data allows one to support the crucial role of the photon energy on the formation and energetic properties of the traps in BHJ of such disordered systems. This parameter obtained from the sites occupied by both localized polarons and fullerene anion radicals in P3HT:PC$_{61}$BM changes extremely with $h\nu_{\mathrm{ph}}$ attaining the minimum at $h\nu_{\mathrm{ph}} \sim 2.4$ eV. The width of energy distribution of the traps in the P3HT:*bis*-PC$_{61}$BM system decreases with growing $h\nu_{\mathrm{ph}}$. On the other hand, mobile charge carriers are characterized by extremal $E_0(h\nu_{\mathrm{ph}})$ dependences with a maximum at $h\nu_{\mathrm{ph}} \sim 2.3$ eV for P3HT:PC$_{61}$BM and a minimum at $h\nu_{\mathrm{ph}} \sim 2.1$ eV for P3HT:*bis*-PC$_{61}$BM (Table 5.2). This indicates that the local structure and ordering govern the depth of spin traps and their distribution in these composites.

It should be noted that the other polymer:fullerene composites are characterized by much shorter-lived centers making it impossible to register them just after their light excitation due to a higher ordering of such systems. In order to increase the number of charge carriers and, therefore, the efficiency of polymer:fullerene systems, the opposite charge carriers must be separated in brief time before their recombination and absorption by the lattice. This means that an initial exciton should rapidly drift to another layer in the device where the charge separation occurs, but the fast reabsorption of excitons before charge separation is the main reason of low efficiency. To solve this problem, one should use organic composites able to generate the exciton and separate the charges, analogously to that as it occurs in more complex systems modified by carbon nanotubes [629] or graphene. While the excitons produced in graphene are quickly reabsorbed back into polymer matrix, the excitons produced in the polymer/graphene interface are skipped across interface, where the charge separation occurred. This should provide a higher probability for an exciton separation. One can expect that the pulse exciton initiation in such a system combined with their resonant handling can create new spintronic devices with better electronic properties. The placement of a charge carrier with aspherical distribution of unpaired electron into higher magnetic fields increases a distance between its characteristic resonant fields (see Equation 2.2). Besides, higher Zeeman splitting between the energies of parallel and antiparallel spins exponentially decreases the probability of cross-relaxation in the system [107] that, in turn, makes excitons to be less interacting and, therefore, decreases a probability of the random singlet–triplet transformation mentioned earlier. In this case, one can expect the imbalance in the exciton formation rate and effective mixing between different exciton states to be more synchronous, and the coherent spin manipulation in organic molecular spin electronics becomes more effective.

5.3 Light-Induced EPR Linewidth

Various static and dynamic factors can affect a linewidth of PC stabilized and/or induced in nanomodified polymer BHJ. One of them is the previously mentioned hyperfine interaction of the spins and the nuclei of heteroatoms included in the composite backbone. Interaction of polarons and methanofullerene ion radicals with other neighboring or guest spins should also accelerate electron relaxation of all spin reservoirs and, therefore, broaden their EPR lines. Translational dynamics of polarons and reorientational diffusion of fullerene cage in BHJ should also be taken into account. If a part of the charge carriers is partly fixed in spin traps formed in polymer matrix, an effective linewidth should also be changed. The linewidths of both charge carriers photoinduced in different polymer:fullerene BHJ at low temperatures are also summarized in Table 5.1. It should be noted that this value obtained from polarons photoinduced in the composites is approximately 0.15–0.18 mT evaluated for respective charge carriers stabilized in different P3AT matrices [105]. However, this value is considerably lower than that determined for undoped polythiophene [105,331], which is an evidence of weaker spin interaction with the P3AT lattice. The LEPR linewidth should reflect different processes occurring in a polymer:fullerene composite. One of them is the association of mobile polarons with the counter charges. Another process realized in the system is exchange interaction between mobile and trapped polarons and fullerenes that broadens the line by $\delta\Delta B_{ex} = \mu_B/R_0^3 = 4/3\pi\mu_B n_p$, where R_0 is the distance between dipoles. Assuming anisotropic character of the main magnetic resonance parameters of polarons discussed earlier, one can evaluate $R_0 \approx 2.3–2.5$ nm for a distance between dipoles, for example, in the P3AT:PC$_{61}$BM systems.

Let us analyze the dependence of the linewidth of charge carriers on the structure of polymer matrix and fullerene derivative as well as on the photon energy $h\nu_{ph}$ and temperature.

Figure 5.15 depicts the dependences of effective LEPR linewidth $\Delta B_{pp}^{(0)}$ of charge carriers photoinduced by different light photons in nanocomposites formed by the chains of P3DDT and PCDTBT with PC$_{x1}$BM and 2-(azahomo[60] fullereno)-5-nitropyrimidine (AFNP) fullerene derivatives. Light illumination of such systems was shown [57] to lead to reversible formation in their

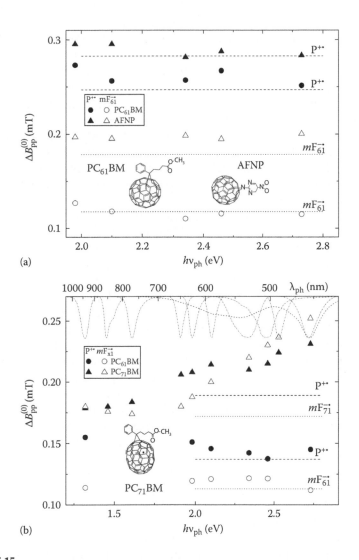

FIGURE 5.15
Dependence of linewidths of charge carriers photoinduced in the P3DDT:PC$_{61}$BM and
P3DDT:AFNP (a) as well as in the PCDTBT:PC$_{61}$BM and PCDTBT:PC$_{71}$BM (b) composites at
$T = 77$ K on the photon energy $h\nu_{ph}$. These values obtained for polarons and fullerene ion radi-
cals initiated in the bulk heterojunctions by achromatic white light are shown by the dashed
and dotted sectors, respectively. The symbol (0) in $\Delta B_{pp}^{(0)}$ implies that the light-induced electron
paramagnetic resonance spectra were measured far from microwave saturation, when $B_1 \rightarrow 0$.
The fullerene derivatives are shown schematically.

polymer matrices of spin traps whose number, depth, and distribution depend on the photon energy hv_{ph}. The analysis of the data presented in Figure 5.15a indicates a rather weak $\Delta B_{pp}^{(0)}$ (hv_{ph}) dependence obtained from both charge carriers photoinitiated in the P3DDT:PC$_{61}$BM and P3DDT:AFNP BHJ. On the other hand, the notable broadening of LEPR linewidths of both charge carriers is registered as PC$_{61}$BM globes are replaced by PC$_{71}$BM cages in their composite with PCDTBT chains. Besides, an additional monotonic line broadening with increasing photon energy of both the charge carriers in the PCDTBT:PC$_{71}$BM BHJ can be concluded.

Line broadening of charge carriers depends not only on the energy of the initiated photons but also on the temperature. Temperature dependences of effective LEPR linewidth $\Delta B_{pp}^{(0)}$ of charge carriers photoinduced in the polymer:fullerene composites mentioned earlier are presented in Figures 5.16 and 5.17.

It is seen from Figure 5.16 that the linewidth obtained from polarons diffunding in the P3HT:PC$_{61}$BM and P3HT:bis-PC$_{61}$BM systems is characterized by ∩-like temperature dependence with the extremum of approximately 140 and 130 K, respectively, which remain almost unchanged at the exchange of the fullerene derivatives. Linewidth of the methanofullerene anion radicals demonstrates more monotonic temperature dependence and decreases with the system heating (Figure 5.16). The data presented were interpreted in terms of exchange interaction of spins with different mobility in polymer matrix. The dependences calculated from Equation 2.20 with and $t_{1D} = 1.18$ eV [159] and E_r summarized in Table 5.2 are also presented in Figure 5.16. The fitting is evidence of the applicability of these approaches for interpretation of electronic processes realized in these polymer composites. The energy E_r obtained from P3HT:PC$_{61}$BM and P3HT:bis-PC$_{61}$BM is close to that evaluated for regioregular P3HT from its *ac* conductometric (0.080 eV [630]) and ^{13}C NMR (0.067–0.085 eV at $T < 250$ K [621]) data. Figure 5.16 shows that the linewidth of the PC$_{61}$BM anion radicals decreases with the system heating. Besides, this value decreases as the PC$_{61}$BM acceptor is replaced by bis-PC$_{61}$BM in such polymer:fullerene composite. The latter fact additionally indicates a more ordered structure of the P3HT:bis-PC$_{61}$BM composite as compared with the P3HT:PC$_{61}$BM one.

Figure 5.17a demonstrates how this parameter changes for both charge carriers photoinduced in the P3DDT modified by PC$_{61}$BM and AFNP under irradiation by photons with $hv_{ph} = 1.98$ eV at various temperatures. It is seen from Figure 5.17a that the linewidth of polarons in the P3DDT:PC$_{61}$BM composite broadens with the temperature; however, this value oppositely depends on the temperature as P3DDT is modified by the AFNP nanoadduct. Besides, this parameter becomes higher under such a replacement. This can be explained by stronger spin–spin interaction due to better ordering of the latter composite. The data obtained were also interpreted in terms of the previously mentioned approach of the collision of localized and mobile spins.

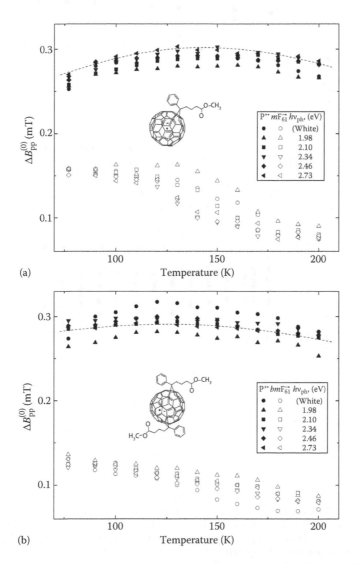

FIGURE 5.16
Linewidth of charge carriers photoinduced in the P3HT:PC$_{61}$BM (a) and P3HT:*bis*-PC$_{61}$BM (b) composites as a function of temperature and photon energy $h\nu_{ph}$. Exemplary dashed lines show the dependences calculated using Equation 2.20 with $E_r = 0.047$ (a) and 0.050 eV (b). These values obtained at other experimental conditions are summarized in Table 5.2. The structures of methanofullerenes are shown schematically. (From Krinichnyi, V.I. and Yudanova, E.I., *Sol. Energy Mater. Sol. Cells*, 95(8), 2302, 2011. With permission.)

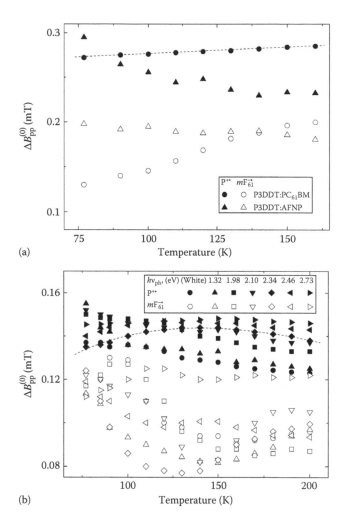

FIGURE 5.17

(a) Temperature dependence of linewidth of charge carriers photoinduced in the P3DDT:PC$_{61}$BM and P3DDT:AFNP bulk heterojunctions by photons with energy $h\nu_{ph} = 1.98$ eV. The exemplary dashed line shows the dependence calculated Equation 2.20 with $E_r = 0.001$ eV. (b) The linewidth of both charge carriers photoinduced in the PCDTBT:PC$_{61}$BM composite by different light sources as function of temperature. Exemplary dashed line shows the dependence calculated using Equation 2.20 with $E_r = 0.017$ eV. These parameter obtained at other experimental conditions are summarized in Table 5.2.

Indeed, the dependences calculated from Equation 5.20 with E_r summarized in Table 5.2 fits well with the experimental data presented.

Respective dependences obtained from PCDTBT:PC$_{61}$BM BHJ under irradiation by different photons at a wide temperature range are presented in Figure 5.17b. It is seen from the figure that the linewidth of polarons and methanofullerene anion radicals also differently depend on the temperature with different temperature sign. It can be due to their different nature and dynamic mechanism in domains inhomogeneously distributed in the composite. Such inhomogeneity seems to be more characteristic for methanofullerene domains than for polaronic phase possibly due to a more ordered layer morphology of the PCDTBT matrix. The energy necessary for activation of both the charge carriers change slightly as $h\nu_{ph}$ excesses the polymer bandgap. One of the dependences calculated from Equation 2.20 and fitting appropriate data are also presented in Figure 5.17b. This is also evidence of the applicability of the approach proposed earlier for interpretation of electronic processes realized in organic composites.

One can conclude that the energy required for initiation of polaron diffusion in the P3HT:PC$_{61}$BM composite exceeds E_r obtained from both the P3DDT:PC$_{61}$BM and PCDTBT:PC$_{61}$BM ones. Such effect can be explained by the better ordering of the latter BHJ. This parameter additionally decreases as PC$_{61}$BM globes are replaced by AFNP nanoadducts in P3DDT matrix [558] and by PC$_{71}$BM cages in PCDTBT matrix [603,604]. Such decrease in E_r indicates electronic properties of appropriate polymer:fullerene composites.

The data presented are additional evidence that relaxation and dynamic processes realized in the composites are governed mainly by the structure of the polymer matrix and fullerene derivative as well as by the nature and dynamics of charge carriers photoinduced in the appropriate BHJ. The number and depth of spin traps are governed by the structure of counterions and the energy of photons initiating their formation. As in the case of spin susceptibility, variation in $\Delta B_{pp}^{(0)}$ ($h\nu_{ph}$) registered for charge carriers can also be attributed to inhomogeneous distribution of domains with different ordering (and, hence, bandgap energies) in the polymer:fullerene BHJ. The photonenergy correlation obtained from the main magnetic resonance parameters of the polymer:fullerene BHJ can be used in the creation of organic molecular devices with spin-assisted (spintronic) properties.

5.4 Electron Relaxation and Dynamics of Spin Charge Carriers

With the increasing magnetic term B_1 of MW irradiation in a polymer:fullerene BHJ, the intensity and width of the LEPR spectra of both spin charge carriers change, as shown in Equations 2.32 and 2.33, respectively. The slope of these dependences is evidently governed by the nature, electron relaxation, and mobility of these spin charge carriers. Figure 5.18 illustrates the change in the intensities of all charge carriers photoinitiated in the P3HT:PC$_{61}$BM composite with B_1 [558]. These data are seen to be indeed well described in Equation 2.32. Since polarons and fullerene anion radicals carrying a charge through BHJ are found to be independent, this allows determining separately both their T_1 and T_2 relaxation times. One should only take

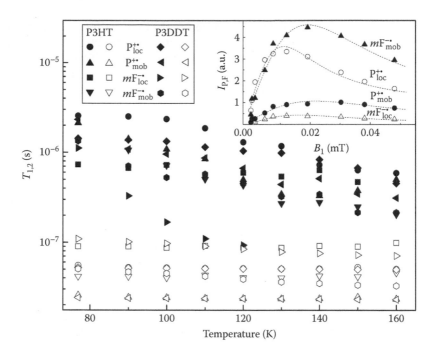

FIGURE 5.18

Temperature dependence of spin–lattice (T_1, filled points) and spin–spin (T_2, open points) relaxation times of charge carriers $P_{loc}^{+\bullet}$ and $mF_{loc}^{-\bullet}$ as well as $P_{mob}^{+\bullet}$ and $mF_{mob}^{-\bullet}$ photoinduced in the P3HT:PC$_{61}$BM and P3DDT:PC$_{61}$BM composites by light with $h\nu_{ph} = 1.98$ eV. The inset shows the changes in intensity of their contributions to the effective light-induced electron paramagnetic resonance spectrum initiated in the P3HT:PC$_{61}$BM bulk heterojunctions by light with $h\nu_{ph} = 1.98$ eV at 90 K as a function of the magnetic term B_1 of microwave field as well as the dependences shown by the dashed lines and calculated using Equation 2.32 with $T_1 = 2.5 \times 10^{-6}$ and $T_2 = 5.3 \times 10^{-8}$ s, $T_1 = 1.4 \times 10^{-6}$ and $T_2 = 2.4 \times 10^{-8}$ s, $T_1 = 1.1 \times 10^{-6}$ and $T_2 = 4.2 \times 10^{-8}$ s, and $T_1 = 6.7 \times 10^{-7}$ and $T_2 = 9.0 \times 10^{-8}$ s, respectively. (Reprinted with permission from Krinichnyi, V.I., Yudanova, E.I., and Spitsina, N.G., Light-induced EPR study of poly(3-alkylthiophene)/fullerene composites, *J. Phys. Chem. C*, 114(39), 16756, 2010. Copyright 2010 American Chemical Society.)

into account the different distribution of spin packets in the LEPR spectra of mobile and localized charge carriers. Figure 5.18 shows an example of these parameters determined for charge carriers stabilized and photoinitiated in the P3HT:PC$_{61}$BM and P3DDT:PC$_{61}$BM composites in a wide temperature range. Both T_1 and T_2 values of charge carriers photoinitiated in some polymer:fullerene composites at $T = 77$ K are summarized in Table 5.3.

The analysis of the data presented shows that the interaction of most charge carriers with the lattice is characterized by monotonic temperature dependences, whereas T_1 of fullerene anion radicals trapped by the P3DDT matrix demonstrates sharper temperature dependence. Spin–spin interaction is seen to be nearly temperature independent. However, it is governed by structural properties of a polymer:fullerene composite (Figure 5.18).

TABLE 5.3

Spin–Lattice T_1 and Spin–Spin T_2 Relaxation Times of Polarons and Methanofullerene Anion Radicals Stabilized $P_{loc}^{+\bullet}$, $mF_{loc}^{-\bullet}$ and Mobile $P_{mob}^{+\bullet}$, $mF_{mob}^{-\bullet}$ Photoinduced in Some Polymer:Fullerene Composites at Different Photon Energy/Wavelength $h\nu_{ph}/\lambda_{ph}$

Parameter		White	$h\nu_{ph}/\lambda_{ph}$ (eV/nm)								
			1.32/ 940	1.46/ 850	1.61/ 770	1.91/ 650	1.98/ 625	2.10/ 590	2.34/ 530	2.46/ 505	2.73/ 455
P3HT:PC₆₁BM											
$P_{loc}^{+\bullet}$	T_1 (10^{-6} s)	2.8	—	—	—	—	2.6	1.9	2.3	1.9	1.4
	T_2 (10^{-8} s)	5.6	—	—	—	—	5.4	5.5	5.3	5.4	5.4
$P_{mob}^{+\bullet}$	T_1 (10^{-6} s)	1.8	—	—	—	—	1.2	1.5	1.7	1.7	2.6
	T_2 (10^{-8} s)	2.6	—	—	—	—	2.6	2.5	2.5	2.5	2.5
$mF_{loc}^{-\bullet}$	T_1 (10^{-6} s)	1.5	—	—	—	—	0.73	0.94	1.3	1.2	1.2
	T_2 (10^{-8} s)	8.9	—	—	—	—	8.9	9.1	8.9	9.4	8.9
$mF_{mob}^{-\bullet}$	T_1 (10^{-6} s)	1.6	—	—	—	—	1.3	1.6	1.4	1.6	1.5
	T_2 (10^{-8} s)	4.1	—	—	—	—	4.2	4.2	4.2	4.3	4.1
P3HT:*bis*-PC₆₁BM											
$P_{loc}^{+\bullet}$	T_1 (10^{-6} s)	3.3	—	—	—	—	2.8	1.3	1.8	1.4	1.3
	T_2 (10^{-8} s)	5.6	—	—	—	—	5.4	4.9	4.7	4.8	4.8
$P_{mob}^{+\bullet}$	T_1 (10^{-6} s)	1.3	—	—	—	—	3.8	3.2	2.4	5.2	1.2
	T_2 (10^{-8} s)	2.6	—	—	—	—	2.5	2.3	2.2	2.2	2.2
$mF_{mob}^{-\bullet}$	T_1 (10^{-6} s)	1.2	—	—	—	—	2.1	1.4	1.1	1.3	1.1
	T_2 (10^{-8} s)	5.4	—	—	—	—	5.3	5.3	5.4	5.3	5.0
$P_{loc}^{+\bullet}$	T_1 (10^{-6} s)	3.3	—	—	—	—	2.8	1.3	1.8	1.4	1.3
	T_2 (10^{-8} s)	5.6	—	—	—	—	5.4	4.9	4.7	4.8	4.8
PCDTBT:PC₆₁BM											
$P_{loc}^{+\bullet}$	T_1 (10^{-6} s)	1.9	2.1	—	—	—	2.9	2.1	2.2	1.6	2.3
	T_2 (10^{-8} s)	4.8	4.4	—	—	—	4.4	4.3	4.9	4.7	4.5
$mF_{mob}^{-\bullet}$	T_1 (10^{-6} s)	1.1	0.89	—	—	—	1.4	1.3	1.3	1.2	0.97
	T_2 (10^{-8} s)	5.8	5.8	—	—	—	5.6	5.4	5.3	5.5	5.9
PCDTBT:PC₇₁BM											
$P_{loc}^{+\bullet}$	T_1 (10^{-6} s)	2.30	2.91	1.93	1.71	1.52	1.51	1.33	1.03	1.16	2.21
	T_2 (10^{-8} s)	3.78	2.44	2.63	2.68	2.89	2.85	2.94	3.09	3.34	2.74
$mF_{mob}^{-\bullet}$	T_1 (10^{-6} s)	0.04	1.29	1.03	0.83	0.83	0.58	0.27	0.31	0.31	0.16
	T_2 (10^{-8} s)	5.72	3.59	3.91	4.02	4.10	4.11	3.86	2.86	2.61	2.84

It is seen that the replacement of $PC_{61}BM$ globes by $PC_{71}BM$ cages accelerates both the spin–lattice and spin–spin relaxation of all spin charge carriers (except spin–spin relaxation of methanofullerene ion radicals) in appropriate $PCDTBT:PC_{x1}BM$ composite. By analyzing the data obtained, one can conclude that the spin–lattice relaxation of charge carriers $P_{loc}^{+\cdot}$ and $mF_{mob}^{-\cdot}$ photoinitiated in the $PCDTBT:PC_{71}BM$ composite accelerates monotonically with the increase of $h\nu_{ph}$. The interaction of immobilized polaron with its own environment, however, decreases at $h\nu_{ph} = 2.73$ eV. The same relaxation parameter of $mF_{loc}^{-\cdot}$ increases remarkably only at $h\nu_{ph} \approx 1.9$–2.1 eV changing weakly at the lower and higher photon energy. All the centers demonstrate weak $T_2(h\nu_{ph})$ dependence within all the energy region used (see Table 5.3). The temperature also slightly affects the spin–spin relaxation of all spin packets (not presented). The T_1 value of immobilized polarons changes weakly within the temperature range 77–160 K. On the other hand, opposite charge carriers are characterized by a much faster energy exchange with their own environment, which accelerates their spin–lattice relaxation. The heating of the composite additionally steps up such exchange of $mF_{loc}^{-\cdot}$.

Diffusion coefficients calculated [612] from Equations 2.51 and 2.52 for both types of charge carriers photoinduced in the $P3HT:PC_{61}BM$ and $P3HT:bis$-$PC_{61}BM$ composites using the relaxation and susceptibility data as well as the appropriate spectral density functions in Equation 2.46 are presented in Figure 5.19 as a function of $h\nu_{ph}$.

The data presented show that spin dynamics in these systems depend on the structure of counterions, photon energy $h\nu_{ph}$, and treatment of their BHJ by temperature annealing as well as MW and US radiation. It is seen, for example, that the coefficients of intrachain D_{1D} and interchain D_{3D} polaron diffusion are governed sufficiently by the energy of initiated photons $h\nu_{ph}$. However, the replacement of $PC_{61}BM$ globes by bis-$PC_{61}BM$ ones suppresses this effect. Besides, such replacement increases anisotropy of polaron diffusion $A = D_{1D}/D_{3D}$ in the initial P3HT matrix. The anisotropy of polaron dynamics decreases by 1.5–2 orders under MW and US treatment of the $P3HT:PC_{61}BM$ BHJ. This effect is explained by the increase of methanofullerene clusters and a well-ordered polymer under a modified system that facilitates polaron diffusion, inhibits fullerene reorientation, and decreases interaction of charge carriers in photoinduced radical pairs. The formation of appropriate crystallites in an amorphous polymer matrix leads to the longer diffusion of charge carriers and higher light conversion efficiency. The A parameter changes slightly under annealing and MW treatment of the $P3HT:bis$-$PC_{61}BM$ composite but, however, decreases by an order in US treatment. This additionally indicates evidences a better ordering of polymer matrix with bis-$PC_{61}BM$ globes embedded due to the fact that the more side-ramified methanofullerenes restrict the number of possible conformations able to be formed by two adjacent thiophene rings rotating near their main C–C bond. Quasi-rotational diffusion coefficient D_{rot} of fullerene anion radicals decreases monotonically with $h\nu_{ph}$ in the initial

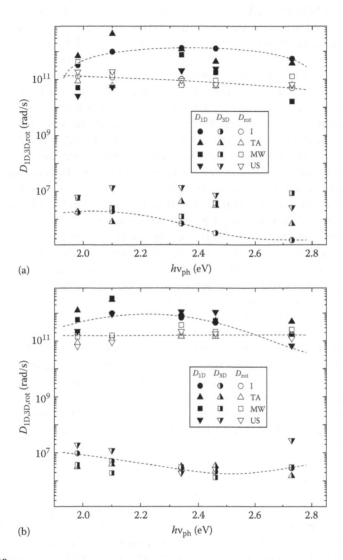

FIGURE 5.19

The dependence of intrachain and interchain translational diffusion coefficients (D_{1D} and D_{3D}, filled and semifilled symbols, respectively) determined for mobile polarons as well as rotational diffusion coefficient (D_{rot}, open points) determined for mobile methanofullerene anion radicals photoinduced in the initial, thermally annealed, microwave, and ultrasonic treated P3HT:PC$_{61}$BM (a) and P3HT:*bis*-PC$_{61}$BM (b) nanocomposites by photons with different energies $h\nu_{ph}$. The registration temperature is $T = 77$ K. The dashed lines connecting experimental points obtained for spin charge carriers photoinitiated in the initial nanocomposites are drawn arbitrarily only for illustration to guide the eye.

and annealed P3HT:*bis*-PC$_{61}$BM composites. This parameter obtained from the US- and MW-treated samples is minimal at $h\nu_{ph} = 2.35$ eV. Besides, the intrachain coefficient of polarons diffusing through BHJ of treated composites is characterized by maximum at $h\nu_{ph} \approx 2.1$–2.3 eV. The change in diffusion coefficients of polarons and methanofullerene anion radicals in a symbate manner obtained from the US- and MW-treated P3HT:PC$_{61}$BM composites may indicate an increase under their modification of crystalline clusters of polymer and fullerene phases with more facilitated intrachain polaron diffusion and inhibited methanofullerene quasi-rotation. As the PC$_{61}$BM molecules are replaced by the *bis*-PC$_{61}$BM ones, the D_{rot} parameter starts to change with $h\nu_{ph}$ in a symbate manner (Figure 5.19). The dependences $D_{rot}(h\nu_{ph})$ obtained from appropriate initial and annealed composites become extreme with a characteristic energy of approximately 2.3 eV. The coefficients of translational polaron and quasi-rotational methanofullerene diffusion in the US- and MW-treated P3HT:*bis*-PC$_{61}$BM composites were obtained to be characterized with extremes at $h\nu_{ph} = 2.10$ and 2.35, respectively. Such effect can be explained by the existence of two equivalent groups in the P3HT:*bis*-PC$_{61}$BM composite increasing spin interaction in polymer:fullerene interfaces. Therefore, one may conclude the sufficient decrease of polaron dynamics in the P3HT:PC$_{61}$BM composite under its US and MW treatment.

Temperature dependences of dynamic parameter of both spin charge carriers photoinitiated by different photons in the initial P3HT:PC$_{61}$BM and P3HT:*bis*-PC$_{61}$BM nanocomposites are shown in Figure 5.20. These parameters are shown to be governed not only by the structure of methanofullerene and the energy of initiated photons $h\nu_{ph}$ but also by the temperature. To account for the LEPR mobility data obtained, different theoretical models can be used. Figure 5.20 indicates that intrachain polaron dynamics in these samples is characterized by strong temperature dependence. Such a behavior can be associated, for example, with the scattering of polarons on the lattice phonons of crystalline domains embedded into an amorphous matrix. Figure 5.20 shows that the D_{1D} obtained from polaron from Equation 2.51 follows well with Equation 2.83 with the phonon energy summarized in Table 5.2. This value is closed to the energy of lattice phonons, 0.09–0.32 eV, determined for other conjugated polymers [118]. E_{ph} obtained from the P3HT:*bis*-PC$_{61}$BM composite appears to be sensitive to the energy of illuminated photons attaining maximum at $h\nu_{ph} = 2.46$ eV.

The interchain spin hopping dynamics can be analyzed, for example, in terms of the previously discussed Hoesterey–Letson formalism of trap-controlled spin mobility. Figure 5.20 shows also exemplary temperature dependences calculated from Equations 2.71 and 2.84 using $T_c = 111$–126 K for P3HT:PC$_{61}$BM and 127–140 K for P3HT:*bis*-PC$_{61}$BM as well as E_0 and E_t summarized in Table 5.2. Figure 5.20 evidences that interchain polaron dynamics can be indeed described in the frames of the theory mentioned earlier.

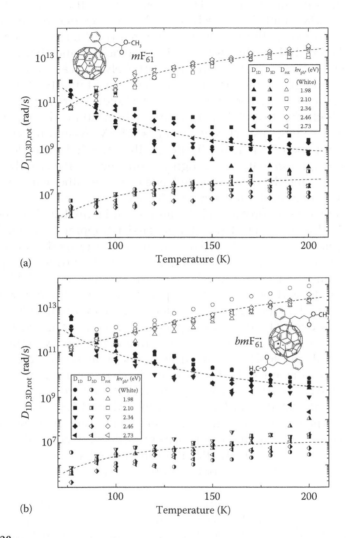

FIGURE 5.20
Temperature dependence of intrachain (D_{1D}, filled points), interchain (D_{3D}, semifilled points), and rotational diffusion (D_{rot}, open points) coefficients of mobile charge carriers $P_{mob}^{+\cdot}$ and $mF_{mob}^{-\cdot}$ (left and right lines of the quasi-pairs 2 in Figure 68, respectively) photoinduced in the P3HT:PC$_{61}$BM (a) and P3HT:*bis*-PC$_{61}$BM (b) composites by the polychromatic white and monochromatic light with different photon energies $h\nu_{ph}$. Exemplary dashed lines show the dependences calculated using Equations 2.71 and 2.77 with $E_a = 0.045$ eV, Equation 2.83 with $E_{ph} = 0.082$ eV, Equation 2.85 with $E_t = 0.140$ eV and $E_0 = 0.018$ eV (a), Equation 2.77 with $E_a = 0.059$ eV, Equation 2.83 with $E_{ph} = 0.077$ eV, and Equation 2.84 with $E_t = 0.128$ eV and $E_0 = 0.017$ eV (b). These values obtained from fitting of other experimental data are summarized in Table 5.2. (From Krinichnyi, V.I. and Yudanova, E.I., *Sol. Energy Mater. Sol. Cells*, 95(8), 2302, 2011. With permission.)

The E_t values obtained from P3HT:PC$_{61}$BM prevail those characteristic of P3HT:*bis*-PC$_{61}$BM (Table 5.2), which is an additional evidence of deeper traps reversibly formed in the former polymer matrix. Moreover, the replacement of the PC$_{61}$BM by the *bis*-PC$_{61}$BM counterions somewhat increases T_c of a polymer:fullerene system. This fact, probably, indicates the decrease in trap concentration due the increase in effective crystallinity of the polymer matrix. The data presented show that the photon energy governs simultaneously both the T_c and E_t parameters, which attain the maximal and the minimal values, respectively, at $h\nu_{ph} \approx 2.5$ eV. Assuming all the electron wave functions exponentially decay, the interchain transfer integral, $t_\perp = 2e^2r/3\varepsilon a^2 \exp(-r/a)$ [631], was roughly estimated for P3HT to be equal to 0.12 eV. Comparing the data presented, one can note rather extremal $E_t(h\nu_{ph})$ dependences with minimum at $h\nu_{ph} \approx 2.3$–2.4 eV.

The fullerene pseudorotational dynamics can be analyzed in the framework of the Elliot model of the charge carrier hopping described earlier. The energies E_a necessary to activate such motion of methanofullerene in the polymer:fullerene composites obtained from the fitting of experimental data in Equations 2.71 and 2.77 are also summarized in Table 5.2. The temperature dependences of dynamic parameters are shown in Figure 5.20 calculated from Equation 2.77 with E_a determined approximately well-fitted experimental data. These values depend on photon energy (see Table 5.2) and lie near those obtained, for example, from molecular dynamics in polycrystalline fullerene [632] and a triphenylamine fullerene complex [633].

Dynamic parameters of both types of spin charge carriers photoinduced in the PCDTBT:PC$_{61}$BM composite in a wide temperature range and photon energy $h\nu_{ph}$ calculated from Equations 2.46, 2.51, and 2.52 are presented in Figure 5.21. The figure shows that the values and frequency dispersion of all diffusion coefficients are characterized by weak dependence on the $h\nu_{ph}$ value, like in the case of the P3HT/*bis*-PC$_{61}$BM composite that appeared to become more ordered than P3HT/PC$_{61}$BM and other known polymer:fullerene systems [578].

The anisotropy of polaron dynamics, $A = D_{1D}/D_{3D}$, in the PCDTBT:PC$_{61}$BM composite, is significantly lower than that obtained from analogous P3DDT:PC$_{61}$BM and P3HT:*bis*-PC$_{61}$BM (see previous text), which is typical for more ordered systems. This value determined at $T = 77$ K is characterized by a U-like dependence on the photon energy $h\nu_{ph}$ (see inset of Figure 5.21). The figure shows that the replacement of PC$_{61}$BM counterions by PC$_{71}$BM ones leads to a significant increase in the anisotropy of polaron dynamics and more selective $A(h\nu_{ph})$ dependence with extremum at $h\nu_{ph} \approx 2.35$ eV. This can be explained by the better ordering of the composite at intermediate $h\nu_{ph}$ and A values due to asymmetrical fullerene cage. One can only note an unusual feature of this dependence, namely, the evident increase of an anisotropy of the polaron diffusion in the former composite at $h\nu_{ph} = 2.10$ eV lying near the polymer bandgap. This effect can be due to a stronger interaction of the polymer matrix with such light photons. Indeed, the layer ordering of

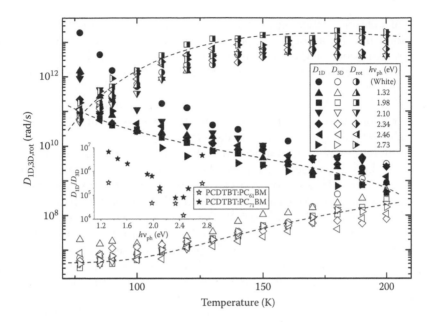

FIGURE 5.21

Temperature dependence of intrachain (D_{1D}, filled points), interchain (D_{3D}, open points), and rotational diffusion (D_{rot}, semifilled points) coefficients of mobile charge carriers $P_{mob}^{+\cdot}$ and $mF_{mob}^{-\cdot}$ photoinduced in the PCDTBT:PC$_{61}$BM composite by the polychromatic white and monochromatic light with different photon energies $h\nu_{ph}$ determined at their appropriate unit concentration n_i. Top–bottom dashed lines show the dependences calculated using Equations 2.20 and 2.71 with $E_r = 0.102$ eV, Equation 2.83 with $E_{ph} = 0.049$ eV, and Equation 2.77 with $E_a = 0.067$ eV as well. The same values obtained from fitting of other experimental data are summarized in Table 5.2. The inset shows the anisotropy of polaron motion, $A = D_{1D}/D_{3D}$, in the PCDTBT:PC$_{61}$BM (open stars) and PCDTBT:PC$_{61}$BM (filled stars) composites as function of $h\nu_{ph}$. (Reprinted with permission from Krinichnyi, V.I., Yudanova, E.I., and Denisov, N.N., The role of spin exchange in charge transfer in low-bandgap polymer:fullerene bulk heterojunctions, *J. Chem. Phys.*, 141(4), 044906/01. Copyright 2014, American Institute of Physics.)

the polymer matrix allows longer polaron diffusion and formation of well-ordered PC$_{61}$BM pools located between these layers [291,552]. This can in principle lead to resonant electronic response on photon energy. If the latter becomes comparable with the polymer bandgap, the stronger polaron interaction with the lattice phonons can initiate the observed change in its *g*-factor and diffusion anisotropy. An analogous sufficient decrease in the anisotropy of polaron dynamics in MW-treated P3HT:PC$_{61}$BM composite due to its better ordering was discussed earlier. So one can conclude better matrix planarity of the PCDTBT:PC$_{x1}$BM composite at illumination by such phonons that accelerates charge transport.

Intrachain polaron dynamics in the PCDTBT:PC$_{61}$BM BHJ, as in the case of other polymer:fullerene composite, is characterized by a strong temperature dependence, so the data presented in Figure 5.21 can also be described in terms of polaron scattering on the phonons of crystalline lattice domains embedded into an amorphous polymer matrix. E_{ph} values obtained from polaron diffusion in the PCDTBT:PC$_{61}$BM composite at different photon energy $h\nu_{ph}$ are also summarized in Table 5.2. Figure 5.21 shows that the D_{1D} obtained from polarons initiated by photons with, for example, $h\nu_{ph} = 1.98$ eV follows well with Equation 2.83 with $E_{ph} = 0.049$ eV. This value lies close to the energy of lattice phonons determined for various conjugated polymers (0.09–0.32 eV) [118] and plastic solar cells [55].

The interchain spin dynamics can also be analyzed in terms of the Hoesterey–Letson concept [247] of the trap-controlled charge hopping between polymer layers. The analysis of the data obtained, however, has showed that such approach cannot be used for the interpretation of the $D_{3D}(T)$ dependences presented in Figure 5.21. These values as well as methanofullerene reorientational diffusion coefficients $D_{rot}(T)$ can rather be explained in the frame of the Pike [634] and Elliott [635] models based on the carrier hopping over the energetic barrier E_a [226]. It may be due to a suggestion that the PCs produced by the influence of light might be expected to have a large effect on the *ac* mobility of charge carriers [616]. The respective energies E_a required to activate polaron transverse diffusion in the PCDTBT:PC$_{61}$BM composite are also listed in Table 5.2.

Reorientational mobility of the methanofullerene cages can rather be described in the framework of the Marcus mechanism mentioned earlier. Reorganization energies E_r obtained from Equation 2.20 with $t_r = 1.18$ eV [159] for charge carriers photoinitiated in the PCDTBT:PC$_{61}$BM composite are summarized in Table 5.2 as well. E_r values obtained exceed the energy required for activation of reorientation of C$_{60}^{\cdot-}$ anions in polymethyl methacrylate (0.026 eV) and C$_{70}$ globes in cyclohexane [636] but, however, lie near those determined for a motion of fullerene derivatives in conjugated polymer matrices [558,578]. It should be noted that the E_r value obtained from methanofullerene photoinitiated by achromatic and monochromic light sources becomes compatible to that (0.224 eV) required for activating fullerene reorientational hopping or rotation in pure C$_{60}$ matrix [637]. The data described consider the dynamics of solitary polarons and fullerene. Undoubtedly, the interaction of these charge carriers with the nearest spins, lattice phonons, etc., may also affect their relaxation and, therefore, should also be taken into account when interpreting the results.

So light excitation of BHJ formed by organic polymer macromolecules with fullerene globes leads to fast formation of two long-living noninteracting PCs with rhombic symmetry, namely, the positively charged polaron P$^+$ (hole) on the polymer backbone and the negatively charged fullerene anion radical F$^-$ located between polymer chains. The main magnetic resonance, relaxation, and dynamic parameters of these charge carriers are governed by the structure, morphology, and ordering of BHJ as well as by the energy of excited photons. The weak interaction of these charge carriers originated from the former radical quasi-pairs stipulates a difference in their dynamics, interaction with their own microenvironment and, hence, the variation in their magnetic resonance parameters. Spatial separation due to charge distribution over full fullerene globule additionally reduces the recombination rate of these charge carriers. This allows one to determine separately all their parameters.

Polarons and fullerene anion radicals photoinitiated tend to recombine, and the probability of such process depends on the rate of polaron diffusion rate and the energy of initiating photons. Light photons reversibly initiate in the amorphous polymer matrix spin traps with different energy depth and distribution. A part of charge carriers is, therefore, trapped in a polymer matrix taking nevertheless indirect part in the charge transport. Such selectivity can be used, for example, in plastic photovoltaic sensors. On the other hand, solar cells should be more homogeneous to be sensitive to all optical photons. The data obtained suggest the impact of the polymer ring-torsion and layer motions on the charge initiation, separation, and diffusion in disordered organic composites.

Optimization of structure of polymer matrix and nanoadditives allows improving electronic properties of appropriate composite. The substitution, for example, $PC_{61}BM$ by *bis*-$PC_{61}BM$ increases planarity and ordering of the P3HT matrix. Besides, it also decreases the number of traps, facilitates local molecular vibrations, and, therefore, accelerates charge transfer through a BHJ. On the other hand, the replacement of P3AT matrix by PCDTBT one also increases the planarity and crystallinity of polymer matrix, suppresses the appearance of spin traps, facilitates local site molecular vibrations that accelerates charge transfer through BHJ, minimizes the energy loss, and, therefore, increases the power conversion efficiency. The illumination of this sample by photons with the energy lying near the polymer bandgap additionally decreases the number of such traps. This reduces significantly the anisotropy of polaron dynamics in such layer-ordered Q2D matrix due to the collective interaction of charge carriers. Besides, this affects more noticeably the splitting of the polarons' σ, π, and σ^* levels, increases the number of initial spin quasi-pairs, broadens the LEPR spectrum, slightly reduces spin interaction with the polymer network, and increases diffusion anisotropy in the polymer network. The replacement of the $PC_{61}BM$ globes by the $PC_{71}BM$ cages in the PCDTBT matrix leads to the change of the E_{ph}, E_a, and E_r parameters of the respective composite from 0.123, 0.079, and 0.218 eV obtained from PCDTBT:$PC_{61}BM$ composite [582] to 0.124, 0.072, and 0.010 eV reported for PCDTBT:$PC_{71}BM$ [603, 604]. Such a sufficient decrease in E_r can evidence more ordered matrix in the latter system. This means that the better charge transport properties of the PCDTBT:$PC_{71}BM$ nanocomposite can be governed by its laminate morphology originating charge separation faster than the localization of the primary excitation to form a bound exciton. The efficiency of energy conversion of polymer:fullerene BHJ grows in the series P3DDT:$PC_{61}BM$ → P3HT:$PC_{61}BM$ → P3HT:*bis*-$PC_{61}BM$ → PCDTBT:$PC_{61}BM$ → PCDTBT:$PC_{71}BM$.

6

Spin-Assisted Charge Transfer in Polymer:Dopant/Polymer:Fullerene Nanocomposites

As discussed in Chapter 5, spin–spin exchange interaction has a significant effect on the main properties of various polymer:fullerene composites. Since charge carriers in conjugated polymers and their composites have spins, this feature can be used for the creation of spintronic devices with spin-assisted electronic properties [36]. In such devices, positive charge on polymer chain (polaron/hole) exchangeable spin-flips negative charge on fullerene, so the further charge recombination becomes dependent on their dynamics, number, polarization, and mutual separation. For large separations, when thermal energy exceeds the interaction potential, the charges are considered as noninteracting. Once the carriers become nearer than the inverted Coulombic interaction potential, their wave functions overlap and exchange interactions become nonnegligible. This can lead to the formation of singlet or triplet excitons in organic semiconductors.

The EPR method was proved [89] to be the most effective direct tool that is able to reveal the underlying nature of spin carriers excited in such systems. This method allows one to study various materials with weak spin–orbit coupling, where the differences in lifetime between the three excited-state triplet sublevels give rise to a spin-dependent buildup of macroscopic polarization [117], including spin charge carriers stabilized in conjugated polymers [105,118] and photoinduced in their fullerene-based photovoltaic nanocompositions [35,55,57].

The interaction between polarons and excitons increases under paramagnetic resonance [119]. Besides, it was proved [638] that the interaction of paramagnetic centers (PC) with excitons increases the spin relaxation rate of PC in comparison with the case of their interaction with free carriers. Thus, singlet excitons are quenched to promote nonradiative decay to the ground state. The study of exchange effects in composites of two or more spin subsystems and their ingredients is expected to provide a good framework for understanding the underlying nature of such interactions among spins in systems with different polaron lattices. However, there is no simple picture that would clarify spin resonance–assisted processes in organic semiconductors governed by spin-dependent exciton–charge interactions and consistent with the spin-dependent polaron pair recombination model [120]. Besides, the data presented earlier prove that such processes are also governed by the energy of initiating photons due mainly to inhomogeneous distribution of polymer and fullerene domains in bulk heterojunction (BHJ). It should also be noted that only a very few data are published on molecular magnetic resonance spectroscopy relating to the actual problems in organic electronics.

In order to study exchange interaction in a multispin composite, P3DDT, which was described earlier as an effective polymer matrix of organic solar cells, can be used as one of the model spin reservoirs [264]. PANI–ES may be chosen as a second suitable spin subsystem for the study of spin-assisted charge transfer in its composite with nanomodified P3DDT. The existence of polarons trapped on chains in amorphous polymer phase and polarons diffusing along and between chains of crystalline polymer clots was demonstrated in typical PANI–ES. Polarons diffusing along polymer chains in such regions appeared to be accessible for triplet excitons injected into the polymer bulk. It was also shown that spin exchange interaction in polymer:fullerene composites leads to collision of domestic and guest spins dramatically changing their magnetic, relaxation, and electronic dynamic parameters. Such effect was registered only in PANI:TSA [474,639,640] due to more accessibility of its spin ensemble for guest spins. In contrast to other PANI–ES, the latter system becomes Fermi glass with high density of states near the Fermi energy level and 3D Mott's VRH charge transfer [641]. This is why PANI:TSA demonstrates better material quality and therefore more metallic behavior with extended states near Fermi level.

The following discussion considers the results of a detailed light-induced electron paramagnetic resonance (LEPR) study of the main magnetic resonance parameters of PC stabilized in highly doped PANI:TSA and photoinduced in PANI:TSA/P3DDT:PC$_{61}$BM composites at a wide temperature range [42,57,264,446]. Such study was expected to design the strategy of spin handling in organic complex nanocomposites for further development of novel molecular devices with spin-assisted electronic transport.

6.1 Composition of Charge Carriers and Spin Precession Frequency

Initial PANI:TSA samples exhibit single X-band EPR spectrum (central spectrum in Figure 6.1) attributed to polarons P_1^{+} with g_{iso} = 2.0028 stabilized in its backbone. This value remains almost constant within a wide temperature range typical for highly conductive crystalline solids [118,184,354]. Another subcomposite P3DDT:PC$_{61}$BM under illumination by white light exhibits superposed lines attributed to positively charged diffusing polarons P_2^{+} with g_{iso} = 2.0018 and negatively charged anion radicals mF_{61}^{-} with g_{iso} = 1.9997 rotating around its own main axis (lower spectrum drawn in Figure 6.1). When combined, these systems form PANI:TSA/P3DDT:PC$_{61}$BM

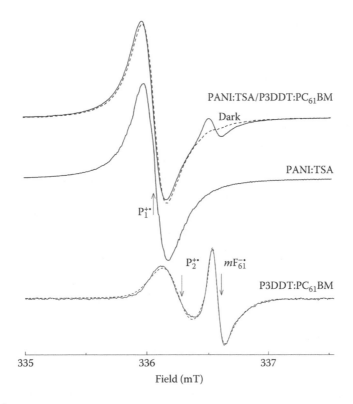

FIGURE 6.1
X-band light-induced electron paramagnetic resonance spectra of the PANI:TSA/
P3DDT:PC$_{61}$BM composite illuminated by white light at $T = 90$ K (top) as well as its contributions due to polarons P$_1^{+\bullet}$ stabilized in PANI:TSA (center) and radical quasi-pairs P$_2^{+\bullet} - m$F$_{61}^{-\bullet}$ (bottom). Top and bottom dashed lines show electron paramagnetic resonance spectra obtained in the absence of illumination and calculated using $\Delta B_{pp}^P = 0.267$ mT, $\Delta B_{pp}^{mF} = 0.117$ mT, and [P$_2^{+}$]/[mF$_{61}^{-}$]= 2.0, respectively. The positions of paramagnetic centers are shown as well. (Reprinted from *Spectroscopy of Polymer Nanocomposites*, Ponnamma, D., Rouxel, D., Thomas, S., and Krinichnyi, V.I., EPR spectroscopy of polymer: Fullerene nanocomposites, pp. 202–275, Chapter 9, Copyright 2016, with permission from Elsevier.)

composite of which dark EPR spectrum mainly demonstrates contribution of polarons P$_1^{+\bullet}$ (upper spectrum is shown in Figure 6.1). Its shape and intensity change under light illumination as shown in Figure 6.1. Such transformation was interpreted as the result of a photoinitiation of quasi-pair P$_2^{+\bullet} - m$F$_{61}^{-\bullet}$ in the P3DDT:PC$_{61}$BM BHJ. This means that two spin subsystems appear in a PANI:TSA/P3DDT:PC$_{61}$BM composite, namely PANI:TSA with polarons P$_1^{+\bullet}$ and P3DDT:PC$_{61}$BM with P$_2^{+\bullet} - m$F$_{61}^{-\bullet}$ spin ensembles. As an illumination is turned off, the spectra originated from the polarons P$_1^{+\bullet}$ stabilized in PANI:TSA composite and polarons P$_2^{+\bullet}$ pinned in P3DDT:PC$_{61}$BM system are only ones detected. In order to study charge-separated states and spin–spin interactions in this composite, its sum spectrum was tentatively deconvoluted [264]. Such procedure allowed to obtain and analyze separately magnetic resonance parameters of all PCs stabilizing in initial polymers and their appropriate composite PANI:TSA/P3DDT:PC$_{61}$BM BHJ.

Figure 6.2a depicts temperature dependences of the linewidth $\Delta B_{pp}^{(0)}$ of polarons $P_1^{\cdot+}$ stabilized in the PANI:TSA and $P_2^{\cdot+}$ photoinitiated in P3DDT:PC$_{61}$BM BHJ, and these values are obtained for darkened and illuminated PANI:TSA/ P3DDT:PC$_{61}$BM composites. It is seen that the EPR linewidth for both polarons stabilized in these systems depends on the structure of polymer matrix. Indeed, the heating of the initial PANI–ES sample is accompanied by a monotonic decrease in $\Delta B_{pp}^{(0)}$ of polarons $P_1^{\cdot+}$ stabilized on its chains. However, this parameter for polarons $P_2^{\cdot+}$ photoinitiated in the P3DDT:PC$_{61}$BM BHJ shows an opposite temperature dependence as compared with that for polarons $P_1^{\cdot+}$ (Figure 6.2). This effect can be explained by different interactions of these polarons with appropriate polymer lattice. The formation of the PANI:TSA/ P3DDT:PC$_{61}$BM composite does not noticeably change the linewidth for PC $P_1^{\cdot+}$. However, this originates the change in the temperature dependence of $P_2^{\cdot+}$ charge carriers photoinitiated in the P3DDT matrix.

It was shown in Section 4.6 that relaxation and dynamic properties of polarons stabilized in the PANI:TSA system are governed by the spin precession frequency ω_e. Figure 4.39 evidences that the exchange interaction between polarons and guest spins also depends on resonant frequency ω_e. On the other hand, the electronic properties of a polymer:fullerene composite determine its quantum efficiency: the higher the concentration ratio of mobile spin charge carriers to localized ones, the faster the charge transfer in such nanocomposite (see, e.g., $[mF_{mob}^{\cdot-}]/[P_{loc}^{\cdot+}]$ in Section 5.2). However, the comparative analysis of multifrequency LEPR spectra of different polymer:fullerene nanocomposites (shown as an example in Figure 5.2) has shown that the number of mobile charge carriers decreases as the spin precession frequency increases. This means that the latter factor may also play an important role in the determination of the effective electronic properties of a multispin nanocomposite and, therefore, should be taken into account in engineering advanced molecular devices.

6.2 Handling of Charge Transfer in Multispin Polymer Systems

Spin properties of both polaronic reservoirs in the multispin PANI:TSA/ P3DDT:PC$_{61}$BM composite are strongly governed by the morphology of PANI chains that determines its main electronic properties [642]. It is seen from Figure 6.2 that once the polymers form a composite, the polarons $P_1^{\cdot+}$ and $P_2^{\cdot+}$ formed on their chains start to demonstrate extremal temperature-dependent linewidths characterized by a corresponding critical point $T_{ex} \approx$ 150 K. A similar effect was observed in the EPR study of exchange interaction

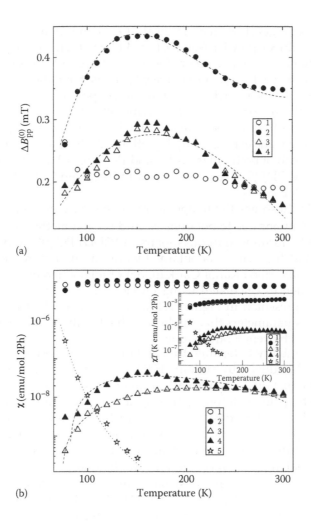

(a)

(b)

FIGURE 6.2

Temperature dependence of peak-to-peak linewidth $\Delta B_{pp}^{(0)}$ (a) and spin susceptibility χ and χT product (insets) (b) determined for domestic polarons $P_1^{+\cdot}$ stabilized in the initial PANI:TSA backbone (1), PANI:TSA/P3DDT:PC$_{61}$BM composite (2), and polarons $P_2^{+\cdot}$ stabilized in the darkened PANI:TSA/P3DDT:PC$_{61}$BM composite (3) and irradiated by white light (4), as well as methanofullerene radical anions $mF_{61}^{-\cdot}$ (5) photoinitiated in the composite. The upper (0) symbol in $\Delta B_{pp}^{(0)}$ implies this parameter to be measured far from the spectrum microwave saturation. Dashed lines in (a) show the dependences calculated using Equation 2.15 with $\omega_{hop}^0 = 1.2 \times 10^9 \mathrm{s}^{-1}$, $\Delta E_{ij} = 0.006$ eV (top line), $\omega_{hop}^0 = 1.3 \times 10^9 \mathrm{s}^{-1}$, $\Delta E_{ij} = 0.012$ eV (bottom line), $J_{ex} = 0.110$ eV, and $n_p = 1.2 \times 10^{-4}$. Top and bottom dashed lines in (b) show the dependences calculated using Equation 2.7 with $C = 1.0 \times 10^{-8}$ emu/mol 2Ph, $a_d = 0.98$, and $J_c = 0.015$ eV and $C = 9.8 \times 10^{-7}$ emu/mol 2Ph, $a_d = 0.98$, and $J_c = 0.010$ eV, respectively. The dotted line shows the dependences calculated using Equation 2.15 with $E_r = 0.050$ eV. (Reprinted from *Spectroscopy of Polymer Nanocomposites*, Ponnamma, D., Rouxel, D., Thomas, S., and Krinichnyi, V. I., EPR spectroscopy of polymer: Fullerene nanocomposites, pp. 202–275, Chapter 9, Copyright 2016, with permission from Elsevier.)

of polarons with guest oxygen biradicals ·O–O· in highly doped PANI:HCl [153] and PANI:TSA powder (see Section 4.6). Such effect was identified as exchange interaction in quasi-pairs formed by the guest spins with domestic polarons hopping across energy barrier E_r. Thus, the data presented can be described in terms of an exchange interaction of polarons hopping along and between the closely located solitary polymer chains.

The collision of both type spins should broaden the absorption term of EPR line on the value expressed by Equation 2.24. Indeed, this equation well fits the linewidth of both polarons $P_1^{\cdot+}$ and $P_2^{\cdot+}$ at $n_p = 1.2 \times 10^{-4}$ obtained for P3DDT:PC$_{61}$BM BHJ [558] and at E_r equal to 0.006 and 0.012 eV, respectively (see Figure 6.2a).

Figure 6.2b illustrates the temperature dependence of spin susceptibility χ with contributions due to polarons $P_1^{\cdot+}$ and $P_2^{\cdot+}$ and methanofullerene radical anions $mF_{61}^{\cdot-}$ forming spin quasi-pairs in P3DDT:PC$_{61}$BM and PANI:TSA/ P3DDT:PC$_{61}$BM BHJ. The data obtained were interpreted in the framework of the earlier described Kahol-Clark model of interaction of $N_s/2$ spin pairs. $\chi_{ST}(T)$ dependences calculated using Equation 2.7 with accurate C, a_d, and J_c values are presented in Figure 6.2b as well. It is evident that the model used provides parameters that can be suitable for all experimental data sets within all possible temperature range.

Spin susceptibility determined for polarons $P_1^{\cdot+}$ is close to that obtained for PANI highly doped by sulfonic [643] and hydrochloric [165] acids. The last term of Equation 2.6 is normally a function of distance. When polymer chains librate, J_c for polarons diffusing along neighboring chains would oscillate and should be described by a stochastic process [644]. However, such effect appears at low temperatures, when $k_B T \ll J_c$. Thus, it can be neglected within all temperature range used. Nevertheless, this constantly increases as polarons $P_1^{\cdot+}$ start to interact with polarons $P_2^{\cdot+}$ in the composite. This is an additional evidence of strong interaction of polarons stabilized in the matrices of both PANI:TSA and P3DDT:PC$_{61}$BM nanocomposites. When Fermi energy ε_F lies close to the mobility edge, the temperature dependence of spin susceptibility gradually changes from Curie law behavior $\chi_C \propto 1/T$ to temperature-independent Pauli-type behavior with increasing temperature. The corresponding density of states $n(\varepsilon_F)$ for both spin directions per monomer unit at ε_F can be determined from the analysis of the $\chi(T)T$ dependence for all polarons stabilized in both polymers (see inset in Figure 6.2b). It was shown [639,640] that the TSA-treated PANI is characterized by higher $n(\varepsilon_F)$ as compared with other PANI:ES. This can be explained by the difference in their metallic properties mentioned earlier and also by on-site electron–electron interaction [422].

The spin susceptibility obtained in methanofullerene radical anions $mF_{61}^{\cdot-}$ photoinduced in P3DDT:PC$_{61}$BM nanocomposite demonstrates sharper temperature dependence (Figure 6.2b). This can be explained by the fast recombination of $P_2^{\cdot+} - mF_{61}^{\cdot-}$ quasi-pairs. Effective paramagnetic susceptibility of this charge carrier should inversely depend on the probability of their recombination, which in turn is governed by polaron Q1D hopping between polymer units [645]. In this case, χ value should be defined by using Equation 2.13. The dependence value calculated by using Equation 2.13 with $E_r = 0.050$ eV is also presented in Figure 6.2b. Therefore, the decay of long-lived charge carriers originated from initial spin pairs photoinduced in the PANI:TSA/ P3DDT:PC$_{61}$BM composite can indeed be described in terms of the model mentioned earlier. This process is also determined by the structure and morphology of radical anion and its environment in a polymer backbone. The use, for example, of PCDTBT, instead of P3DDT, and PC$_{71}$BM, instead of PC$_{61}$BM, should facilitate the excitation to reach the polymer:fullerene interface for charge separation before it becomes spatially self-localized and bound within an exciton [552]. Therefore, the main properties of an exciton are irrelevant to ultrafast charge transfer and do not limit effective charge transfer in such nanocomposite.

The data shown in Figure 6.2b indicate that spin susceptibility of polarons P$_1^{\cdot+}$ stabilized in the initial PANI:TSA sample is characterized by weak temperature dependence without any anomaly. Interaction between neighboring polarons provokes extremal χ versus T dependence obtained for both polarons P$_1^{\cdot+}$ and P$_2^{\cdot+}$ (see Figure 6.2b). Such interaction increases the overlapping of their wave functions and the energy barrier that overcomes the polaron crossing BHJ. This affects the polaron intrachain mobility and, therefore, probability of its recombination with fullerene anion.

It was noted earlier that several relaxation and dynamic processes, for example, dipole–dipole, hyperfine, exchange interactions between PC of different spin packets, can shorten spin relaxation times and hence change a shape of a multispin EPR line. So the study of spin relaxation can provide us with important information about spin-assisted electronic processes applied in the PANI:TSA/P3DDT:PC$_{61}$BM composite. It was demonstrated earlier that the initial EPR linewidth is proportional to the spin–spin relaxation rate. Besides, spin–lattice relaxation also shortens the lifetime of a spin state and broadens the line. It was also shown earlier that spin relaxation of spin charge carriers stabilized in polymer matrices is strongly defined by the structural, conformational, and electronic properties of their microenvironment. So it is important to analyze also how spin exchange affects spin–lattice relaxation of polarons in polymer matrix.

Figure 6.3 exhibits temperature dependencies of T_1 and T_2 values for polarons P$_1^{\cdot+}$ stabilized in PANI:TSA and PANI:TSA/P3DDT:PC$_{61}$BM samples. Spin–spin relaxation was shown earlier to be governed by spin–spin exchange interaction. The T_1 value of the samples was measured at RT to be 4.5×10^{-8} and 3.3×10^{-8} s, respectively. These values are in good agreement with $T_1 = 9.8 \times 10^{-8}$ s obtained by Wang et al. [166] for highly doped PANI:HCl. It is determined that spin–lattice relaxation of P$_1^{\cdot+}$ stabilized initially in a PANI:TSA composite changes weakly as the temperature increases

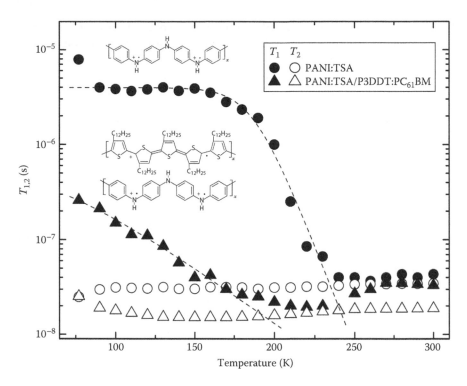

FIGURE 6.3

Temperature dependence of spin–lattice, T_1, and spin–spin, T_2, relaxation times determined for polarons P_1^{+} stabilized in the PANI:TSA backbone and respective PANI:TSA/P3DDT:PC$_{61}$BM composite without light illumination. (From Krinichnyi, V. et al., *Synth. Met.*, 179(1), 67, 2013. With permission.)

up to $T \sim 180$ K, which is typical for organic ordered systems. This process accelerates suddenly at $T \sim 210$ K possibly to a phase transition and then plateaus at higher temperatures. As P_1^{+} starts to interact with P_2^{+} in the PANI:TSA/P3DDT:PC$_{61}$BM nanocomposite, their spin–lattice relaxation strongly accelerates and becomes more temperature dependent as shown in Figure 6.3. This is an additional evidence of the exchange between polarons stabilized in different neighboring polymer chains. Figure 6.3 demonstrates that at high temperatures, T_1 tends to T_2. This is typical for PC stabilized in organic systems of low dimensionality and can be explained by the defrosting of macromolecular dynamics.

Therefore, light excitation of P3DDT:PC$_{61}$BM BHJ in the PANI:TSA/P3DDT:PC$_{61}$BM composite leads to charge separation and transfer from P3DDT chains to methanofullerene globes. This is accompanied by the appearance of polarons P_2^{+} on the P3DDT backbone and anion radicals mF_{61}^{-} located between its chains. Polarons P_2^{+} moving in P3DDT solitary chains interact with polarons P_1^{+} stabilized on neighboring PANI:TSA chains due to overlapping of their wave functions. Such interaction is governed mainly by nanomorphology of conjugated PANI:TSA subdomains. Exchange interaction and polaron relaxation are governed by Q1D activation hopping of P_2^+ along domestic polymer chains. Paramagnetic susceptibility of both polarons is described in the framework design of the model of exchange-coupled spin pairs differently distributed in appropriate polymer matrices. This deepens overlapping of wave functions of these charge carriers and increases the energy barrier which overcomes the polaron under its crossing through a BHJ. It is evident that the EPR investigation of domestic and photoexcited PC in complex PANI:TSA/P3DDT:PC$_{61}$BM composite and its spin subsystems allows controlling its texture and other structural properties.

Subsequent multifrequency EPR investigations of spin-assisted handling of domestic and photoinduced spin dynamics can open new horizons in creating novel spintronic devices with spin- and field-handling electronic properties based on conjugated polymer nanocomposites in which hands-on quantum mechanics with electron spins is performed. The flexibility, processability, and scalability of organic electronics have great impacts on the development of novel devices with specific properties. The significance of resonant photon absorption under high local magnetic fields could also induce the development of sensitive selective detectors of THz wave band attracting interest for, for example, security applications. The study of anisotropic spin dynamics in organic semiconductors may help us to understand more clearly the electronic processes in more conventional mesoscopic systems such as quantum dots. Indeed, the spin pairs transferring a charge in organic semiconductors are often created in coupled quantum dots also being a coherent quantum mechanical superposition of spin states [646], so that one can expect the arise of such entangled excitations naturally in molecular materials. The controlling and exploiting the interaction of polarons with the spin ensembles in organic semiconductors with weak dielectric screening takes a possibility to regulate the charge separation by means of electric fields [647]. In such electrostatic control of carrier separation, one may expect that analogous interactions forced by spin-resonant perturbation of the spin packets could form the basis of controlled programming of quantum dots by a spin excitation in organic semiconductors, as was obtained in phosphorus-doped silicon quantum dots [648]. It can be realized during two-stage separation and storage of photoinduced charge carriers under an electric field, their coherent manipulation using a spin-resonant excitation with following recombination in a controlled spin ensemble by removal of the electric field. This could then provide the quantum computing requiring logic operations of two or more well-organized spin qubits.

7

Concluding Remarks

The results presented in this book show that multifrequency EPR spectroscopy is a method that allows one to obtain qualitatively new information on molecular dynamics, transfer mechanism of spin carriers, magnetic and relaxation properties of conjugated polymers and their nanocomposites, and other low-dimensional condensed systems. This facilitates solving various practical problems in physics, chemistry, molecular biology, and interdisciplinary sciences. A higher spectral resolution of a millimeter wave band EPR provides enhanced accuracy in the measurement of magnetic resonance parameters and makes g-factor of organic free radicals an important characteristic. The method enables to resolve individual contributions of organic radicals with close magnetic resonant characteristics, to provide their reliable identification, and to establish the correlation between structural, dynamic, and magnetic parameters of such systems. Besides, the interaction between different spin packets is weakened significantly in high magnetic fields, so they may be considered as noninteracting and their parameters can be determined more precessionally. It allows obtaining qualitatively new information on the metrology of free radicals, molecular dynamics and electron mobility, electron and dimensional structure of paramagnetic centers, matrix's morphology, polarity, etc. Therefore, it can be reliably and efficiently used in the study of different molecular and electron processes realized in various polymer systems. This provides the unique possibility of the registration of fine structural and morphologic transitions, as well as electronic processes in these compounds with their following interpretation in the frames of appropriate theories. It should be noted that EPR spectroscopy is still developing and not all its abilities are exhausted so far. All these allow one to hope that the novel millimeter wave band technique combined, for example, with the multiresonant and pulse approaches, will be very beneficial in the future for the increase in informativity and the advancement of EPR methods.

The variety of electronic processes taking place in conjugated polymers and their nanocomposites is determined by the structure, morphology, ordering degree of their matrices, and the number and origin of nanoadducts. Among the general relationships, peculiar to these compounds, one can distinguish the following.

Solitons and polarons may be stabilized in conjugated polymers by dimerization distortion of their chains. They possess spins and elemental charges and are characterized by the high mobility of polymer chains. The main properties of such charge carriers depend on various factors, for example, method of synthesis of a polymer, its structure and morphology. The collapse of polaron pairs into diamagnetic (spinless) bipolarons with slower dynamics and higher effective mass can occur with the increase of doping level of a polymer. Besides, the doping process leads to the change in charge-transfer mechanism. The conductivity of neutral and weakly nanomodified conjugated polymers is governed mainly by interchain charge tunneling in the frames of *small polarons* or Kivelson theories based on high interaction of spin carriers with phonons of polymer lattice. Once the doping level of a polymer increases and its matrix becomes more ordered, these mechanisms cease to dominate, and the charge can be transferred as a result of its thermal activation from widely separated localized states in the bandgap to close localized states in the tails of both the valence and conjugated bands. This stipulates the formation in some matrices of complex quasi-particles, namely, the molecular-lattice polarons, due to libron–phonon interactions analogous to that as it is realized in organic molecular crystals. Because conjugated polymers have *a priori* a lower dimensionality as compared with more ordered systems mentioned earlier, the dynamics of their charge carriers is appeared to be more anisotropic. In heavily doped samples, crystalline clusters are formed in bulk of polymer matrix. The charge transport in such systems is realized in terms of Q1D charge hopping between clusters and its Q3D charge hopping between the cluster's chains characterized by its strong interaction with lattice phonons.

Polarons diffusing along polymer chains interact with other polarons stabilized on neighboring chains and/or with radicals embedded into polymer matrix as counteranions. Fullerene cages are reoriented by small angles between polymer Q1D and Q2D units according to the Marcus mechanism and can act as a nanoscopic probe of polaron dynamics. Such exchange interaction complicates charge transport in polymer:fullerene and polymer:dopant/polymer:fullerene multispin nanocomposites. Charge dynamics in these systems is also realized in the mechanisms mentioned earlier which add up due to spin exchange interaction. These spin-assisted processes are governed mainly by the structure of the ingredients of a composite as well as by the nature and dynamics of photoinduced charge carriers. The specific morphology of such polymer nanocomposites changes their energy levels and shifts the competition between excited states in their BHJ. LEPR data presented earlier contribute to a better understanding of the correlations of structural, magnetic, and electronic properties of polymer:fullerene and polymer:dopant/polymer:fullerene composites and their ingredients. Such correlations obtained by direct EPR method can be used for further development and optimization of nanomodified polymer photovoltaic devices. The fact that the polaron possesses a spin plays a key role in interchain coupling of different spin charge carriers that can be used for the handling of charge transfer through BHJ of multispin composites. Initiation of spins by different photons allows making such handling more delicate that is a crucial strategy in creating optimal systems with spin-assisted charge transfer. The correlations established between dynamic electronic, and structural parameters of these systems at wide region of spin precession frequency can be used for further controllable synthesis of various organic spintronic devices with optimal properties. This will open new horizon in the creation of flexible and scalable organic molecular devices with spin-assisted electronic properties. Solitary spin carriers trapped in the bulk of a darkened polymer matrix can in principle be used as elemental dots for quantum computing. Since coherent spin dynamics in such organic BHJ is anisotropic, our strategy allow us to obtain complex correlations of anisotropic electron transport and spin dynamics for the fast-growing evolution of molecular electronics and spintronics.

References

1. Williams, J. M., Ferraro, J. R., Thorn, R. J., Carlson, K. D., Geiser, U., Wang, H. H., Kini, A. M., and Whangboo, M. H., *Organic Superconductors (Including Fullerenes): Synthesis, Structure, Properties, and Theory.* Englewood Cliffs, NJ: Prentice-Hall, Inc., 1992, 400p.
2. Wright, J. D., *Molecular Crystals.* Cambridge, U.K.: Cambridge University Press, 1995, 236p.
3. Nalwa, H. S., Ed., *Handbook of Organic Conductive Molecules and Polymers.* Chichester, U.K.: John Wiley & Sons, 1997, 3334p.
4. Roth, S. and Carroll, D., *One-Dimensional Metals: Conjugated Polymers, Organic Crystals, Carbon Nanotubes.* Weinheim, Germany: Wiley-VCH, 2004, 264p.
5. Todres, Z. V., *Ion-Radical Organic Chemistry: Principles and Applications.* Boca Raton, FL: CRC Press, 2008, 496p.
6. Kadish, K. M. and Ruoff, R. S., Eds., *Fullerenes: Chemistry, Physics, and Technology.* New York: Wiley-Interscience, 2000, 978p.
7. Nalwa, H., Ed., *Handbook of Advanced Electronic and Photonic Materials and Devices*, vols. 1–10. Maryland Heights, MO: Academic Press, 2001, 3366p.
8. Prassides, K., Ed., *Fullerene-Based Materials: Structures and Properties*, Structure and Bonding, vol. 109. Berlin, Germany: Springer, 2004, 285p.
9. Sorokin, A. B., Phthalocyanine metal complexes in catalysis, *Chemical Review* **113**(10), 8152–8191, 2013.
10. Kalyanasundaram, K., Ed., *Dye-Sensitized Solar Cells.* Lausanne, Switzerland: EFPL Press, 2010, 320p.
11. Bhattacharya, S. K., *Metal Filled Polymers (Plastics Engineering)*, vol. 11. Boca Raton, FL: CRC Press, 1986, 376p.
12. Zanardi, C., Terzi, F., Pigani, L., and Seeber, R., Electrode coatings consisting of polythiophene-based composites containing metal centres, in *Encyclopedia of Polymer Composites: Properties, Performance and Applications*, Lechkov, M. and Prandzheva, S., Eds. Hauppauge, NY: Nova Science Publishers, 2009, pp. 1–74.
13. Mark, J. E., Ed., *Physical Properties of Polymers Handbook.* New York: Springer, 2007, 1076p.
14. Petrova-Koch, V., Hezel, R., and Goetzberger, A., Eds., *High-Efficient Low-Cost Photovoltaics.* Berlin, Germany: Springer, 2009, 228p.
15. Veca, L. M., Wang, W., Lin, Y., Meziani, M. J., Tian, L., Connel, J. W., Ghose, S., Kong, C. Y., and Sun, Y.-P., Thermal conductive materials based on carbon nanotubes and graphene nanosheets, in *Handbook of Carbon Nano Materials*, vol. 2, D'Souza, F. and Kadish, K. M., Eds. Singapore: World Scientific Publishing, 2011, Chapter 10, 972p.
16. Alvi, F., Ram, M. K., Basnayaka, P. A., Stefanakos, E., Goswami, Y., and Kumar, A., Graphene-polyethylenedioxythiophene conducting polymer nanocomposite based supercapacitor, *Electrochimica Acta* **56**(25), 9406–9412, 2011.
17. Sadasivuni, K. K., Ponnamma, D., Kim, J., and Thomas, S., Eds., *Graphene-Based Polymer Nanocomposites in Electronics*, Springer Series on Polymer and Composite Materials. Heidelberg, Germany: Springer, 2015, 382p.

18. Mohanty, S., Nayak, S. K., Kaith, B. S., and Kaila, S., Eds., *Polymer Nanocomposites Based on Inorganic and Organic Nanomaterials*. Hoboken, NJ: Wiley, 2015, 300p.
19. Brédas, J. L. and Silbey, R., Eds., *Conjugated Polymers: The Novel Science and Technology of Highly Conducting and Nonlinear Optically Active Materials*. Dordrecht, the Netherlands: Kluwer Academic, 1991, 624p.
20. Menon, R., Charge transport in conducting polymers, in *Handbook of Organic Conductive Molecules and Polymers*, vol. 4, Nalwa, H. S., Ed. Chichester, U.K.: John Wiley & Sons, 1997, pp. 47–145.
21. Skotheim, T., Elsenbaumer, R., and Reynolds, J., Eds., *Handbook of Conducting Polymers*. New York: Marcel Dekker, Inc., 1998, 1105p.
22. Chandrasekhar, P., Ed., *Conducting Polymers, Fundamentals and Applications: A Practical Approach*. Boston, MA: Kluwer Academic Publishers, 1999, 718p.
23. Hadziioannou, G. and van Hutten, P. F., Eds., *Semiconducting Polymers: Chemistry, Physics and Engineering*. Weinheim, Germany: Wiley, 2000, 768p.
24. Malhotra, B. D., Ed., *Handbook of Polymers in Electronics*. Shawbury, U.K.: Rapra Technology Ltd., 2002, 488p.
25. Skotheim, T. A. and Reynolds, J. R., *Conjugated Polymers: Processing and Applications*, vol. 3d. Boca Raton, FL: CRC Press, 2006, 656p.
26. Skotheim, T. E. and Reynolds, J. R., Eds., *Handbook of Conducting Polymers*. Boca Raton, FL: CRC Press, 2007, 1680p.
27. Wan, M. X., *Conducting Polymers with Micro or Nanometer Structure*. Berlin, Germany: Springer, 2008, 170p.
28. Heeger, A. J., Sariciftci, N. S., and Namdas, E. B., *Semiconducting and Metallic Polymers*. London, U.K.: Oxford Univertsity Press, 2010, 288p.
29. Inzelt, G., *Conducting Polymers: A New Era in Electrochemistry*. Berlin, Germany: Springer, 2012, 309p.
30. Launay, J.-P. and Verdaguer, M., *Electrons in Molecules: From Basic Principles to Molecular Electronics*. Oxford, U.K.: Oxford University Press, 2013, 512p.
31. Kobayashi, S. and Müllen, K., Eds., *Encyclopedia of Polymeric Nanomaterials*. Berlin, Germany: Springer, 2015, 2640p.
32. Little, W. A., Possibility of synthesizing an organic superconductor, *Physical Review* **134**(6A), 1416–1424, 1964.
33. Saxena, V. and Malhotra, B. D., Prospects of conducting polymers in molecular electronics, *Current Applied Physics* 3(2–3), 293–305, 2003.
34. Zhang, G. and Manjooran, N., Eds., *Nanofabrication and Its Application in Renewable Energy*, RSC Nanoscience & Nanotechnology. Cambridge, U.K.: The Royal Society of Chemistry, 2014, 225p.
35. Brabec, C., Scherf, U., and Dyakonov, V., Eds., *Organic Photovoltaics: Materials, Device Physics, and Manufacturing Technologies*. Weinheim, Germnay: Wiley-VCH, 2014, 642p.
36. Xu, Y., Awschalom, D. D., and Nitta, J., Eds., *Handbook of Spintronics*. Berlin, Germany: Springer-Verlag, 2015, 1700p.
37. Krinichnyi, V. I., Eremenko, O. N., Rukhman, G. G., Letuchii, Y. A., and Geskin, V. M., Sensors based on conducting organic polymers, *Polyaniline, Polymer Science U.S.S.R.* **31**(8), 1819–1825, 1989.
38. Krinichnyi, V. I., Eremenko, O. N., Rukhman, G. G., Geskin, V. M., and Letuchy, Y. A., Polyaniline based sensors for solution components, *Synthetic Metals* **41**(3), 1137–1137, 1991.

39. Goldenberg, L. M. and Krinichnyi, V. I., Water-sensitive sensor based on modified poly(vinylchloride), *Synthetic Metals* **53**(3), 403–407, 1993.
40. Harsanyi, G., *Polymer Films in Sensor Applications*. Boca Raton, FL: CRC Press, 1995, 465p.
41. ElKaoutit, M., Application of conducting polymers in electroanalysis, in *Aspects on Fundaments and Applications of Conducting Polymers*, Motheo, A. J., Ed. Rijeka, Croatia: InTech, 2011, pp. 43–66.
42. Krinichnyi, V. I., Dynamics of spin charge carriers in polyaniline, *Applied Physics Reviews* **1**(2), 021305/01–021305/40, 2014.
43. Goldenberg, L. M., Krinichnyi, V. I., and Nazarova, I. B., The Schottky device based on doped poly(para-phenylene), *Synthetic Metals* **44**(2), 199–203, 1991.
44. Nazarova, I. B., Krinichnyi, V. I., and Goldenberg, L. M., Schottky diodes based on poly(*para*-phenylene) and poly(1,4-dipyrrolobenzene), *Synthetic Metals* **53**(3), 399–402, 1993.
45. Schuhmann, W., Conducting polymers and their application in amperometric biosensors, in *Diagnostic Biosensor Polymers*, Usmani, A. M. and Akmal, N., Eds. Washington, DC: American Chemical Society, 1994, pp. 110–123.
46. Arshak, K., Velusamy, V., Korostynska, O., Oliwa-Stasiak, K., and Adley, C., Conducting polymers and their applications to biosensors: Emphasizing on foodborne pathogen detection, *IEEE Sensors Journal* **9**(12), 1942–1951, 2009.
47. Cosnier, S. and Karyakin, A., Eds., *Electropolymerization: Concepts, Materials and Applications*. Weinheim, Germany: Wiley, 2010, 296p.
48. Wang, C., Jiang, L., and Hu, W., Organic/polymeric field-effect transistors, in *Organic Optoelectronics*, Hu, W., Ed. Weinheim, Germnay: Wiley-VCH Verlag, 2013, pp. 43–94.
49. Klauk, H., Organic thin-film transistors, *Chemical Society Reviews* **39**(7), 2643–2666, 2010.
50. Bittle, E. G., Basham, J. I., Jackson, T. N., Jurchescu, O. D., and Gundlach, D. J., Mobility overestimation due to gated contacts in organic field-effect transistors, *Nat Commun* **7**, 10908/01-10908/07, 2016.
51. Sun, S. S. and Sariciftci, N. S., Eds., *Organic Photovoltaics: Mechanisms, Materials, and Devices (Optical Engineering)*, Optical Science and Engineering Series. Boca Raton, FL: CRC Press, 2005, 664p.
52. Poortmans, J. and Arkhipov, V., Eds., *Thin Film Solar Cells: Fabrication, Characterization and Applications*. West Sussex, U.K.: Wiley, 2006, 502p.
53. Jeffries-El, M. and McCullough, R. D., Regioregular polythiophenes, in *Handbook of Conducting Polymers*, vol. 3d, Scotheim, T. E. and Reynolds, J. R., Eds. Boca Raton, FL: CRC Press, 2007, Chapter 9, pp. 9.1–9.49.
54. Pagliaro, M., Palmisano, G., and Ciriminna, R., *Flexible Solar Cells*. Weinheim, Germnay: Wiley-WCH, 2008, 202p.
55. Krinichnyi, V. I., LEPR spectroscopy of charge carriers photoinduced in polymer/fullerene composites, in *Encyclopedia of Polymer Composites: Properties, Performance and Applications*, Lechkov, M. and Prandzheva, S., Eds. Hauppauge, NY: Nova Science Publishers, 2009, pp. 417–446.
56. Hu, W., Ed., *Organic Optoelectronics*. Weinheim, Germany: Wiley-VCH Verlag, 2013, 507p.

57. Krinichnyi, V. I., EPR spectroscopy of polymer: Fullerene nanocomposites, in *Spectroscopy of Polymer Nanocomposites*, Ponnamma, D., Rouxel, D., and Thomas, S., Eds. Amsterdam, the Netherlands: Elsevier, 2016, pp. 202–275.

58. Salamone, J. C., Ed., *Polymeric Material Encyclopedia*. Boca Raton, FL: CRC Press, 1996, 9600p.

59. Scotheim, T. E., Elsenbaumer, R. L., and Reynolds, J. R., Eds., *Handbook of Conducting Polymers*. New York: Marcel Dekker, 1997, 1097p.

60. Menon, R., Yoon, C. O., Moses, D., and Heeger, A. J., Metal-Insulator transition in doped conducting polymers, in *Handbook of Conducting Polymers*, Skotheim, T. A., Elsenbaumer, R. L., and Reynolds, J. R., Eds. New York: Marcel Dekker, 1997, pp. 27–84.

61. Wessling, B., Metallic properties of conductive polymers due to dispersion, in *Handbook of Organic Conductive Molecules and Polymers*, vol. 3, Nalwa, H. S., Ed. Chichester, U.K.: John Wiley & Sons, 1997, pp. 497–632.

62. Blakemore, J. S., *Solid State Physics*. Cambridge, U.K.: Cambridge University Press, 1985, 520p.

63. Parker, M. A., *Solid State and Quantum Theory for Optoelectronics*. Boca Raton, FL: CRC Press, 2009, 848p.

64. Yu, P. Y. and Cardona, M., *Fundametals of Semiconductors: Physics and Materials Properties*. Heidelberg, Germany: Springer, 2010, 775p.

65. Chance, R. R., Boudreaux, D. S., Brédas, J. L., and Silbey, R., Solitons, polarons and bipolarons in conjugated polymers, in *Handbook of Conducting Polymers*, vol. 2, Scotheim, T. E., Ed. New York: Marcel Dekker, Inc., 1986, pp. 825–857.

66. Lu, Y., Ed., *Solitons and Polarons in Conducting Polymers*. River Edge, NJ: World Scientific, 1988, 772p.

67. Peierls, R. E., *Quantum Theory of Solids*. London, U.K.: Oxford University Press, 1996, 238p.

68. Chien, J. C. W., *Polyacetylene: Chemistry, Physics and Material Science*. Orlando, FL: Academic Press, 1984, 634p.

69. Curran, S., Stark-Hauser, A., and Roth, S., Polyacetylene, in *Handbook of Organic Conductive Molecules and Polymers*, vol. 2, Nalwa, H. S., Ed. New York: John Wiley & Sons, 1997, Chapter 1, pp. 1–59.

70. Su, W. P., Schrieffer, J. R., and Heeger, A. J., Solitons in polyacetylene, *Physical Review Letters* **42**, 1698–1701, 1979.

71. Su, W. P., Schrieffer, J. R., and Heeger, A. J., Soliton excitations in polyacetylene, *Physical Review B* **22**(4), 2099–2111, 1980.

72. Brédas, J. L., Electronic structure of highly conducting polymers, in *Handbook of Conducting Polymers*, vol. 2, Scotheim, T. E., Ed. New York: Marcel Dekker, Inc., 1986, pp. 859–913.

73. Brédas, J. L., Chance, R. R., Baughman, R. H., and Silbey, R., Abinitio effective Hamiltonian study of the electronic properties of conjugated polymers, *Journal of Chemical Physics* **76**(7), 3673–3678, 1982.

74. Elsenbaumer, R. L. and Shacklette, L. W., Phenylene-based conducting polymers, in *Handbook of Conducting Polymers*, vol. 1, Scotheim, T. E., Ed. New York: Marcel Dekker, Inc., 1986, pp. 213–263.

75. Devreux, F., Genoud, F., Nechtschein, M., and Villeret, B., On polaron and bipolaron formation in conducting polymers, in *Electronic Properties of Conjugated Polymers*, vol. 76, Kuzmany, H., Mehring, M., and Roth, S., Eds. Berlin, Germany: Springer-Verlag, 1987, pp. 270–276.

76. Westerling, M., Osterbacka, R., and Stubb, H., Recombination of long-lived photoexcitations in regioregular polyalkylthiophenes, *Physical Review B* **66**(16), 165220/01–165220/07, 2002.

77. Shaheen, S. E., Brabec, C. J., Sariciftci, N. S., Padinger, F., Fromherz, T., and Hummelen, J. C., 2.5% Efficient organic plastic solar cells, *Applied Physics Letters* **78**(6), 841–843, 2001.

78. Hwang, I. W., Soci, C., Moses, D., Zhu, Z. G., Waller, D., Gaudiana, R., Brabec, C. J., and Heeger, A. J., Ultrafast electron transfer and decay dynamics in a small band gap bulk heterojunction material, *Advanced Materials* **19**(17), 2307–2312, 2007.

79. Kraabel, B., McBranch, D., Sariciftci, N. S., Moses, D., and Heeger, A. J., Ultrafast spectroscopic studied of photoinduced electron-transfer from semiconducting polymers to C-60, *Physical Review B* **50**(24), 18543–18552, 1994.

80. Brabec, C. J., Zerza, G., Cerullo, G., DeSilvestri, S., Luzatti, S., Hummelen, J. C., and Sariciftci, N. S., Tracing photoinduced electron transfer process in conjugated polymer/fullerene bulk heterojunctions in real time, *Chemical Physics Letters* **340**(3–4), 232–236, 2001.

81. Banerji, N., Cowan, S., Leclerc, M., Vauthey, E., and Heeger, A. J., Exciton formation, relaxation, and decay in PCDTBT, *Journal of the American Chemical Society* **132**(49), 17459–17470, 2010.

82. Dediu, V. A., Hueso, L. E., Bergenti, I., and Taliani, C., Spin routes in organic semiconductors, *Nature Materials* **8**, 707–716, 2009.

83. Garnica, M., Stradi, D., Barja, S., Calleja, F., Diaz, C., Alcami, M., Martin, N. et al., Long-range magnetic order in a purely organic 2D layer adsorbed on epitaxial graphene, *Nature Physics* **9**(6), 368–374, 2013.

84. Shen, L., Zeng, M., Wu, Q., Bai, Z., and Feng, Y. P., Graphene spintronics: Spin generation and manipulation in graphene, in *Graphene Optoelectronics*, Yusoff, R. b. M., Ed. Weinheim, Germany: Wiley, 2014, 167–188.

85. Lloyd, S., A potentially realizable quantum computer, *Science* **261**, 1569–1571, 1993.

86. Blank, A., Kastner, R., and Lebanon, H., Exploring new active materials for low noise, room temperature, microwave amplifiers and other devices, *IEEE Transactions of Microwave Theory and Techniques* **46**(12), 2137–2144, 1998.

87. Wohlgenannt, M., Tandon, K., Mazumdar, S., Ramasesha, S., and Vardeny, Z. V., Formation cross-sections of singlet and triplet excitons in p-conjugated polymers, *Nature* **409**, 494–497, 2001.

88. Baunsgaard, D., Larsen, M., Harrit, N., Frederiksen, J., Wilbrandt, R., and Stapelfeldt, H., Photophysical properties of 2,3,6,7,10,11-hexakis(n-hexylsulfanyl)triphenylene and 2,3,6,7,10,11-hexakis(n-hexylsulfonyl)triphenylene insolution, *Journal of Chemical Society, Faraday Transactions*, **93**(10), 1893–1901, 1997.

89. Lupton, J. M., McCamey, D. R., and Boehme, C., Coherent spin manipulation in molecular semiconductors: Getting a handle on organic spintronics, *Chemical Physics Chemistry* **11**(14), 3040–3058, 2010.

90. Lloyd, M. T., Lim, Y. F., and Malliaras, G. G., Two-step exciton dissociation in poly(3-hexylthiophene)/fullerene heterojunctions, *Applied Physics Letters* **92**(14), 143308–143310, 2008.

91. Im, C., Lupton, J. M., Schouwink, P., Heun, S., Becker, H., and Bassler, H., Fluorescence dynamics of phenyl-substituted polyphenylenevinylene–trinitrofluorenone blend systems, *Journal of Chemical Physics* **117**, 1395–1402, 2002.

92. Rouxel, D., Thomas, S., and Ponnamma, D., Eds., *Spectroscopy of Polymer Nanocomposites*. Amsterdam, the Netherlands: Elsevier, 2016, 482p.

93. List, E. J. W., Scherf, U., Mullen, K., Graupner, W., Kim, C. H., and Shinar, J., Direct evidence for singlet-triplet exciton annihilation in π-conjugated polymers, *Physical Review B* **66**(23), 235203–235207, 2002.

94. Reufer, M., Lagoudakis, P. G., Walter, M. J., Lupton, J. M., Feldmann, J., and Scherf, U., Evidence for temperature-independent triplet diffusion in a ladder-type conjugated polymer, *Physical Review B* **74**(24), 241201–241204, 2006.

95. Schlick, S., Ed., *Advanced ESR Methods in Polymer Research*. New York: John Wiley & Sons Inc., 2006, 368p.

96. Eaton, G. R., Eaton, S. S., Barr, D. P., and Weber, R. T., *Quantitative EPR*. Wien, Austria: Springer, 2010, 185p.

97. Ranby, B. and Rabek, J. F., *ESR Spectroscopy in Polymer Research*. New York: Springer, 2011, 410p.

98. Misra, S. K., Ed., *Multifrequency Electron Paramagnetic Resonance—Theory and Applications*. Weinheim, Germany: Wiley-VCH, 2011, 1056p.

99. Lund, A., Shiotani, M., and Shimada, S., *Principles and Applications of ESR Spectroscopy*. New York: Springer, 2011, 430p.

100. Lund, A. and Shiotan, M., Eds., *EPR of Free Radicals in Solids I: Trends in Methods and Applications*, Progress in Theoretical Chemistry and Physics 24. Dordrecht, the Netherlands: Springer, 2013, 414p.

101. Lund, A. and Shiotan, M., Eds., *EPR of Free Radicals in Solids II: Progress in Theoretical Chemistry and Physics*, Progress in Theoretical Chemistry and Physics 24. Dordrecht, the Netherlands: Springer, 2013, 387p.

102. Misra, S. K., Ed., *Handbook of Multifrequency Electron Paramagnetic Resonance: Data and Techniques*. Weinheim, Germany: Wiley-VCH, 2014, 320p.

103. Baranov, P., *Magnetic Resonance of Semiconductors and Semiconductor Nanostructures*. Vienna, Austria: Springer-Verlag, 2016, 400p.

104. Bernier, P., The magnetic properties of conjugated polymers: ESR studies of undoped and doped systems, in *Handbook of Conducting Polymers*, vol. 2, Scotheim, T. E., Ed. New York: Marcel Dekker, Inc., 1986, pp. 1099–1125.

105. Mizoguchi, K. and Kuroda, S., Magnetic properties of conducting polymers, in *Handbook of Organic Conductive Molecules and Polymers*, vol. 3, Nalwa, H. S., Ed. Chichester, U.K.: John Wiley & Sons, 1997, pp. 251–317.

106. Kuroda, S., ESR and ENDOR spectroscopy of solitons and polarons in conjugated polymers, in *EPR in the 21st Century: Basics and Applications to Material, Life and Earth Sciences*, Kawamori, A., Yamauchi, J., and Ohta, H., Eds. Amsterdam, the Netherlands: Elsevier Science B.V., 2002, pp. 113–124.

107. Altshuler, S. A. and Kozirev, B. M., *Electron Paramagnetic Resonance*. New York: Academic Press, 1972, 372p.

108. Jeschke, G. and Schlick, S., Eds., Continuous-Wave and pulsed ESR methods, in *Advanced ESR Methods in Polymer Research*, Hoboken, NJ: Wiley, 2006, 1–24, Chapter 1.

109. Salikhov, K. M., Semenov, A. G., and Tsvetkov, Y. D., *Electron Spin Echo and Its Utilization (Russ)*. Novosibirsk, Russia: Nauka, 1976, 342p.

110. Tanaka, T., *Experimental Methods in Polymer Science: Modern Methods in Polymer Research and Technology*. New York: Academic Press, 2000, 693p.

111. Kohmoto, T., Fukuda, Y., and Kunitomo, M., Fourier-transform ESR spectroscopy and observation of ultrafast spin-lattice relaxation by optical means, in *EPR in the 21st Century*, Ohta, A. K. Y., Ed. Amsterdam, the Netherlands: Elsevier Science B.V., 2002, pp. 700–705.

112. Poole, C. P., *Electron Spine Resonance*. London, U.K.: *Int. Sci. Publ.*, 1967, 959p.
113. Kevan, L. and Kispert, L. D., *Electron Spin Double Resonance Spectroscopy*. New York: Wiley Interscience, 1976, 438p.
114. Hyde, J. S. and Dalton, L. R., Saturation-transfer spectroscopy, in *Spin Labeling II: Theory and Application*, vol. 2, Berliner, L. J., Ed. New York: Academic, 1979, pp. 1–70.
115. Wolf, S. A., Awschalom, D. D., Buhrman, R. A., Daughton, J. M., von Molnar, S., Roukes, M. L., Chtchelkanova, A. Y., and Treger, D. M., Spintronics: A spin-based electronics vision for the future, *Science* **294**(5546), 1488–1495, 2001.
116. Cinchetti, M., Heimer, K., Wustenberg, J. P., Andreyev, O., Bauer, M., Lach, S., Ziegler, C., Gao, Y. L., and Aeschlimann, M., Determination of spin injection and transport in a ferromagnet/organic semiconductor heterojunction by two-photon photoemission, *Nature Materials* **8**, 115–119, 2009.
117. Kinoshita, M., Iwasaki, N., and Nishi, N., Molecular spectroscopy of the triplet state through optical detection of zero-field magnetic resonance, *Applied Spectroscopy Review* **17**(1), 1–94, 1981.
118. Krinichnyi, V. I., 2-mm Waveband electron paramagnetic resonance spectroscopy of conducting polymers (review), *Synthetic Metals* **108**(3), 173–222, 2000.
119. Lee, M. K., Segal, M., Soos, Z. G., Shinar, J., and Baldo, M. A., Yield of singlet excitons in organic light-emitting devices: A double modulation photoluminescence-detected magnetic resonance study, *Physical Review Letters* **94**(13), 137403–137406, 2005.
120. McCamey, D. R., van Schooten, K. J., Baker, W. J., Lee, S. Y., Paik, S.-Y., Lupton, J. M., and Boehme, C., Hyperfine-field-mediated spin beating in electrostatically bound charge carrier pairs, *Physical Review Letters* **104**, 017601–017604, 2010.
121. Grinberg, O. Y., Dubinskii, A. A., and Lebedev, Y. S., Electron-paramagnetic resonance of free radicals in a 2-millimeter range of wave-length, *Uspekhi Khimii* **52**(9), 1490–1513, 1983.
122. Krinichnyi, V. I., *2-mm Wave Band EPR Spectroscopy of Condensed Systems*. Boca Raton, FL: CRC Press, 1995, 223p.
123. Grinberg, O. Y. and Berliner, L. J., Eds., *Very High Frequency (VHF) ESR/EPR*, Biological Magnetic Resonance. New York: Kluwer Academic Plenum Publishers, 2004, 592p.
124. Krinichnyi, V. I., High field ESR spectroscopy of conductive polymers, in *Advanced ESR Methods in Polymer Research*, Schlick, S., Ed. Hoboken, NJ: Wiley, 2006, pp. 307–338.
125. Krinichnyi, V. I., Investigation of biological systems by high-resolution 2-mm wave band, *ESR Journal of Biochemical and Biophysical Methods* **23**(1), 1–30, 1991.
126. Krinichnyi, V. I., Investigation of biological systems by high resolution 2-mm wave band EPR, *Applied Magnetic Resonance* **2**(1), 29–60, 1991.
127. Lebedev, Y. S., High-field ESR, in *Electron Spin Resonance*, vol. 14, Atherton, N. M., Davies, M. J., and Gilbert, B. C., Eds. Cambridge, U.K.: Royal Society of Chemistry, 1994, pp. 63–87.
128. Eaton, S. S. and Eaton, G. R., High magnetic fields and high frequrncies, in *Handbook of Electron Spin Resonance*, vol. 2, Poole, C. P. and Farach, H. A., Eds. New York: Springer-Verlag, 1999, pp. 344–367.
129. Hagen, W. R., High-frequency EPR of transition ion complexes and metalloproteins, *Coordination Chemistry Reviews* **192**, 209–229, 1999.

130. Smith, G. M. and Riedi, P. C., *Progress in High Field EPR*. Cambridge, U.K.: RSC, 2000, 164p.
131. Marsh, D., Kurad, D., and Livshits, V. A., High-field electron spin resonance of spin labels in membranes, *Chemistry and Physics of Lipids* **116**(1–2), 93–114, 2002.
132. Hustedt, E. J. and Beth, A. H., High field/high frequency saturation transfer electron paramagnetic resonance spectroscopy: Increased sensitivity to very slow rotational motions, *Biophysical Journal* **86**, 3940–3950, 2004.
133. Grinberg, O. and Berliner, L., Eds., *Very High Frequency (VHF) ESR/EPR (Biological Magnetic Resonance)*, Biological Magnetic Resonance (Book 22). New York: Springer-Verlag, 2011, 569p.
134. Carrington, F. and McLachlan, A. D., *Introduction to Magnetic Resonance with Application to Chemistry and Chemical Physics*. New York: Harrer & Row, Publishers, 1967, 266p.
135. Buchachenko, A. L. and Vasserman, A. M., *Stable Radicals (Russ)*. Moscow, Russia: Khimija, 1973, 408p.
136. Buchachenko, A. L., Turton, C. N., and Turton, T. I., *Stable Radicals*. New York: Consultants Bureau, 1995, 180p.
137. Kim, Y., Cook, S., Tuladhar, S. M., Choulis, S. A., Nelson, J., Durrant, J. R., Bradley, D. D. C. et al., A strong regioregularity effect in self-organizing conjugated polymer films and high-efficiency polythiophene: Fullerene solar cells, *Nature Materials* **5**(3), 197–203, 2006.
138. Dinse, K.-P. and Kato, T., Multi-frequency EPR study of metallo-endofullerenes, in *Novel NMR and EPR Techniques*, Dolinšek, J., Vilfan, M., and Žumer, S., Eds. Heidelberg, Germany: Springer, 2006, pp. 185–207.
139. Möbius, K., Lubitz, W., and Savitsky, A., High-field EPR on membrane proteins— Crossing the gap to NMR, *Progress in Nuclear Magnetic Resonance Spectroscopy* **75**(1), 1–49, 2013.
140. Silinsh, E. A., Kurik, M. V., and Chapek, V., *Electronic Processes in Organic Molecular Crystals: The Phenomena of Localization and Polarization (Russian)*. Riga, Latvia: Zinatne, 1988, 329p.
141. Bobbert, P. A., Organic semiconductors: What makes the spin relax?, *Nature Materials* **9**, 288–290, 2010.
142. Glenn, R., Baker, W. J., Boehme, C., and Raikh, M. E., Analytical description of spin-Rabi oscillation controlled electronic transitions rates between weakly coupled pairs of paramagnetic states with S = 1/2, *Physical Review B* **87**(15), 155208/01–155208/10, 2013.
143. Poole, C. P., *Electron Spin Resonance: A Comprehensive Treatise on Experimental Techniques*. New York: John Wiley & Sons, 1983, 780p.
144. Vonsovskii, S. V., *Magnetism. Magnetic Properties of Dia-, Para-, Ferro-, Antiferro-, and Ferrimagnetics*, vol. 1, 2. New York: John Wiley & Sons, 1974, 1256p.
145. Salavagione, H., Morales, G. M., Miras, M. C., and Barbero, C., Synthesis of a self-doped polyaniline by nucleophilic addition, *Acta Polymerica* **50**(1), 40–44, 1999.
146. Kronmüller, H. and Parkin, S., Eds., *Handbook of Magnetism and Advanced Magnetic Materials*. Chichester, U.K.: Wiley, 2007, 3064p.
147. Clark, W. G. and Tippie, L. C., Exchange-coupled pair model for the random-exchange Heisenberg antiferromagnetic chain, *Physical Review B* **20**(7), 2914–2923, 1979.

148. Kahol, P. K. and Mehring, M., Exchange-coupled pair model for the non-curie-like susceptibility in conducting polymers, *Synthetic Metals* **16**(2), 257–264, 1986.
149. Nelson, J., Diffusion-limited recombination in polymer-fullerene blends and its influence on photocurrent collection, *Physical Review B* **67**(15), 155209/01–155209/10, 2003.
150. Schultz, N. A., Scharber, M. C., Brabec, C. J., and Sariciftci, N. S., Low-temperature recombination kinetics of photoexcited persistent charge carriers in conjugated polymer/fullerene composite films, *Physical Review B* **64**(24), 245210/01–245210/07, 2001.
151. Tachiya, M. and Seki, K., Theory of bulk electron-hole recombination in a medium with energetic disorder, *Physical Review B* **82**(8), 085201/01–085201/08, 2010.
152. Molin, Y. N., Salikhov, K. M., and Zamaraev, K. I., *Spin Exchange*. Berlin, Germany: Springer, 1980, 260p.
153. Houze, E. and Nechtschein, M., ESR in conducting polymers: Oxygen-induced contribution to the linewidth, *Physical Review B* **53**(21), 14309–14318, 1996.
154. Roth, H. K., Keller, F., and Schneider, H., *Hochfrequenzspectroskopie in der Polymerforschung*. Berlin, Germany: Academie Verlag, 1984, 374p.
155. Blumenfeld, L. A., Voevodski, V. V., and Semenov, A. G., *Application of Electron Paramagnetic Resonance in Chemistry (Russ)*. Novosibirsk, Russia: Izdat. SO AN SSSR, 1962, 216p.
156. Lebedev, Y. S. and Muromtsev, V. I., *EPR and Relaxation of Stabilized Radicals (Russ)*. Moscow, Russia: Khimija, 1972, 256p.
157. Van Vooren, A., Kim, J.-S., and Cornil, J., Intrachain versus interchain electron transport in poly(fluorene-alt-benzothiadiazole): A quantum-chemical insight, *European Journal of Chemical Physics and Physical Chemistry* **9**(7), 989–993, 2008.
158. Lan, Y.-K. and Huang, C.-I., A theoretical study of the charge transfer behavior of the highly regioregular poly-3-hexylthiophene in the ordered state, *Journal of Physical Chemistry B* **112**(47), 14857–14862, 2008.
159. Cheung, D. L., McMahon, D. P., and Troisi, A., Computational study of the structure and charge-transfer parameters in low-molecular-mass P3HT, *Journal of Physical Chemistry B* **113**(28), 9393–9401, 2009.
160. Troisi, A. and Orlandi, G. J., Dynamics of the intermolecular transfer integral in crystalline organic semiconductors, *Journal of Physical Chemistry A* **110**(11), 4065–4070, 2006.
161. Kirkpatrick, J., Marcon, V., Nelson, J., Kremer, K., and Andrienko, D., Charge mobility of discotic mesophases: A multiscale quantum and classical study, *Physical Review Letters* **98**(22), 227402/01–227402/04, 2007.
162. Butler, M. A., Walker, L. R., and Soos, Z. G., Dimensionality of spin fluctuations in highly anisotropic TCNQ salts, *Journal of Chemical Physics* **64**(9), 3592–3601, 1976.
163. Krasicky, P. D., Silsbee, R. H., and Scott, J. C., Studies of a polymeric chromium phosphinate—Electron-spin resonance and spin dynamics, *Physical Review B* **25**(9), 5607–5626, 1982.
164. Hennessy, M. J., McElwee, C. D., and Richards, P. M., Effect of interchain coupling on electron-spin resonance in nearly one-dimensional systems, *Physical Review B* **7**(3), 930–947, 1973.

165. Wang, Z. H., Ray, A., MacDiarmid, A. G., and Epstein, A. J., Electron localization and charge transport in poly(o-toluidine)—A model polyaniline derivative, *Physical Review B* **43**(5), 4373–4384, 1991.

166. Wang, Z. H., Scherr, E. M., MacDiarmid, A. G., and Epstein, A. J., Transport and EPR studies of polyaniline—A quasi-one-dimensional conductor with 3-dimensional metallic states, *Physical Review B* **45**(8), 4190–4202, 1992.

167. Kahol, P. K., Kumar, K. K. S., Geetha, S., and Trivedi, D. C., Effect of dopants on electron localization length in polyaniline, *Synthetic Metals* **139**(2), 191–200, 2003.

168. Weinberger, B. R., Ehrenfreund, E., Pron, A., Heeger, A. J., and MacDiarmid, A. G., Linewidth of cis-PA, *Journal of Chemical Physics* **72**, 4749–4755, 1980.

169. Heeger, A. J. and Schrieffer, J. R., Neutral solitons in polyacetylene—Implication of the ENDOR results, *Solid State Communications* **48**(3), 207–210, 1983.

170. Nechtschein, M., Devreux, F., Genoud, F., Guglielmi, M., and Holczer, K., Magnetic-resonance studies in undoped *trans*-polyacetylene (CH)$_X$. 2, *Physical Review B* **27**(1), 61–78, 1983.

171. Abragam, A., *Principles of Nuclear Magnetism*. Oxford, U.K.: Clarendon Press, 1983, 614p.

172. Slichter, C. P., *Principles of Magnetic Resonance*, vol. 3. Berlin, Germnay: Springer, 1989, 319p.

173. Khazanovich, T. N., Theory of nuclear magnetic relaxation in liquid-phase polymers, *Polymer Science U.S.S.R.* **4**(4), 727–736, 1963.

174. Dyson, F. J., Electron spin resonance absorbtion in metals. II. Theory of electron diffusion and the skin effect, *Physical Review B* **98**(2), 349–359, 1955.

175. Wilamowski, Z., Oczkiewicz, B., Kacman, P., and Blinowski, J., Asymmetry of the EPR absorption line in CdF$_2$ with Y, *Physica Status Solidi B* **134**(1), 303–306, 1986.

176. Goldberg, I. B., Crowe, H. R., Newman, P. R., Heeger, A. J., and MacDiarmid, A. G., Electron spine resonance of polyacetylene and AsF$_5$-doped polyacetylene, *Journal of Chemical Physics* **70**, 1132, 1979.

177. Kaeriyama, K., Synthesis and properties of processable polythiophenes, in *Handbook of Organic Conductive Molecules and Polymers*, vol. 2, Nalwa, H. S., Ed. Chichester, NY: Wiley, 1997, Ch. 7, pp. 271–308.

178. Goldenberg, L. M., Pelekh, A. E., Krinichnyi, V. I., Roshchupkina, O. S., Zueva, A. F., Lyubovskaja, R. N., and Efimiv, O. N., Investigation of poly(*p*-phenylene) obtained by electrochemical oxidation of benzene in a BuPyCl-AlCl$_3$ melt, *Synthetic Metals* **36**(2), 217–228, 1990.

179. Monkman, A. P., Bloor, D., and Stevens, G. C., Effects of oxidation-state on the measured g-value in polyaniline, *Journal of Physics D* **23**(5), 627–629, 1990.

180. Iida, M., Asaji, T., Inoue, M. B., and Inoue, M., EPR study of polyaniline perchlorates—Spin species-related to charge-transport, *Synthetic Metals* **55**(1), 607–612, 1993.

181. Krinichnyi, V. I., Nazarova, I. B., Goldenberg, L. M., and Roth, H. K., Spin dynamics in conducting poly(aniline), *Polymer Science, Series A* **40**(8), 835–843, 1998.

182. Krinichnyi, V. I., Chemerisov, S. D., and Lebedev, Y. S., EPR and charge-transport studies of polyaniline, *Physical Review B* **55**(24), 16233–16244, 1997.

183. Krinichnyi, V. I., Chemerisov, S. D., and Lebedev, Y. S., Mechanism of spin and charge transport in poly(aniline), *Polymer Science, Series A* **40**(8), 826–834, 1998.

184. Krinichnyi, V. I., The nature and dynamics of nonlinear excitations in conducting polymers, *Polyaniline, Russian Chemical Bulletin* **49**(2), 207–233, 2000.
185. Krinichnyi, V. I., Roth, H. K., Hinrichsen, G., Lux, F., and Lüders, K., EPR and charge transfer in H_2SO_4-doped polyaniline, *Physical Review B* **65**(15), 155205/01–155205/14, 2002.
186. Kon'kin, A. L., Shtyrlin, V. G., Garipov, R. R., Aganov, A. V., Zakharov, A. V., Krinichnyi, V. I., Adams, P. N., and Monkman, A. P., EPR, charge transport, and spin dynamics in doped polyanilines, *Physical Review B* **66**(7), 075203/01–075203/11, 2002.
187. Krinichnyi, V. I., Konkin, A. L., and Monkman, A., Electron paramagnetic resonance study of spin centers related to charge transport in metallic polyaniline, *Synthetic Metals* **162**(13–14), 1147–1155, 2012.
188. Kahol, P. K., Clark, G. C., and Hehring, M., in *Conjugated Conducting Polymers*, vol. 102, Kiess, H., Ed. Berlin, Germany: Springer-Verlag, 1992, pp. 217–304.
189. Chapman, A. C., Rhodes, P., and Seymour, E. F. W., The effect of eddy currents on nuclear magnetic resonance in metals, *Proceedings of the Physical Society B* **70**(4), 345–360, 1957.
190. Abragam, A. and Bleany, B., *Electron Paramagnetic Resonance of Transition Ions.* Oxford, U.K.: Oxford University Press, 2012, 944p.
191. Lösche, A., *Kerninduction.* Berlin, Germany: VEB Deutscher Verlag, 1957, 684p.
192. Poole, C. P. and Farach, H. A., Eds., *Handbook of Electron Spin Resonance.* New York: Springer-Verlag, 1999, 375p.
193. Bugai, A. A., Passing effects for inhomoneously broadened EPR lines at high-frequency-modulation of magnetic field, *Physics of the Solid State* **4**, 3027–3030, 1962.
194. Salpeter, E. E., Nuclear induction signals for long relaxation times, *Proceedings of the Physical Society A* **63**, 337–349, 1950.
195. Feher, G., Electron spin resonance experiments on donors in silicon. I. Electronic structure of donors by electron nuclear double resonance technique, *Physical Review B* **114**(5), 1219, 1959.
196. Weger, M., Passage effects in paramagnetic resonance experiments, *Bell System Technical Journal* **39**(4), 1013–1112, 1960.
197. Ammerlaan, C. A. J. and van der Wiel, A., The divacancy in silicon: Spin-lattice relaxation and passage effects in electron paramagnetic resonance, *Journal of Magnetic Resonance* **21**(3), 387–513, 1976.
198. Krinichnyi, V. I., Investigation of biological systems by EPR method of high spectral resolution at 2-mm wave band, *Journal of Applied Spectroscopy* **52**(6), 575–591, 1990.
199. Cullis, P. R., EPR in inhomogeneously broadened systems: A spin temperature approach, *Journal of Magnetic Resonance* **21**(3), 397–418, 1976.
200. Pelekh, A. E., Krinichnyi, V. I., Brezgunov, A. Y., Tkachenko, L. I., and Kozub, G. I., ESR study of relaxational parameters of paramagnetic centers in polyacetylene, *Polymer Science U.S.S.R.* **33**(8), 1615–1623, 1991.
201. Krinichnyi, V. I., Pelekh, A. E., Brezgunov, A. Y., Tkachenko, L. I., and Kozub, G. I., The EPR study of undoped polyacetylene, *Materials Science & Engineering* **17**(1), 25–29, 1991.
202. Torrey, H. C., Nuclear spin relaxation by translation diffusion, *Physical Review B* **92**, 962–969, 1953.

203. Dobrynin, A. V., Molecular simulation of charged polymers, in *Simulation Methods for Polymers*, Kotelyanskii, M. and Theodorou, D. N., Eds. New York: Marcel Dekker, 2004, pp. 259–312.

204. Mizoguchi, K., Spin dynamics in conducting polymers, *Makromolekulare Chemie-Macromolecular Symposia* **37**, 53–65, 1990.

205. Nechtschein, M., Electron spin dynamics, in *Handbook of Conducting Polymers*, Skotheim, T. A., Elsenbaumer, R. L., and Reynolds, J. R., Eds. New York: Marcel Dekker, 1997, pp. 141–163.

206. Berliner, L. J., Ed., *Spin Labeling I: Theory and Application*. New York: Academic Press, 1976, 583p.

207. Kuznetsov, A. N., *Spin Probe Method (Russ)*. Moscow, Russia: Nauka, 1976, 210p.

208. Berliner, L. J. and Reuben, J., Eds., *Spin Labeling—Theory and Applications*, Biological Magnetic Resonance 8. New York: Plenum Press, 1989, 670p.

209. Bravaya, N. M. and Pomogailo, A. D., Spin labels as the instrument for analysis of topochemistry of polymer-immobilized Ziegler catalytic systems, *Journal of Inorganic and Organometallic Polymers* **10**(1), 1–22, 2000.

210. Winter, H., Sachs, G., Dormann, E., Cosmo, R., and Naarmann, H., Magnetic-properties of spin-labeled polyacetylene, *Synthetic Metals* **36**(3), 353–365, 1990.

211. Goldman, G. A., Bruno, G. V., and Freed, J. H., Estimating slow-motional rotational correlation times for nitroxides by electron spin resonance, *Journal of Physical Chemistry* **76**, 1858–1860, 1972.

212. Wasserman, A. M. and Kovarski, A. L., *Spin Labels and Probes in Physics and Chemistry of Polymers (Russian)*. Moscow, Russia: Nauka, 1986, 246p.

213. Robinson, B. H. and Dalton, L. R., Anisotropic rotational diffusion studied by passage saturation transfer electron paramagnetic resonance, *Journal of Chemical Physics* **72**, 1312–1324, 1980.

214. Dubinski, A. A., Grinberg, O. Y., Kurochkin, V. I., Oransky, L. G., Poluektov, O. G., and Lebedev, Y. S., Investigation of anisotropy of nitroxide radicals rotation using 2-mm wave band EPR spectra, *Theoretical and Experimental Chemistry* **17**(2), 231–236, 1981.

215. Krinichnyi, V. I., Grinberg, O. Y., Dubinskii, A. A., Livshits, V. A., Bobrov, Y. A., and Lebedev, Y. S., Study of anisotropic molecular rotations by saturation transfer EPR spectroscopy at 2-mm wave band, *Biofizika* **32**(3), 534–535, 1987.

216. Krinichnyi, V. I., Lebedev, Y. S., and Grinberg, O. Y., Study of viscous liquids by spin micro- and macroprobe methods at 2-mm wave band EPR, *Applied Magnetic Resonance* **13**(1–2), 259–267, 1997.

217. Klinger, M. I., *Problems of Linear Electron (Polaron) Transport Theory in Semiconductors*. Oxford, U.K.: Pergamon Press, 1979, 962p.

218. Pope, M. and Swenberg, C. E., *Electronic Processes in Organic Crystals and Polymers*, Oxford, U.K.: Oxford University Press, 1999, 1360p.

219. Kivelson, S., Electron hopping in a soliton band—Conduction in lightly doped $(CH)_x$, *Physical Review B* **25**(6), 3798–3821, 1982.

220. Kivelson, S., Electron hopping conduction in the soliton model of polyacetylene, *Molecular Crystals and Liquid Crystals* **77**(1–4), 65–79, 1981.

221. Kivelson, S., Electron hopping conduction in the soliton model of polyacetylene, *Physical Review Letters* **46**, 1344–1346, 1981.

222. Zuppiroli, L., Paschen, S., and Bussac, M. N., Role of the dopant counterions in the transport and magnetic-properties of disordered conducting polymers, *Synthetic Metals* **69**(1–3), 621–624, 1995.

223. Tawansi, A., Oraby, A. H., Zidan, H. M., and Dorgham, M. E., Effect of one-dimensional phenomena on electrical, magnetic and ESR properties of $MnCl_2$-filled PVA films, *Physica B* **254**(1–2), 126–133, 1998.

224. Long, A. R. and Balkan, N., AC loss in amorphous germanium, *Philosophical Magazine B* **41**(3), 287–305, 1980.

225. El Kadiri, M. and Parneix, J. P., Frequency- and temperature-dependent complex conductivity of some conducting polymers, in *Electronic Properties of Polymers and Related Compounds*, vol. 63, Kuzmany, H., Mehring, M., and Roth, S., Eds. Berlin, Germany: Springer-Verlag, 1985, pp. 183–186.

226. Parneix, J. P. and El Kadiri, M., Frequency- and temperature-dependent dielectric losses in lightly doped conducting polymers, in *Electronic Properties of Conjugated Polymers*, vol. 76, Kuzmany, H., Mehring, M., and Roth, S., Eds. Berlin, Germany: Springer-Verlag, 1987, pp. 23–26.

227. Schafer-Siebert, D., Budrowski, C., Kuzmany, H., and Roth, S., Influence of the conjugation length of polyacetylene chains on the DC-conductivity, in *Electronic Properties of Conjugated Polymers*, vol. 76, Kuzmany, H., Mehring, M., and Roth, S., Eds. Berlin: Springer-Verlag, 1987, pp. 38–42.

228. Schafer-Siebert, D. and Roth, S., Limitation of the conductivity of polyacetylene by conjugational defects, *Synthetic Metals* **28**(3), D369–D374, 1989.

229. Meier, H., Application of the semiconductor properties of dyes: Possibilities and problems, in *Topics in Current Chemistry*, vol. 61, Bayley, H., Houk, K. N., Hughes, G., Hunter, C. A., Ishihara, K., Krische, M. J., Lehn, J.-M. et al., Eds. Berlin, Germnay: Springer-Verlag, 1976, pp. 85–131.

230. Jozefowicz, M. E., Laversanne, R., Javadi, H. H. S., Epstein, A. J., Pouget, J. P., Tang, X., and MacDiarmid, A. G., Multiple lattice phases and polaron-lattice spinless-defect competition in polyaniline, *Physical Review B* **39**(17), 12958–12961, 1989.

231. Dos Santos, D. A., Galvao, D. S., Laks, B., and dos Santos, M. C., Poly(alkylthiophenes): Chain conformation and thermochromism, *Synthetic Metals* **51**, 203–209, 1992.

232. Hoogmartens, I., Adreansens, P., Carleer, R., Vanderzande, D., Martens, H., and Gelan, J., An investigation into the electronic structure of poly(isothianaphthene), *Synthetic Metals* **51**(1–3), 219–228, 1992.

233. Kwezera, P., Dresselhaus, M. S., and Adler, D., Electrical properties of $NiS_{(2-x)}Se_{(x)}$, *Physical Review B* **21**(6), 2328–2335, 1980.

234. Masse, M. A., Schlenoff, J. B., Karasz, F. E., and Thomas, E. L., Crystalline phases of electrically conductive poly(p-phenylene vinylene), *Journal of Polymer Science—Polymer Physics Edition* **27**(10), 2045–2059, 1989.

235. Kuntscher, S. A., van der Marel, D., Dressel, M., Lichtenberg, F., and Mannhart, J., Signatures of polaronic excitations in quasi-one-dimensional $LaTiO_{3.41}$, *Physical Review B* **67**(15), 035105/01–035105/05, 2003.

236. Friedman, L. and Holstein, T., Studies of polaron motion. Part III: The Hall mobility of the small polaron, *Annals of Physics* **21**(3), 494–549, 1963.

237. Nagels, P., Experimental Hall effect data for a small-polaron semiconductor, in *The Hall Effect and Its Application*, Chien, C. L. and Westgate, C. R., Eds. New York: Plenum Press, 1980, pp. 253–280.

238. Mott, N. F., *Conduction in Non-Crystalline Materials*. Oxford, U.K.: Clarendon Press, 1987, 150p.

239. Mott, N. F. and Davis, E. A., *Electronic Processes in Non-Crystalline Materials*. Oxford, U.K.: Clarendon Press, 2012, 608p.

240. Austin, I. G. and Mott, N. F., Polarons in crystalline and non-crystalline materials, *Advances in Physics* **18**(71), 41–102, 1969.

241. Sheng, P., Abeles, B., and Arie, Y., Hopping conductivity in granular metals, *Physical Review Letters* **31**(1), 44–47, 1973.

242. Sheng, P., Fluctuation-induced tunneling conduction in disordered materials, *Physical Review B* **21**, 2180, 1980.

243. Sheng, P. and Klafter, J., Hopping conductivity in granular disordered systems, *Physical Review B* **27**(4), 2583, 1983.

244. Sheng, P., Sichel, E. K., and Gittleman, J. I., Fluctuation-induced tunneling conduction in carbon-polyvinylchloride composites, *Physical Review Letters* **40**(18), 1197–1200, 1978.

245. Pietronero, L., Ideal conductivity of carbon π-polymers and intercalation compounds, *Synthetic Metals* **8**, 225–231, 1983.

246. Kivelson, S. and Heeger, A. J., Intrinsic conductivity of conducting polymers, *Synthetic Metals* **22**(4), 371–384, 1988.

247. Fishchuk, I. I., Kadashchuk, A. K., Bässler, H., and Weiss, D. S., Nondispersive charge-carrier transport in disordered organic materials containing traps, *Physical Review B* **66**(20), 205208/01–205208/12, 2002.

248. Elliott, R. J., Effect of spin-orbit coupling on paramagnetic resonance in semiconductors, *Physical Review* **96**, 266–279, 1954.

249. Beuneu, F. and Monod, P., The Elliott relation in pure metals, *Physical Review B* **18**, 2422–2425, 1978.

250. Zuo, F., Angelopoulos, M., MacDiarmid, A. G., and Epstein, A. J., Transport studies of protonated emeraldine polymer—A antigranulocytes polymeric metal system, *Physical Review B* **36**(6), 3475–3478, 1987.

251. MacDiarmid, A. G. and Epstein, A. J., Polyanilines—A novel class of conducting polymers, *Faraday Discussions* **88**, 317–332, 1989.

252. Epstein, A. J. and MacDiarmid, A. G., Novel concepts in electronic polymers—Polyaniline and its derivatives, *Makromolekulare Chemie-Macromolecular Symposia* **51**, 217–234, 1991.

253. Pouget, J. P., Jozefowicz, M. E., Epstein, A. J., Tang, X., and MacDiarmid, A. G., X-ray structure of polyaniline, *Macromolecules* **24**(3), 779–789, 1991.

254. Lee, K. and Heeger, A. J., Optical reflectance studies of conducting polymers on the metal-insulator boundary, *Synthetic Metals* **84**(1–3), 715–718, 1997.

255. Anderson, P. W., The Fermi glass: Theory and experiment, *Comments on Solid State Physics* **2**, 193–198, 1970.

256. Brustolon, M. and Giamello, E., Eds., *Electron Paramagnetic Resonance: A Practitioners Toolkit*. Hoboken, NJ: Wiley, 2009, 560p.

257. Herlach, F., Ed., *Strong and Ultrastrong Magnetic Fields and Their Applications*, Topics in Applied Physics. Berlin, Germany: Springer-Verlag, 1985, 362p.

258. Grinberg, O. Y., Dubinskii, A. A., Shuvalov, V. F., Oransky, L. G., Kurochkin, V. I., and Lebedev, Y. S., Submillimeter EPR spectroscopy of free radicals, *Doklady Akademii Nauk SSSR* **230**, 884–886, 1976.

259. Kawamori, A., Yamauchi, J., and Ohta, H., Eds., *EPR in the 21st Century: Basics and Applications to Material, Life and Earth Sciences.* Amsterdam, the Netherlands: Elsevier Science B.V., 2002, 830p.

260. Galkin, A. A., Grinberg, O. Y., Dubinskii, A. A., Kabdin, N. N., Krymov, V. N., Kurochkin, V. I., Lebedev, Y. S., Oransky, L. G., and Shuvalov, V. F., EPR spectrometer in 2-mm range for chemical research, *Instruments and Experimental Techniques* **20**(4), 1229–1229, 1977.

261. Barra, A. L., Grund, L. C., and Robert, J. B., EPR spectroscopy at very high field, *Chemical Physics Letters* **165**(1), 107–109, 1990.

262. Krinichnyi, V. I., Pelekh, A. E., Lebedev, Y. S., Tkachenko, L. I., Kozub, G. I., Barra, A. L., Brunel, L. C., and Robert, J. B., Very high field EPR study of cis-polyacetylene and trans-polyacetylene, *Applied Magnetic Resonance* **7**(4), 459–467, 1994.

263. Kuska, H. A. and Rogers, M. T., Electron spin resonance of first row transition metal complex ions, in *Radical Ions*, Kaiser, E. T. and Kevan, L., Eds. New York: John Wiley & Sons, 1968, pp. 579–745.

264. Krinichnyi, V. I., Yudanova, E. I., and Wessling, B., Influence of spin–spin exchange on charge transfer in PANI-ES/P3DDT/PCBM composite, *Synthetic Metals* **179**, 67–73, 2013.

265. Kolaczkowski, S. V., Cardin, J. T., and Budil, D. E., Some remarks on reported inconsistencies in the high-field EPR spectrum of DPPH, *Applied Magnetic Resonance* **16**(2), 293–298, 1999.

266. Chen, O., Zhuang, J., Guzzetta, F., Lynch, J., Angerhofer, A., and Cao, Y. C., Synthesis of Water-soluble 2,2′-diphenyl-1-picrylhydrazyl nanoparticles: A new standard for electron paramagnetic resonance spectroscopy, *Journal of the American Chemical Society* **131**(35), 12542–12543, 2009.

267. Orlinski, S. B., Schmidt, J., Mokhov, E. N., and Baranov, P. G., Silicon and carbon vacancies in neutron-irradiated SiC: A high-field electron paramagnetic resonance study, *Physical Review B* **67**(12), 125207/01–125207/08, 2003.

268. Wind, R. A., Bai, S., Hu, J. Z., Solum, M. S., Ellis, P. D., Grant, D. M., Pugmire, R. J., Taylor, C. M. V., and Yonker, C. R., H¹ dynamic nuclear polarization in supercritical ethylene at 1.4 T, *Journal of Magnetic Resonance* **143**, 233–239, 2000.

269. Surján, P. R., Charge vs. spin density waves in the fullerene polymer, *International Journal of Quantum Chemistry* **63**(2), 425–435, 1997.

270. Harneit, W., Fullerene-based electron-spin quantum computer, *Physical Review A* **65**(3), 032322/01–032322/06, 2002.

271. Stesmans, A. and van Gorp, G., Novel method for accurate g measurements in electron-spin resonance, *Review of Scientific Instruments* **60**, 2949–2952, 1989.

272. Riekel, C., Hasslin, H. W., Menke, K., and Roth, S., Crystalline features in AsF₅ doped PA, *Journal of Chemical Physics* **77**(8), 4254–4255, 1982.

273. Hasslin, H. W., Riekel, C., Menke, K., and Roth, S., A neutron diffraction study on the doping of polyacetylene by AsF₅, *Macromolecular Chemistry and Physics* **185**(2), 397–416, 1984.

274. Pradere, P. and Boudet, A., Influence of the mode of synthesis on the morphology and structure of polyparaphenylene, *Journal of Materials Science* **22**(12), 4240–4246, 1987.

275. Kovacic, P., Feldman, M. B., Kovacic, J. P., and Lando, J. B., Polymerization of aromatic nuclei. XIII. X-ray analysis and properties of poly(p-phenylene) pellets, *Journal of Applied Polymer Science* **12**(7), 1735–1743, 1968.
276. Tabor, B. J., Magre, E. P., and Boom, J., The crystal structure of poly-*p*-phenylene sulphide, *European Polymer Journal* **7**, 1127–1133, 1971.
277. Giess, R. H., Street, G. B., Volksen, W., and Economy, J., Polymer structure determination using electron diffraction techniques, *IBM Journal of Research and Development* **27**(4), 321–329, 1983.
278. Street, G. B., Clarke, T. C., Giess, R. H., Lee, V. Y., Nazzal, A., Pfluger, P., and Scott, J. C., Characterization of polypyrrole, *Journal de Physique Colloques* **44**, C3-599–C3-606, 1983.
279. Lee, K., Heeger, A. J., and Cao, Y., Reflectance spectra of polyaniline, *Synthetic Metals* **72**(1), 25–34, 1995.
280. Yang, C. Y., Smith, P., Heeger, A. J., Cao, Y., and Osterholm, J. E., Electron-diffraction studies of the structure of polyaniline dodecylbenzenesulfonate, *Polymer* **35**(6), 1142–1147, 1994.
281. Wessling, B., Srinivasan, D., Rangarajan, G., Mietzner, T., and Lennartz, W., Dispersion-induced insulator-to-metal transition in polyaniline, *European Physical Journal E* **2**(3), 207–210, 2000.
282. Jozefowicz, M. E., Epstein, A. J., Pouget, J. P., Masters, J. G., Ray, A., and MacDiarmid, A. G., X-ray structure of the polyaniline derivative poly(ortho-toluidine)—The structural origin of charge localization, *Macromolecules* **24**(21), 5863–5866, 1991.
283. Jozefowicz, M. E., Epstein, A. J., Pouget, J. P., Masters, J. G., Ray, A., Sun, Y., Tang, X., and MacDiarmid, A. G., X-ray structure of polyanilines, *Synthetic Metals* **41**(1–2), 723–726, 1991.
284. Gupta, M. C. and Umare, S. S., Studies on poly(ortho-methoxyaniline), *Macromolecules* **25**(1), 138–142, 1992.
285. Roy, B. C., Gupta, M. D., and Ray, J. K., Studies on conducting polymers. 1. Aniline-initiated polymerization of nitroanilines, *Macromolecules* **28**(6), 1727–1732, 1995.
286. Patzsch, J., Elektrische Eigenschaften von polymeren Tetrathiafulvalen (PTTF), PhD thesis, Technische Hochschule Leipzig, Leipzig, Germany, 1991.
287. Łuźny, W., Trznadel, M., and Proń, A., X-ray diffraction study of reoregular poly(3-alkylthiophenes), *Synthetic Metals* **81** 71–74, 1996.
288. Tashiro, K., Ono, K., Minagawa, Y., Kobayashi, M., Kawai, T., and Yoshino, K., Structure and thermochromic solid-state phase-transition of poly(3-alkylthiophene), *Journal of Polymer Science Part B—Polymer Physics* **29**(10), 1223–1233, 1991.
289. Tashiro, K., Kobayashi, M., Kawai, T., and Yoshino, K., Crystal structural change in poly(3-alkyl thiophene)s induced by iodine doping as studied by an organized combination of x-ray diffraction, infrared/Raman spectroscopy and computer simulation techniques, *Polymer* **38**(12), 2867–2879, 1997.
290. Samuelsen, E. J. and Mardalen, J., Structure of polythiophenes, in *Handbook of Organic Conductive Molecules and Polymers*, vol. 3, Nalwa, H. S., Ed. Chichester, U.K.: John Wiley, 1997, pp. 87–120.
291. Lu, X. H., Hlaing, H., Germack, D. S., Peet, J., Jo, W. H., Andrienko, D., Kremer, K., and Ocko, B. M., Bilayer order in a polycarbazole-conjugated polymer, *Nature Communications* **3**, 1290/01–1290/07, 2012.

292. Curran, S., Stark-Hauser, A., and Roth, S., Polyacetylene, in *Handbook of Organic Conductive Molecules and Polymers*, vol. 2, Nalwa, H. S., Ed. New York: John Wiley & Sons, 1997, pp. 1–59.

293. Kuroda, S., ESR and ENDOR studies of solitons and polarons in conjugated polymers, *Applied Magnetic Resonance* **23**(3–4), 455–468, 2003.

294. Trhlikova, O., Zednik, J., Matejicek, P., Horacek, M., and Sedlacek, J., Degradation and cis-to-trans isomerization of poly(2,4-difluorophenyl)acetylene s of various initial molecular weight: SEC, NMR, DLS and EPR study, *Polymer Degradation and Stability* **98**(9), 1814–1826, 2013.

295. Aoki, Y., Fujita, S., Haramizu, S., Akagi, K., and Shirakawa, H., Current progress in synthesis of polyacetylene films, *Synthetic Metals* **84**(1–3), 307–310, 1997.

296. Zuravleva, T. S., Studies of polyacetylene by magnetic resonance methods (Russ.), *Russian Chemical Reviews* **56**(1), 128–147, 1987.

297. Bartl, A., Frohner, J., Zuzok, R., and Roth, S., Characterization of segmented and highly oriented polyacetylene by electron spin resonance, *Synthetic Metals* **51**(1–3), 197–201, 1992.

298. Rachdi, F. and Bernier, P., ESR study of metallic complexes of alkali-doped polyacetylene, in *Electronic Properties of Conjugated Polymers*, vol. 76, Kuzmany, H., Mehring, M., and Roth, S., Eds. Berlin, Germany: Springer-Verlag, 1987, pp. 160–164.

299. Leising, G., Kahlert, H., and Leitner, O., Intrinsic anisotropic properties of *trans*-(CH)$_x$, in *Electronic Properties of Polymers and Related Compounds*, vol. 63, Kuzmany, H., Mehring, M., and Roth, S., Eds. Berlin, Germany: Springer-Verlag, 1985, pp. 56–62.

300. Mizoguchi, K., Kume, K., Masubuchi, S., and Shirakawa, H., Characterization of neutral soliton dynamics in pristine *trans*-polyacetymene by means of anisotropic electron-spin-resonance T_1 and linewidth, *Solid State Communications* **59**(7), 465–468, 1986.

301. Mizoguchi, K., Kume, K., Masubuchi, S., and Shirakawa, H., Neutral soliton diffusion and anisotropic T_1 and T_2 of electron-spin-resonance and NMR in stretch oriented tert-polyacetylene, *Synthetic Metals* **17**(1–3), 405–411, 1987.

302. Mizoguchi, K., Komukai, S., Tsukamoto, T., Kume, K., Suezaki, M., Akagi, K., and Shirakawa, H., Spin dynamics in undoped *trans*-(CH)$_x$, *Synthetic Metals* **28**(3), D393–D398, 1989.

303. Brazovskii, S. A., Electron excitations in the Peierls-Frohlich state, *JETP Letters* **28**(10), 606–610, 1978.

304. Brazovskii, S. A. and Kirova, N. N., Excitons, polarons, and bipolarons in conducting polymers, *JETP Letters* **33**(1), 4–8, 1981.

305. Tomkiewicz, Y., Schultz, T. D., Broom, H. B., Clarke, T. C., and Street, G. B., Evidence against solitons in PA: Magnetic measurements, *Physical Review Letters* **43**, 1532–1536, 1979.

306. Nechtschein, M., Devreux, F., Greene, R. L., Clarke, T. C., and Street, G. B., One-Dimensional spin diffusion in polyacetylene, (CH)$_x$, *Physical Review Letters* **44**(5), 356–359, 1980.

307. Nechtschein, M., NMR and electron-spin-resonance studies of polyacetylene, *Bulletin of the American Physical Society* **26**(3), 231–232, 1981.

308. Clarke, T. C. and Scott, J. C., NMR studies of polyacetylene, in *Handbook of Conducting Polymers*, vol. 2, Skotheim, T. A., Ed. New York: Marcel Dekker, Inc., 1986, pp. 1127–1156.

309. Masin, F., Gusman, G., and Deltour, R., NMR spin-lattice relaxation of H^1 in iodine doped polyacetylene, *Solid State Communications* **40**(4), 513–515, 1981.

310. Ziliox, M., Spegt, P., Mathis, C., Francois, B., and Weill, G., Spin diffusion in the nuclear-magnetic relaxation of isotopically labeled $(CH)_x$, *Solid State Communications* **51**(6), 393–396, 1984.

311. Mizoguchi, K., Kume, K., and Shirakawa, H., Frequency-dependence of electron spin-lattice relaxation rate at 5–450 MHz in pristine *trans*-polyacetylene— New evidence of one dimensional motion of electron-spin (neutral soliton), *Solid State Communications* **50**(3), 213–218, 1984.

312. Mizoguchi, K., Masubuchi, S., Kume, K., Akagi, K., and Shirakawa, H., Neutral soliton dynamics in t-polyacetylene—Consistent picture for magnetic resonances, *Synthetic Metals* **84**(1–3), 865–866, 1997.

313. Shiren, N. S., Tomkiewicz, Y., Kazyaka, T. G., Taranko, A. R., Thomann, H., Dalton, L., and Clarke, T. C., Spin dynamics in *trans*-polyacetylene, *Solid State Communications* **44**(8), 1157–1160, 1982.

314. Tomkiewicz, Y., Shiren, N. S., Schultz, T. D., Thomann, H., Dalton, L. R., Zettl, A., Gruner, G., and Clarke, T. C., Spin and charge excitation in polyacetylene, *Molecular Crystals and Liquid Crystals* **83**(1–4), 1049–1063, 1982.

315. Dalton, L. R., Thomann, H., Morrobelsosa, A., Chiu, C., Galvin, M. E., Wnek, G. E., Tomkiewicz, Y., Shiren, N. S., Robinson, B. H., and Kwiram, A. L., Study of polyacetylene and composites of polyacetylene polyethylene by Electron Nuclear Double-Resonance, Electron Nuclear Nuclear Triple Resonance, and Electron-Spin Echo Spectroscopies, *Journal of Applied Physics* **54**(10), 5583–5591, 1983.

316. Chien, J. C. W., Wnek, G. E., Karasz, F. E., Warakomski, J. M., Dickinson, L. C., Heeger, A. J., and MacDiarmid, A. G., Electron-paramagnetic resonance saturation characteristics of pristine and doped polyacetylenes, *Macromolecules* **15**(2), 614–621, 1982.

317. Reichenbach, J., Kaiser, M., Anders, J., Burne, H., and Roth, S., Picosecond photoconductivity in $(CH)_x$, *Synthetic Metals* **51**(1–3), 245–250, 1992.

318. Holczer, K., Boucher, J. P., Devreux, F., and Nechtschein, M., Magnetic-resonance studies in undoped trans-polyacetylene, $(CH)_x$, *Physical Review B* **23**(3), 1051–1063, 1981.

319. Wang, Z. H., Theophilou, N., Swanson, D. B., MacDiarmid, A. G., and Epstein, A. J., EPR of Naarmann-Theophilou polyacetylene—Critical role of interchain interactions, *Physical Review B* **44**(21), 12070–12073, 1991.

320. Mizoguchi, K., Sakurai, H., Shimizu, F., Masubuchi, S., and Kume, K., Comparative-study on Shirakawa polyacetylene and Naarmann-Theophilou-type polyacetylene—Neutral soliton dynamics, *Synthetic Metals* **68**(3), 239–242, 1995.

321. Wada, Y. and Schrieffer, J. R., Brownian motion of a domain wall and the diffusion constants, *Physical Review B* **18**(8), 3897–3912, 1978.

322. Maki, K., The soliton diffusion in polyacetylene. I, II, *Physical Review B*, **26**(4), 2181–2191, 1982.

323. Terai, A. and Ono, Y., Motion of solitons in one-dimensional psin-density-wave systems, *Synthetic Metals* **57**(2–3), 4672–4677, 1993.

324. Epstein, A. J., AC conductivity of polyacetylene: Distinguishing mechanisms of charge transport, in *Handbook of Conducting Polymers*, vol. 2, Scotheim, T. E., Ed. New York: Marcel Dekker, 1986, pp. 1041–1097.

325. Foxonet, N., Bernier, P., and Voit, J., Conductivity of N-doped polyacetylene—Dependence of doping level and temperature, *Journal de Chimie Physique et de Physico-Chimie Biologique* **89**(5), 977–986, 1992.

326. Chen, X. and Deng, W. G., Carrier transport in polyacetylene with intermediate doping, *Physical Review B* **49**(5), 3155–3160, 1994.

327. Anglada, M. C., Ferreranglada, N., Ribo, J. M., and Movaghar, B., Fluctuation-assisted transport in polymers, *Synthetic Metals* **78**(2), 169–176, 1996.

328. Graeff, C. F. O., Brandt, M. S., Faria, R. M., and Leising, G., Electrically detected magnetic resonance in undoped polyacetylene and polyaniline, *Physica Status Solidi A* **162**(2), 713–721, 1997.

329. Tang, J., Norris, J. R., and Shirakawa, H., EPR lineshape analysis of one-dimensional soliton diffusion in trans-polyacetylene, *Journal of Physics and Chemistry of Solids* **58**(3), 475–480, 1997.

330. Kirtman, B., Hasan, M., and Chipman, D. M., Solitons in polyacetylene—Magnetic hyperfine constants from abinitio calculations, *Journal of Chemical Physics* **95**(10), 7698–7716, 1991.

331. Krinichnyi, V. I., Grinberg, O. Y., Nazarova, I. B., Kozub, G. I., Tkachenko, L. I., Khidekel, M. L., and Lebedev, Y. S., Polythiophene and polyacetylene conductors in the 3-cm and 2-mm electron spin resonance bands, *Bulletin of the Academy of Sciences of the USSR, Division of Chemical Science* **34**(2), 425–427, 1985.

332. Pelekh, A. E., Krinichnyi, V. I., Tkachenko, L. I., and Kozub, G. I., The study of undoped polyacetylene by means of fast running at the 2-mm wave band of ESR, *Synthetic Metals* **41**(1–2), 111–111, 1991.

333. Shirakawa, H. and Ikeda, S., Infrared spectra of poly(acetylene), *Polymer Journal* **2**, 231–244, 1971.

334. Ito, T., Shirakawa, H., and Ikeda, S., Simultaneous polymerization and formation of polyacetylene film on the surface of concentrated soluble Ziegler-type catalyst solution, *Journal of Polymer Science Part A—Polymer Chemistry* **12**, 11–20, 1974.

335. Traven', V. F., *Electronic Structure and Properties of Organic Molecules (Russian)*. Moscow, Russia: Khimija, 1989, 384p.

336. Poole, C. P., *Electron Spin Resonance: A Comprehensive Treatise on Experimental Techniques*. New York: Dover Publications, 1997, 810p.

337. Kivshar, Y. S. and Malomed, B. A., Dynamics of solitons in nearly integrable systems, *Reviews of Modern Physics* **61**(4), 763–915, 1989.

338. Tkachenko, L. I., Makova, M. K., Krinichnyi, V. I., Roshchupkina, O. S., Kozub, G. I., Khrapova, I. M., and Belov, G. P., Synthesis and properties of polyacetylene formed using Ti(OBu) 4-polyalumoxane catalysts, *Vysokomolekulyarnye Soedineniya, Seriya A & Seriya B* **35**(9), 1409–1413, 1993.

339. Kurzin, S. P., Tarasov, B. G., Fatkullin, N. F., and Aseeva, R. M., The electron spin-lattice relaxation in pyrolyzed poly-2-methyl-5-ethinylpyridine, *Polymer Science U.S.S.R, Series A* **24**(1), 134–142, 1982.

340. Krinichnyi, V. I., Tkachenko, L. I., and Kozub, G. I., Investigation of irradiated cis-polyacetylene by 2-mm wave band EPR method, *Khimicheskaya Fizika* **8**(11), 1283–1286, 1989.

341. Krinichnyi, V. I., Pelekh, A. E., Tkachenko, L. I., and Kozub, G. I., Study of spin dynamics in trans-polyacetylene at 2-mm waveband EPR, *Synthetic Metals* **46**(1), 1–12, 1992.

342. Krinichnyi, V. I., Pelekh, A. E., Tkachenko, L. I., and Kozub, G. I., Study of anisotropic spin dynamics in pristine *trans*-polyacetylene by means of 2-mm EPR spectroscopy, *Synthetic Metals* **46**(1), 13–22, 1992.

343. Mank, V. V. and Lebovka, N. I., *NMR Spectroscopy of Water in Heterogenic Systems (Russian).* Kiev, Ukraine: Naukova Dumka, 1988, 204p.

344. Bartenev, G. M. and Frenkel, S. Y., *Physics of Polymers (Russ).* Leningrad, Russia: Khimija, 1990, 432p.

345. Davydov, A. S., *Solitons in Molecular Systems (Russ.).* Kiev, Ukraine: Naukova Dumka, 1984, 288p.

346. Richter, A. M., Richter, J. M., Beye, N., and Fanghänel, E., Organische Elektronenleiter und Vorstufen. V. Synthese von poly(organylthio-acetylenen), *Journal Praktische Chemie* **329**(5), 811–816, 1987.

347. Hempel, G., Richter, A. M., Fanghänel, E., and Schneider, H., NMR-Untersuchungen zu Struktur und Beweglichkeit von substituierten Polyacetylenen, *Acta Polymerica* **41**(10), 522–527, 1990.

348. Roth, H. K., Gruber, H., Fanghänel, E., Richter, A. M., and Horig, W., Laser-induced generation of highly conducting areas in poly(*bis*-alkylthioacetylenes), *Synthetic Metals* **37**, 151–164, 1990.

349. Bleier, H., Roth, S., Shen, Y. Q., and Schafer-Siebert, D., Photoconductivity in *trans*-polyacetylene: Transport and recombination of photogenerated charged excitations, *Physical Review B* **38**, 6031–6040, 1988.

350. Roth, H. K., Krinichnyi, V. I., Schrödner, M., and Stohn, R. I., Electronic properties of laser modified poly(*bis*-alkylthio-acetylene), *Synthetic Metals* **101**(1–3), 832–833, 1999.

351. Krinichnyi, V. I., Roth, H. K., and Schrödner, M., Spin charge carrier dynamics in poly(*bis*-alkylthioacetylene), *Applied Magnetic Resonance* **23**, 1–17, 2002.

352. Krinichnyi, V. I., The nature and dynamics of non-linear excitations in conducting polymers, *Polyacetylene, Russian Chemical Reviews* **65**(1), 81–93, 1996.

353. Krinichnyi, V. I., Relaxation and dynamics of charge carriers in organic polymer semiconductors: Polyacetylene (review), *Physics of the Solid State* **39**(1), 1–13, 1997.

354. Krinichnyi, V. I., The nature and dynamics of nonlinear excitations in conducting polymers. Heteroaromatic polymers, *Russian Chemical Reviews* **65**(6), 521–536, 1996.

355. Roth, H. K., Brunner, W., Volkel, G., Schrödner, M., and Gruber, H., ESR and ESE studies on polymer semiconductors of weakly doped poly(tetrathiafulvalene), *Makromolekulare Chemie-Macromolecular Symposia* **34**, 293–307, 1990.

356. Roth, H. K. and Krinichnyi, V. I., ESR studies on polymers with particular electronic and magnetic properties, *Makromolekulare Chemie-Macromolecular Symposia* **72**, 143–159, 1993.

357. Cameron, T. S., Haddon, R. C., Mattar, S. M., Parsons, S., Passmore, J., and Ramirez, A. P., The synthsis, characterization, x-ray crystal-structure and solution ESR-spectrum of the paramagnetic solid, 4,5-bis(trifluoromethyl)-1,2,3-trithiolium hexafluorarsenate—Implication for the identity of 1,2-dithiete cations, *Journal of the Chemical Society-Chemical Communications* (6), 358–360, 1991. DOI: 10.1039/C39910000358.

358. Cameron, T. S., Haddon, R. C., Mattar, S. M., Parsons, S., Passmore, J., and Ramirez, A. P., Preparation and crystal-structure of the paramagnetic solid $F_3CCSSSCCF_3ASF_6$—Implication for the identity of RCSSCR-bullet$^+$, *Journal of the Chemical Society-Dalton Transactions* (9), 1563–1572, 1992. DOI: 10.1039/DT9920001563.

359. Krinichnyi, V. I., Herrmann, R., Fanghänel, E., Mörke, W., and Lüders, K., Spin relaxation and magnetic properties of benzo-1,2,3-trithiolium radical cations, *Applied Magnetic Resonance* 12(2–3), 317–327, 1997.

360. Masters, J. G., Ginder, J. M., MacDiarmid, A. G., and Epstein, A. J., Thermochromism in the insulating forms of polyaniline—Role of ring-torsional conformation, *Journal of Chemical Physics* 96(6), 4768–4778, 1992.

361. Bock, H., Rittmeyer, P., Krebs, A., Schultz, K., Voss, J., and Kopke, B., Radical ions. 56. One-electron oxidation of 1,2-dithiete derivatives, *Phosphorus Sulfur and Silicon and the Related Elements* 19, 131–136, 1984.

362. Brédas, J. L., Quattrocchi, C., Libert, J., MacDiarmid, A. G., Ginder, J. M., and Epstein, A. J., Influence of ring-torsion dimerization on the band-gap of aromatic conjugated polymers, *Physical Review B* 44(12), 6002–6010, 1991.

363. Iida, M., Asaji, T., Ikeda, R., Inoue, M. B., Inoue, M., and Nakamura, D., Electron-paramagnetic resonance study of intrinsic paramagnetism in poly(aniline trifluoromethanesulfonate), $[(-C_6H_4NH-)(CF_3SO_3)_{0.5}{\cdot}0.5H_2O]_X$, *Journal of Materials Chemistry* 2(3), 357–360, 1992.

364. Barta, P., Niziol, S., Leguennec, P., and Pron, A., Doping-induced magnetic phase-transition in poly(3-alkylthiophenes), *Physical Review B* 50(5), 3016–3024, 1994.

365. Cik, G., Sersen, F., Dlhan, L., Szabo, L., and Bartus, J., Anomaly in magnetic properties of poly(3-alkylthiophene)s depending on alkyl chain length, *Synthetic Metals* 75(1), 43–48, 1995.

366. Kawai, T., Mizobuchi, H., Okazaki, S., Araki, H., and Yoshino, K., Ferromagnetic tendency of TDAE-fullerene-conducting polymer system, *Japanese Journal of Applied Physics Part 2—Letters* 35(5B), L640–L643, 1996.

367. Cik, G., Sersen, F., Dlhan, L., Cerven, I., Stasko, A., and Vegh, D., Study of magnetic properties of copolymer of 3-dodecylthiophene and 2,3-R,R-thieno[3,4-b]pyrazine, *Synthetic Metals* 130(2), 213–220, 2002.

368. Stafstrom, S. and Brédas, J. L., Band-structure calculations for the polaron lattice in the highly doped regime of polyacetylene, polythiophene, and polyaniline, *Molecular Crystals and Liquid Crystals* 160, 405–420, 1988.

369. Stafström, S. and Brédas, J. L., Evolution of the electronic-structure of polyacetylene and polythiophene as a function of doping level and lattice conformation, *Physical Review B* 38(6), 4180–4191, 1988.

370. Ahlskog, M., Menon, R., Heeger, A. J., Noguchi, T., and Ohnishi, T., Metal-insulator transition in oriented poly(p-phenylenevinylene), *Physical Review B* 55(11), 6777–6787, 1997.

371. Chang, Y. H., Lee, K., Kiebooms, R., Aleshin, A., and Heeger, A. J., Reflectance of conducting poly(3,4-ethylenedioxythiophene), *Synthetic Metals* 105(3), 203–206, 1999.

372. Ahlskog, M., Reghu, M., and Heeger, A. J., The temperature dependence of the conductivity in the critical regime of the metal-insulator transition in conducting polymers, *Journal of Physics—Condensed Matter* 9(20), 4145–4156, 1997.

373. Krinichnyi, V. I., Pelekh, A. E., Roth, H. K., and Lüders, K., Spin relaxation studies on conducting poly(tetrathiafulvalene), *Applied Magnetic Resonance* **4**(3), 345–356, 1993.

374. Brédas, J. L. and Chance, R. R., Eds., *Conjugated Polymeric Materials: Opportunities in Electronics, Optoelectronics, and Molecular Electronics*, NATO Advanced Study Series. Dordrecht, the Netherlands: Kluwer Academic Publishers, 1990, 608p.

375. Nalwa, H. S., Ed., *Advanced Functional Molecules and Polymers: Electronic and Photonic Properties*, vol. 3. Boca Raton, FL: CRC Press, 2001, 386p.

376. Zhou, W. G., Xu, J. K., Du, Y. K., and Yang, P., Electrochemical polymerization of p-terphenyl in mixed electrolyte of boron trifluoride diethyl etherate and CH_2Cl_2, *Journal of Applied Polymer Science* **117**(5), 2688–2694, 2010.

377. Matthews, M. J., Dresselhaus, M. S., Kobayashi, N., Enoki, T., Endo, M., and Nishimura, K., Localized spins in partially carbonized polyparaphenylene, *Physical Review B* **60**(7), 4749–4757, 1999.

378. Xie, S. J., Mei, L. M., and Lin, D. L., Transitio between bipolaron and polaron states in doped heterocycle polymers, *Physical Review B* **50**(18), 13364–13370, 1994.

379. Lacaze, P. C., Aeiyach, S., and Lacroix, J. C., Poly(p-phenylenes): Preparation, techniques and properties, in *Handbook of Organic Conductive Molecules and Polymers*, vol. 2, Nalwa, H. S., Ed. Chichester, U.K.: Wiley, 1997, pp. 205–270.

380. Goldenberg, L. M., Pelekh, A. E., Krinichnyi, V. I., Roshchupkina, O. S., Zueva, A. F., Lyubovskaja, R. N., and Efimiv, O. N., Investigation of poly(*para*-phenylene) obtained in electrochemical oxidation of benzene in the BuPyCl-$AlCl_3$ melt and in organic solvents with oleum, *Synthetic Metals* **43**(1–2), 3071–3074, 1991.

381. Dubois, M., Merlin, A., and Billaud, D., Electron spin resonance in lithium and sodium electrochemically intercalated poly(paraphenylene), *Solid State Communications* **111**(10), 571–576, 1999.

382. Kuivalainen, P., Stubb, H., Raatikainen, P., and Holmstrom, C., Electron spin resonance studies of magnetic defects in $FeCl_3$-doped polyparaphenylene $(p\text{-}C_6H_4)_x$, *Journal de Physique Colloques* **44**, C3-757–C753-760, 1983.

383. Kivelson, S., Hopping conduction and the continuous-time random-walk model, *Physical Review B* **21**(12), 5755–5767, 1980.

384. Rodriguez, J., Grande, H. J., and Otero, T. F., Polypyrroles: From basic research to technological applications, in *Handbook of Organic Conductive Molecules and Polymers*, vol. 2, Nalwa, H. S., Ed. Chichester, U.K.: Wiley, 1997, pp. 415–468.

385. Epstein, A. J., Conducting polymers: Electrical conductivity, in *Physical Properties of Polymers Handbook*, Mark, J. E., Ed. Berlin, Germany: Springer, 2007, pp. 725–744.

386. Saunders, B. R., Fleming, R. J., and Murray, K. S., Recent advances in the physical and spectroscopic properties of polypyrrole films, particularly those containing transition-metal complexes as counteranions, *Chemistry of Materials* **7**(6), 1082–1094, 1995.

387. Schmeisser, D., Bartl, A., Dunsch, L., Naarmann, H., and Gopel, W., Electronic and magnetic properties of polypyrrole films depending on their one-dimensional and two-dimensional microstructures, *Synthetic Metals* **93**(1), 43–58, 1998.

388. Kanemoto, K. and Yamauchi, J., Electron-spin dynamics of polarons in lightly doped polypyrroles, *Physical Review B* **61**(2), 1075–1082, 2000.
389. Taunk, M. and Chand, S., Variable range hopping transport in polypyrrole composite films, in *Physics of Semiconductor Devices*, Jain, V. K. and Verma, A., Eds. Heidelberg, Germany: Springer, 2014, pp. 903–904.
390. Sakamoto, H., Kachi, N., Mizoguchi, K., Yoshioka, K., Masubuchi, S., and Kazama, S., Origin of ESR linewidth for polypyrrole, *Synthetic Metals* **101**(1–3), 481–481, 1999.
391. Kanemoto, K. and Yamauchi, J., ESR broadening in conducting polypyrrole because of oxygen: Application to the study of oxygen adsorption, *Journal of Physical Chemistry B* **105**(11), 2117–2121, 2001.
392. Kanemoto, K. and Yamauchi, J., Doping-induced variation of electron spin relaxation behavior in polypyrroles, *Synthetic Metals* **114**(1), 79–84, 2000.
393. Scott, J. C., Pfluger, P., Kroumbi, M. T., and Street, G. B., Electron-spin-resonanse studies of pyrrole polymers: Evidence for bipolarons, *Physical Review B* **28**(4), 2140–2145, 1983.
394. Chakrabarti, S., Das, B., Banerji, P., Banerjee, D., and Bhattacharya, R., Bipolaron saturation in polypyrrole, *Physical Review B* **60**(11), 7691–7694, 1999.
395. Audebert, P., Binan, G., Lapkowski, M., and Limosin, Grafting, ionomer composites, and auto-doping of conductive polymers, in *Electronic Properties of Conjugated Polymers*, Springer Series in Solid State Sciences, vol. 76, Kuzmany, H., Mehring, M., and Roth, S., Eds. Berlin, Germany: Springer-Verlag, 1987, pp. 366–384.
396. Shchegolikhin, A. N., Yakovleva, I. V., and Motyakin, M., Optical-properties of poly(diacetylene) block-copoly(ether-urethanes), containing covalently bound nitroxyl spin labels in the main-chain, *Synthetic Metals* **71**(1–3), 2091–2092, 1995.
397. Sersen, F., Cik, G., and Veis, P., Study of conformational changes in poly (3-dodecylthiophene) dependent on backbone stereoregularity using a spin-probe technique, *Journal of Applied Polymer Science* **88**(9), 2215–2223, 2003.
398. Pelekh, A. E., Goldenberg, L. M., and Krinichnyi, V. I., Study of dopped poly-pyrrole by the spin probe method at 3-cm and 2-mm waveband EPR, *Synthetic Metals* **44**(2), 205–211, 1991.
399. Buchachenko, A. L., *Complexes of Radicals and Molecular Oxygen with Organic Molecules (Russ)*. Moscow, Russia: Nauka, 1984, 157p.
400. Reddoch, A. and Konishi, S., The solvent effect on di-tert-butyl nitroxide. A dipole-dipole model for polar solutes in polar solvents, *Journal of Chemical Physics* **70**(5), 2121–2130, 1979.
401. Hinh, L. V., Schukat, G., and Fanghänel, E., Tetrathiafulvalene. VII [1]. Arylenverbrückte polymere Tetrathiafulvalene, *Journal fur Praktische Chemie* **321**(2), 299–307, 1979.
402. Trinh, V. Q., Hinh, L. V., Shukat, G., and Fanghänel, E., Tetrathiafulvalene. XXV [1]. Konjugativ verknüpfte polymere Tetrathiafulvalene (TTF), *Journal fur Praktische Chemie* **331**(5), 826–834, 1989.
403. Quang, T., The synthesis of PTTF-THA from 4,5,10,11-tetrahydroantra[1,2-di (5,6-di)]bis(1,3-dithioliumperchlorate), PhD thesis, Carl Schorlemmer Technical University, Merseburg, Germany, 1987.
404. Faridbod, F., Ganjali, M. R., Dinarvand, R., and Norouzi, P., Developments in the field of conducting and non-conducting polymer based potentiometric membrane sensors for ions over the past decade, *Sensors* **8**(4), 2331–2412, 2008.

405. Roth, H. K., Gruber, H., Fanghänel, E., and Quang, T. v., ESR on polymer semi-conductors of poly(tetrathiafulvalene), *Progress in Colloid Polymer Science* **78**, 75–78, 1988.

406. Gruber, H., Roth, H. K., Patzsch, J., and Fanghänel, E., Electrical-properties of poly(tetrathiafulvalenes), *Makromolekulare Chemie-Macromolecular Symposia* **37**, 99–113, 1990.

407. Roth, H. K., Gruber, H., Voelkel, G., Brunner, W., and Fanghänel, E., Electron spin resonance and relaxation studies on conducting poly(tetrathiafulvalenes), *Progress in Colloid and Polymer Science* **80**(1), 254–263, 1989.

408. Patzsch, J. and Gruber, H., Small polarons in polymeric polytetrathiafulvalenes (PTTF), in *Electronic Properties of Polymers*, vol. 107, Kuzmany, H., Mehring, M., and Roth, S., Eds. Berlin, Germany: Springer-Verlag, 1992, pp. 121–124.

409. Krinichnyi, V. I., Denisov, N. N., Roth, H. K., Fanghänel, E., and Lüders, K., Dynamics of paramagnetic charge carriers in poly(tetrathiafulvalene), *Polymer Science, Series A* **40**(12), 1259–1269, 1998.

410. Rånby, B. and Rabek, J. F., *EPR Spectroscopy in Polymer Research*. Berlin, Germany: Springer-Verlag, 1977, 410p.

411. Devreux, F. and Lecavelier, H., Evidence for anomalous diffusion in a conducting polymer, *Physical Review Letters* **59**(22), 2585–2587, 1987.

412. Madhukar, A. and Post, W., Exact solution for the diffusion of a particle in a medium with site diagonal and off-diagonal dynamic disorder, *Physical Review Letters* **39**(22), 1424–1427, 1977.

413. Bhadra, S., *Polyaniline: Preparation, Properties, Processing and Applications*. Saarbrücken, Germany: Lap Lambert Academic Publishing, 2010, 92p.

414. Wiederrecht, G. P., Ed., *Handbook of Nanoscale Optics and Electronics*. Amsterdam, the Netherlands: Elsevier Academic Press, 2010, 401p.

415. Syed, A. A. and Dinesan, M. K., Polyaniline—A nowel polymeric material (review), *Talanta* **38**(8), 815–837, 1991.

416. Bhadra, S., Chattopadhyay, S., Singha, N. K., and Khastgir, D., Improvement of conductivity of electrochemically synthesized polyaniline, *Journal of Applied Polymer Science* **108**(1), 57–64, 2008.

417. Stafstrom, S., Brédas, J. L., Epstein, A. J., Woo, H. S., Tanner, D. B., Huang, W. S., and MacDiarmid, A. G., Polaron lattice in highly conducting polyaniline—Theoretical and optical studies, *Physical Review Letters* **59**(13), 1464–1467, 1987.

418. McCall, R. P., Ginder, J. M., Roe, M. G., Asturias, G. E., Scherr, E. M., MacDiarmid, A. G., and Epstein, A. J., Massive polarons on large-energy gap polymers, *Physical Review B* **39**(14), 10174–10178, 1989.

419. Vignolo, P., Farchioni, R., and Grosso, G., Tight-binding effective Hamiltonians for the electronic states of polyaniline chains, *Physica Status Solidi B—Basic Solid State Physics* **223**(3), 853–866, 2001.

420. Epstein, A. J., MacDiarmid, A. G., and Pouget, J. P., Spin dynamics and conductivity in polyaniline, *Physical Review Letters* **65**(5), 664–664, 1990.

421. Trivedi, D. C., Polyanilines, in *Handbook of Organic Conductive Molecules and Polymers*, vol. 2, Nalwa, H. S., Ed. Chichester, U.K.: John Wiley, 1997, pp. 505–572.

422. Ginder, J. M., Richter, A. F., MacDiarmid, A. G., and Epstein, A. J., Insulator-to-metal transition in polyaniline, *Solid State Communications* **63**(2), 97–101, 1987.

423. Epstein, A. J. and MacDiarmid, A. G., Polaron and bipolaron defects in polymers—Polyaniline, *Journal of Molecular Electronics* **4**(3), 161–165, 1988.

424. Dai, L. M., Lu, J. P., Matthews, B., and Mau, A. W. H., Doping of conducting polymers by sulfonated fullerene derivatives and dendrimers, *Journal of Physical Chemistry B* **102**(21), 4049–4053, 1998.
425. Saprigin, A. V., Brenneman, K. R., Lee, W. P., Long, S. M., Kohlman, R. S., and Epstein, A. J., Li$^+$ doping-induced localization in polyaniline, *Synthetic Metals* **100**(1), 55–59, 1999.
426. Reghu, M., Cao, Y., Moses, D., and Heeger, A. J., Counterion-induced processibility of polyaniline—Transport of the metal-insulator boundary, *Physical Review B* **47**(4), 1758–1764, 1993.
427. Yoon, C. O., Reghu, M., Moses, D., Heeger, A. J., and Cao, Y., Counterion-induced processibility of polyaniline—Thermoelectric power, *Physical Review B* **48**(19), 14080–14084, 1993.
428. Lee, K. H., Heeger, A. J., and Cao, Y., Reflectance of polyaniline protonated with camphor sulfonic acid—Disordered metal on the metal-insulator boundary, *Physical Review B* **48**(20), 14884–14891, 1993.
429. Lee, K., Heeger, A. J., and Cao, Y., Reflectance of conducting polyaniline near the metal-insulator-transition, *Synthetic Metals* **69**(1–3), 261–262, 1995.
430. Holland, E. R., Pomfret, S. J., Adams, P. N., and Monkman, A. P., Conductivity studies of polyaniline doped with CSA, *Journal of Physics—Condensed Matter* **8**(17), 2991–3002, 1996.
431. Abell, L., Adams, P. N., and Monkman, A. P., Electrical conductivity enhancement of predoped polyaniline by stretch orientation, *Polymer* **37**(26), 5927–5931, 1996.
432. Abell, L., Pomfret, S. J., Adams, P. N., Middleton, A. C., and Monkman, A. P., Studies of stretched predoped polyaniline films, *Synthetic Metals* **84**(1–3), 803–804, 1997.
433. Wang, Z. H., Li, C., Scherr, E. M., MacDiarmid, A. G., and Epstein, A. J., 3 Dimensionality of metallic states in conducting polymers—Polyaniline, *Physical Review Letters* **66**(13), 1745–1748, 1991.
434. Monkman, A. P. and Adams, P., Observed anisotropies in stretch oriented polyaniline, *Synthetic Metals* **41**(1–2), 627–633, 1991.
435. Reghu, M., Cao, Y., Moses, D., and Heeger, A. J., Metal-insulator-transition in polyaniline doped with surfactant counterion, *Synthetic Metals* **57**(2–3), 5020–5025, 1993.
436. Adams, P. N., Laughlin, P. J., Monkman, A. P., and Bernhoeft, N., A further step toward stable organic metals—Oriented films of polyaniline with high elrctrical-conductivity and anisotropy, *Solid State Communications* **91**(11), 875–878, 1994.
437. Joo, J., Chung, Y. C., Song, H. G., Baeck, J. S., Lee, W. P., Epstein, A. J., MacDiarmid, A. G., and Jeong, S. K., Charge transport studies of doped polyanilines with various dopants and their mixtures, *Synthetic Metals* **84**(1–3), 739–740, 1997.
438. Adams, P. N., Laughlin, P. J., and Monkman, A. P., Synthesis of high molecular weight polyaniline at low temperatures, *Synthetic Metals* **76**(1–3), 157–160, 1996.
439. Epstein, A. J., Ginder, J. M., Zuo, F., Bigelow, R. W., Woo, H. S., Tanner, D. B., Richter, A. F., Huang, W. S., and MacDiarmid, A. G., Insulator-to-metal transition in polyaniline, *Synthetic Metals* **18**(1–3), 303–309, 1987.
440. Fite, C., Cao, Y., and Heeger, A. J., Magnetic-susceptibility of crystalline polyaniline, *Solid State Communications* **70**(3), 245–247, 1989.

441. Epstein, A. J. and MacDiarmid, A. G., Structure, order, and the metallic state in polyaniline and its derivatives, *Synthetic Metals* **41**(1–2), 601–606, 1991.

442. Cromack, K. R., Jozefowicz, M. E., Ginder, J. M., Epstein, A. J., McCall, R. P., Du, G., Leng, J. M. et al., Thermal-process for orientation of polyaniline films, *Macromolecules* **24**(14), 4157–4161, 1991.

443. Wang, Z. H., Javadi, H. H. S., Ray, A., MacDiarmid, A. G., and Epstein, A. J., Electron localization in polyaniline derivatives, *Physical Review B* **42**(8), 5411–5413, 1990.

444. Grosse, P., *Freie Elektronen in Festkörpern*. Berlin, Germany: Springer-Verlag, 1979, 296p.

445. Joo, J., Oh, E. J., Min, G., MacDiarmid, A. G., and Epstein, A. J., Evolution of the conducting state of polyaniline from localized to mesoscopic metallic to intrinsic metallic regimes, *Synthetic Metals* **69**(1–3), 251–254, 1995.

446. Krinichnyi, V. I., Relaxation and dynamics of spin charge carriers in polyaniline, in *Advances in Materials Science Research*, vol. 17, Wythers, M. C., Ed. Hauppauge, NY: Nova Science Publishers, 2014, pp. 109–160.

447. Menardo, C., Genoud, F., Nechtschein, M., Travers, J. P., and Hani, P., On the acidic functions of polyaniline, in *Electronic Properties of Conjugated Polymers*, Springer Series in Solid State Sciences, vol. 76, Kuzmany, H., Mehring, M., and Roth, S., Eds. Berlin, Germany: Springer-Verlag, 1987, pp. 244–248.

448. Lapkowski, M. and Genies, E. M., Evidence of 2 kinds of spin in polyaniline from insitu EPR and electrochemistry—Influence of the electrolyte-composition, *Journal of Electroanalytical Chemistry* **279**(1–2), 157–168, 1990.

449. MacDiarmid, A. G. and Epstein, A. J., The polyanilines: Potential technology based on new chemistry and new properties, in *Science and Application of Conducting Polymers*, Salaneck, W. R., Clark, D. T., and Samuelsen, E. J., Eds. Bristol, U.K.: Adam Hilger, 1991, pp. 117–128.

450. Sariciftci, N. S., Heeger, A. J., and Cao, Y., Paramagnetic susceptibility of highly conducting polyaniline—Disordered metal with weak electron-electron interactions (Fermi glass), *Physical Review B* **49**(9), 5988–5992, 1994.

451. Sariciftci, N. S., Kolbert, A. C., Cao, Y., Heeger, A. J., and Pines, A., Magnetic resonance evidence for metallic state in highly conducting polyaniline, *Synthetic Metals* **69**(1–3), 243–244, 1995.

452. Mizoguchi, K., Nechtschein, M., Travers, J. P., and Menardo, C., Spin dynamics in the conducting polymer, *Polyaniline, Physical Review Letters* **63**(1), 66–69, 1989.

453. Mizoguchi, K., Nechtschein, M., and Travers, J. P., Spin dynamics and conductivity in polyaniline—Temperature-dependence, *Synthetic Metals* **41**(1–2), 113–116, 1991.

454. Mizoguchi, K., Nechtschein, M., Travers, J. P., and Menardo, C., Spin dynamics study in polyaniline, *Synthetic Metals* **29**(1), E417–E424, 1989.

455. Mizoguchi, K. and Kume, K., Metallic temperature-dependence in the conducting polymer, Polyaniline—Spin dynamics study by ESR, *Solid State Communications* **89**(12), 971–975, 1994.

456. Mizoguchi, K. and Kume, K., Metallic temperature-dependence of microscopic electrical-conductivity in HCl-doped polyaniline studied by ESR, *Synthetic Metals* **69**(1–3), 241–242, 1995.

457. Nechtschein, M., Genoud, F., Menardo, C., Mizoguchi, K., Travers, J. P., and Villeret, B., On the nature of the conducting state of polyaniline, *Synthetic Metals* **29**(1), E211–E218, 1989.

458. Inoue, M., Inoue, M. B., Castillo-Ortega, M. M., Mizuno, M., Asaji, T., and Nakamura, D., Intrinsic paramagnetism of doped polypyrroles and polythiophenes: Electron spin resonance of the polymers prepared by the use of copper(II) compounds as oxidative coupling agents, *Synthetic Metals* **33**, 355–364, 1989.

459. Iida, M., Asaji, T., Inoue, M., Grijalva, H., Inoue, M. B., and Nakamura, D., Electron-spin-resonance study of intrinsic paramagnetism of soluble polyaniline perchlorates, *Bulletin of the Chemical Society of Japan* **64**(5), 1509–1513, 1991.

460. Ohsawa, T., Kimura, O., Onoda, M., and Yoshino, K., An ESR study on polyaniline in nonaqueous electrolyte, *Synthetic Metals* **47**(2), 151–156, 1992.

461. Bartle, A., Dunsch, L., Naarmann, H., Schmeisser, D., and Gopel, W., ESR studies of polypyrrole films with a two-dimensional microstructure, *Synthetic Metals* **61**(1–2), 167–170, 1993.

462. Nechtschein, M. and Genoud, F., On the broadening of the ESR line in presence of air or oxygen in conducting polymers, *Solid State Communications* **91**(6), 471–473, 1994.

463. Aasmundtveit, K., Genoud, F., Houze, E., and Nechtschein, M., Oxygen-induced ESR line broadening in conducting polymers, *Synthetic Metals* **69**(1–3), 193–196, 1995.

464. Kahol, P. K., Dyakonov, A. J., and McCormick, B. J., An electron-spin-resonance study of polymer interactions with moisture in polyaniline and its derivatives, *Synthetic Metals* **89**(1), 17–28, 1997.

465. Kahol, P. K., Dyakonov, A. J., and McCormick, B. J., An Electron-spin-resonance study of polyaniline and its derivatives—Polymer interactions with moisture, *Synthetic Metals* **84**(1–3), 691–694, 1997.

466. Kang, Y. S., Lee, H. J., Namgoong, J., Jung, B., and Lee, H., Decrease in electrical conductivity upon oxygen exposure in polyanilines doped with HCl, *Polymer* **40**(9), 2209–2213, 1999.

467. Mizoguchi, K., Kachi, N., Sakamoto, H., Yoshioka, K., Masubuchi, S., and Kazama, S., The effect of oxygen on the ESR linewidth in polypyrrole doped by PF6, *Solid State Communications* **105**(2), 81–84, 1998.

468. Lubentsov, B. Z., Timofeeva, O. N., Saratovskikh, S. L., Krinichnyi, V. I., Pelekh, A. E., Dmitrenko, V. V., and Khidekel, M. L., The study of conducting polymer interaction with gaseous substances. 4. The water-content influence on polyaniline crystal structure and conductivity, *Synthetic Metals* **47**(2), 187–192, 1992.

469. Lux, F., Hinrichsen, G., Krinichnyi, V. I., Nazarova, I. B., Chemerisov, S. D., and Pohl, M. M., Conducting islands concept for highly conductive polyaniline— Recent results of TEM-measurements, x-ray-diffraction-measuremants, EPR-measurements, DC conductivity-measurements and magnetic susceptibility-measurements, *Synthetic Metals* **55**(1), 347–352, 1993.

470. Krinichnyi, V. I., Chemerisov, S. D., and Lebedev, Y. S., Charge transport in slightly doped polyaniline, *Synthetic Metals* **84**(1–3), 819–820, 1997.

471. Krinichnyi, V. I., Konkin, A. L., Devasagayam, P., and Monkman, A. P., Multifrequency EPR study of charge transport in doped polyaniline, *Synthetic Metals* **119**, 281–282, 2001.

472. Krinichnyi, V. I., Roth, H. K., and Hinrichsen, G., Charge transfer in heavily H_2SO_4-doped polyaniline, *Synthetic Metals* **135**(1–3), 431–432, 2003.

473. Krinichnyi, V. I. and Tokarev, S. V., Charge transport in polyaniline heavily doped with p-toluenesulfonic acid, *Polymer Science, Series A* **47**(3), 261–269, 2005.
474. Krinichnyi, V. I., Tokarev, S. V., Roth, H. K., Schrödner, M., and Wessling, B., Multifrequency EPR study of metal-like domains in polyaniline, *Synthetic Metals* **152**(1–3), 165–168, 2005.
475. Long, S. M., Cromack, K. R., Epstein, A. J., Sun, Y., and MacDiarmid, A. G., ESR of pernigraniline base solutions revisited, *Synthetic Metals* **62**(3), 287–289, 1994.
476. Cao, Y., Li, S. Z., Xue, Z. J., and Guo, D., Spectroscopic and electrical characterization of some aniline oligomers and polyaniline, *Synthetic Metals* **16**(3), 305–315, 1986.
477. Javadi, H. H. S., Cromack, K. R., MacDiarmid, A. G., and Epstein, A. J., Microwave transport in the emeraldine form of polyaniline, *Physical Review B* **39**(6), 3579–3584, 1989.
478. Kahol, P. K., Pinto, N. J., and McCormick, B. J., Charge-transport and electron localization in alkyl ring-substituted polyanilines, *Solid State Communications* **91**(1), 21–24, 1994.
479. Kahol, P. K., Pinto, N. J., Berndtsson, E. J., and McCormick, B. J., Electron localization effects on the conducting state in polyaniline, *Journal of Physics—Condensed Matter* **6**(29), 5631–5638, 1994.
480. Pinto, N. J., Kahol, P. K., McCormick, B. J., Dalal, N. S., and Wan, H., Charge-transport and electron localization in polyaniline derivatives, *Physical Review B* **49**(19), 13983–13986, 1994.
481. Lim, H. Y., Jeong, S. K., Suh, J. S., Oh, E. J., Park, Y. W., Ryu, K. S., and Yo, C. H., Preparation and properties of fullerene doped polyaniline, *Synthetic Metals* **70**(1–3), 1463–1464, 1995.
482. Beau, B., Travers, J. P., and Banka, E., NMR evidence for heterogeneous disorder and quasi-1D metallic state in polyaniline CSA, *Synthetic Metals* **101**(1–2), 772–775, 1999.
483. Beau, B., Travers, J. P., Genoud, F., and Rannou, P., NMR study of aging effects in polyaniline CSA, *Synthetic Metals* **101**(1–2), 778–779, 1999.
484. Pratt, F. L., Blundell, S. J., Hayes, W., Nagamine, K., Ishida, K., and Monkman, A. P., Anisotropic polaron motion in polyaniline studied by muon spin relaxation, *Physical Review Letters* **79**(15), 2855–2858, 1997.
485. Krinichnyi, V. I., The 140-GHz (D-band) saturation transfer electron paramagnetic resonance studies of macromolecular dynamics in conducting polymers, *Journal of Physical Chemistry B* **112**(32), 9746–9752, 2008.
486. Krinichnyi, V. I., 2-mm Waveband saturation transfer electron paramagnetic resonance of conducting polymers, *Journal of Chemical Physics* **129**(13), 134510–134518, 2008.
487. Raghunathan, A., Kahol, P. K., Ho, J. C., Chen, Y. Y., Yao, Y. D., Lin, Y. S., and Wessling, B., Low-temperature heat capacities of polyaniline and polyaniline polymethylmethacrylate blends, *Physical Review B* **58**(24), R15955–R15958, 1998.
488. Pelster, R., Nimtz, G., and Wessling, B., Fully protonated polyaniline—Hopping transport on a mesoscopic scale, *Physical Review B* **49**(18), 12718–12723, 1994.
489. Li, Y., Song, Y., Zhang, X., Wu, X., Wang, F., and Wang, Z., Programmable polymer memory device based on hydrophilic polythiophene and poly(ionic liquid) electrolyte, *Macromolecular Chemistry and Physics* **216**(1), 113–121, 2015.

490. Springborg, M., The electronic-properties of polythiophene, *Journal of Physics—Condensed Matter* **4**(1), 101–120, 1992.
491. Kaneto, K., Hayashi, S., Ura, S., and Yoshino, K., Electron-spin-resonance and transport studies in electrochemically doped polythiophene film, *Journal of the Physical Society of Japan* **54**(3), 1146–1153, 1985.
492. Moraes, F., Davidov, D., Kobayashi, M., Chung, T. C., Chen, J., Heeger, A. J., and Wudl, F., Doped poly(thiophene)—Electron-spin resonance determination of the magnetic susceptibility, *Synthetic Metals* **10**(3), 169–179, 1985.
493. Hayashi, S., Kaneto, K., Yoshino, K., Matsushita, R., and Matsuyama, T., Electrical-Conductivity and Electron-spin-resonance studies in iodine-doped polythiophene from semiconductor to metallic regime, *Journal of the Physical Society of Japan* **55**(6), 1971–1980, 1986.
494. Chen, J., Heeger, A. J., and Wudl, F., Confined soliton pairs (bipolarons) in polythiophene—*In situ* magnetic-resonance measurements, *Solid State Communications* **58**(4), 251–257, 1986.
495. Mizoguchi, K., Honda, M., Masubuchi, S., Kazama, S., and Kume, K., Study of spin dynamics and electronic-structure in polythiophene heavily-doped with ClO_4, *Japanese Journal of Applied Physics Part 2—Letters* **33**(9A), L1239–L1241, 1994.
496. Masubuchi, S., Kazama, S., Mizoguchi, K., Honda, M., Kume, K., Matsushita, R., and Matsuyama, T., Metallic transport-properties in electrochemically as-grown and heavily-doped polythiophene and poly(3-methylthiophene), *Synthetic Metals* **57**(2–3), 4962–4967, 1993.
497. Demirboga, B. and Onal, A. M., ESR and conductivity investigations on electrochemically synthesized polyfuran and polythiophene, *Journal of Physics and Chemistry of Solids* **61**(6), 907–913, 2000.
498. Kaeriyama, K., Synthesis and properties of processable polythiophenes, in *Handbook of Organic Conductive Molecules and Polymers*, vol. 2, Nalwa, H. S., Ed. Chichester, U.K.: John Wiley & Sons, 1997, pp. 271–308.
499. Hotta, S., Molecular conductive materials: polythiophenes and oligothiophenes, in *Handbook of Organic Conductive Molecules and Polymers*, Nalwa, H. S., Ed. Chichester, U.K.: John Wiley & Sons, 1997, pp. 309–387.
500. Gronowitz, S., Ed., *Thiophene and Its Derivatives*. The Series of Heterocyclic Compounds, Vol. 44, New York: Wiley, 1991, p. 517.
501. McCullough, R. D., Lowe, R. D., Jayaraman, M., Ewbank, P. C., and Anderson, D. L., Synthesis and physical-properties of regiochemically well-defined, head-to-tail coupled poly(3-alkylthiophenes), *Synthetic Metals* **55**(2–3), 1198–1203, 1993.
502. Conwell, E. M., Transport in conducting polymers, in *Handbook of Organic Conductive Molecules and Polymers*, vol. 4, Nalwa, H. S., Ed. Chichester, U.K.: John Wiley & Sons, 1997, pp. 1–45.
503. Heeger, A. J., Kivelson, S., Schrieffer, J. R., and Su, W. P., Solitons in conducting polymers, *Reviews of Modern Physics* **60**(3), 781–850, 1988.
504. Kunugi, Y., Harima, Y., Yamashita, K., Ohta, N., and Ito, S., Charge transport in a regioregular poly(3-octylthiophene) film, *Journal of Materials Chemistry* **10**(12), 2673–2677, 2000.
505. Masubuchi, S., Imai, R., Yamazaki, K., Kazama, S., Takada, J., and Matsuyama, T., Structure and electrical transport property of poly(3-octylthiophene), *Synthetic Metals* **101**(1–3), 594–595, 1999.

506. Scharli, M., Kiess, H., Harbeke, G., Berlinger, W., Blazey, K. W., and Muller, K. A., ESR of BF_4-doped poly(3-methylthiophene), in *Electronic Properties of Conjugated Polymers*, vol. 76, Kuzmany, H., Mehring, M., and Roth, S., Eds. Berlin, Germany: Springer-Verlag, 1987, pp. 277–280.

507. Tourillon, G., Gourier, D., Garnier, F., and Vivien, D., Electron spin resonance study of electrochemically generated poly(thiophene) and derivatives, *Journal of Physical Chemistry* **88**(6), 1049–1051, 1984.

508. Carter, F. L., Ed., *Molecular Electronic Devices II*, vols. 1, 2. New York: Marcel Dekker, 1982, p. 827.

509. Salaneck, W. R., Clark, D. T., and Samuelsen, E. J., Eds., *Science and Applications of Conducting Polymers, Papers from the Sixth European Industrial Workshop.* New York (Bristol): Adam Hilger, 1991, 196p.

510. Ashwell, G. J., Ed., *Molecular Electronics.* New York: John Wiley, 1992, 362p.

511. Scrosati, B., Ed., *Application of Electroactive Polymers.* London, U.K.: Chapman & Hall, 1993, 354p.

512. Wong, C. P., Ed., *Polymers for Electronic & Photonic Application.* Boston, MA: Elsevier – Academic Press, 1993, 661p.

513. Bobacka, J., Ivaska, A., and Lewenstam, A., Plasticizer-free all-solid-state potassium-selective electrode based on poly(3-octylthiophene) and valinomycin, *Analytica Chimica Acta* **385**(1–3), 195–202, 1999.

514. Sariciftci, N. S. and Heeger, A. J., Photophysics of semiconducting polymer C-60 composites—A comparative-study, *Synthetic Metals* **70**(1–3), 1349–1352, 1995.

515. Lee, C. H., Yu, G., Moses, D., Pakbaz, K., Zhang, C., Sariciftci, N. S., Heeger, A. J., and Wudl, F., Sensitization of the photoconductivity of conducting polymers by C-60—Photoinduced electron-transfer, *Physical Review B* **48**(20), 15425–15433, 1993.

516. Lee, K. H., Janssen, R. A. J., Sariciftci, N. S., and Heeger, A. J., Direct evidence of photoinduced electron-transfer in conducting-polymer-C60 composites Infrared photoexcitation spectroscopy, *Physical Review B* **49**(8), 5781–5784, 1994.

517. Gebeyehu, D., Padinger, F., Fromherz, T., Hummelen, J. C., and Sariciftci, N. S., Photovoltaic properties of conjugated polymer/fullerene composites on large area flexible substrates, *Bulletin of the Chemical Society of Ethiopia* **14**(1), 57–68, 2000.

518. Gebeyehu, D., Brabec, C. J., Padinger, F., Fromherz, T., Hummelen, J. C., Badt, D., Schindler, H., and Sariciftci, N. S., The interplay of efficiency and morphology in photovoltaic devices based on interpenetrating networks of conjugated polymers with fullerenes, *Synthetic Metals* **118**(1–3), 1–9, 2001.

519. Taka, T., Jylha, O., Root, A., Silvasti, E., and Osterholm, H., Characterization of undoped poly(3-octylthiophene), *Synthetic Metals* **55**(1), 414–419, 1993.

520. Chen, T. A., Wu, X. M., and Rieke, R. D., Regiocontrolled synthesis of poly(3-alkylthiophenes) mediated by Rieke zink—Their characterization and solid-state properties, *Journal of the American Chemical Society* **117**(1), 233–244, 1995.

521. Kaniowski, T., Niziol, S., Sanetra, J., Trznadel, M., and Proń, A., Optical studies of regioregular poly(3-octylthiophene)s under pressure, *Synthetic Metals* **94**(1), 111–114, 1998.

522. Roth, H. K. and Krinichnyi, V. I., Spin and charge transport in poly(3-octylthiophene), *Synthetic Metals* **137**(1–3), 1431–1432, 2003.

523. Krinichnyi, V. I. and Roth, H. K., EPR study of spin and charge dynamics in slightly doped poly(3-octylthiophene), *Applied Magnetic Resonance* **26**, 395–415, 2004.

524. Marumoto, K., Takeuchi, N., Ozaki, T., and Kuroda, S., ESR studies of photogenerated polarons in regioregular poly(3-alkylthiophene)-fullerene composite, *Synthetic Metals* **129**, 239–247, 2002.

525. Kahol, P. K., Raghunathan, A., McCormick, B. J., and Epstein, A. J., High temperature magnetic susceptibility studies of sulfonated polyanilines, *Synthetic Metals* **101**(1–3), 815–816, 1999.

526. Owens, J., Evidence for zero-field fluctuations in Cr^{3+} near the phase transition in $NH_4Al(SO_4)_2$ $12H_2O$, *Physica Status Solidi B* **79**(2), 623–628, 1977.

527. Mardalen, J., Samuelsen, E. J., Konestabo, O. R., Hanfland, M., and Lorenzen, M., Conducting polymers under pressure: Synchrotron x-ray determined structure and structure related properties of two forms of poly(octyl-thiophene), *Journal of Physics—Condensed Matter* **10**, 7145–7154, 1998.

528. Sauvajol, J. L., Bormann, D., Palpacuer, M., Lere-Porte, J. P., Moreau, J. J. E., and Dianoux, A. J., Low and high-frequency vibrational dynamics as a function of structural order in polythiophene, *Synthetic Metals* **84**(1–3), 569–570, 1997.

529. Österbacka, R., An, C. P., Jiang, X. M., and Vardeny, Z. V., Delocalized polarons in self-assembled poly(3-hexylthiophene) nanocrystals, *Synthetic Metals* **116**(1–3), 317–320, 2001.

530. Yamauchi, T., Najib, H. M., Liu, Y. W., Shimomura, M., and Miyauchi, S., Positive temperature coefficient characteristics of poly(3-alkylthiophene)s, *Synthetic Metals* **84**(1–3), 581–582, 1997.

531. Pivrikas, A., Sariciftci, N. S., Juska, G., and Osterbacka, R., A review of charge transport and recombination in polymer/fullerene organic solar cells, *Progress in Photovoltaics* **15**(8), 677–696, 2007.

532. Kubis, P., Lucera, L., Machui, F., Spyropoulos, G., Cordero, J., Frey, A., Kaschta, J. et al., High precision processing of flexible P3HT/PCBM modules with geometric fill factor over 95%, *Organic Electronics* **15**(10), 2256–2263, 2014.

533. Green, M. A., Emery, K., Hishikawa, Y., Warta, W., and Dunlop, E. D., Solar cell efficiency tables (version 44), *Progress in Photovoltaics: Research and Applications* **22**(7), 701–710, 2014.

534. Halls, J. J. M., Cornil, J., Dos Santos, D. A., Silbey, R., Hwang, D. H., Holmes, A. B., Brédas, J. L., and Friend, R. H., Charge- and energy-transfer processes at polymer/polymer interfaces: A joint experimental and theoretical study, *Physical Review B* **60**(8), 5721–5727, 1999.

535. Brabec, C. J., Winder, C., Sariciftci, N. S., Hummelen, J. C., Dhanabalan, A., van Hal, P. A., and Janssen, R. A. J., A low-bandgap semiconducting polymer for photovoltaic devices and infrared emitting diodes, *Advanced Functional Materials* **12**(10), 709–712, 2002.

536. Brabec, C. J., Cravino, A., Meissner, D., Sariciftci, N. S., Fromherz, T., Rispens, M. T., Sanchez, L., and Hummelen, J. C., Origin of the open circuit voltage of plastic solar cells, *Advanced Functional Materials* **11**(5), 374–380, 2001.

537. Koster, L. J. A., Mihailetchi, V. D., Ramaker, R., and Blom, P. W. M., Light intensity dependence of open-circuit voltage of polymer:fullerene solar cells, *Applied Physics Letters* **86**(12), 123509–123511, 2005.

538. Koster, L. J. A., Mihailetchi, V. D., and Blom, P. W. M., Ultimate efficiency of polymer/fullerene bulk heterojunction solar cells, *Applied Physics Letters* **88**(9), 093511–093513, 2006.

539. Wu, P.-T., Ren, G., and Jenekhe, S. A., Crystalline random conjugated copolymers with multiple side chains: Tunable intermolecular interactions and enhanced charge transport and photovoltaic properties, *Macromolecules* **43**(7), 3306–3313, 2010.

540. Lenes, M., Wetzelaer, G. J. A. H., Kooistra, F. B., Veenstra, S. C., Hummelen, J. C., and Blom, P. W. M., Fullerene bisadducts for enhanced open-circuit voltages and efficiencies in polymer solar cells, *Advanced Materials* **20**, 2116–2119, 2008.

541. Li, G., Shrotriya, V., Huang, J., Yao, Y., Moriarty, T., Emery, K., and Yang, Y., High-efficiency solution processable polymer photovoltaic cells by self-organization of polymer blends, *Nature Materials* **4**(11), 864–868, 2005.

542. Huang, J. H., Li, K. C., Chien, F. C., Hsiao, Y. S., Kekuda, D., Chen, P. L., Lin, H. C., Ho, K. C., and Chu, C. W., Correlation between exciton lifetime distribution and morphology of bulk heterojunction films after solvent annealing, *Journal of Physical Chemistry C* **114**(19), 9062–9069, 2010.

543. Jarzab, D., Cordella, F., Lenes, M., Kooistra, F. B., Blom, P. W. M., Hummelen, J. C., and Loi, M. A., Charge transfer dynamics in polymer-fullerene blends for efficient solar cells, *Journal of Physical Chemistry B* **113**(52), 16513–16517, 2009.

544. Dyer-Smith, C., Reynolds, L. X., Bruno, A., Bradley, D. D. C., Haque, S. A., and Nelson, J., Triplet formation in fullerene multi-adduct blends for organic solar cells and its influence on device performance, *Advanced Functional Materials* **20**(16), 2701–2708, 2010.

545. Berton, N., Ottone, C., Labet, V., de Bettignies, R., Bailly, S., Grand, A., Morell, C., Sadki, S., and Chandezon, F., New alternating copolymers of 3,6-carbazoles and dithienylbenzothiadiazoles: Synthesis, characterization, and application in photovoltaics, *Macromolecular Chemistry and Physics* **212**(19), 2127–2141, 2011.

546. Kim, J., Yun, M. H., Kim, G. H., Kim, J. Y., and Yang, C., Replacing 2,1,3-benzothiadiazole with 2,1,3-naphthothiadiazole in PCDTBT: Towards a low bandgap polymer with deep HOMO energy level, *Polymer Chemistry* **3**(12), 3276–3281, 2012.

547. Blouin, N., Michaud, A., and Leclerc, M., A low-bandgap poly(2,7-carbazole) derivative for use in high-performance solar cells, *Advanced Materials* **19**(17), 2295–2300, 2007.

548. Choy, W. C. H., Ed., *Organic Solar Cells: Materials and Device Physics*. London, U.K.: Springer, 2013, 265p.

549. Ostroverkhova, O., Ed., *Handbook of Organic Materials for Optical and (Opto) Electronic Devices: Properties and Applications*, Woodhead Publishing Series in Electronic and Optical Materials. Philadelphia, PA: Woodhead Publishing, 2013, 832p.

550. Huh, Y. H. and Park, B., Interface-engineering additives of poly(oxyethylene tridecyl ether) for low-band gap polymer solar cells consisting of PCDTBT:PCBM70 bulk-heterojunction layers, *Optics Express* **21**(1), A146–A156, 2013.

551. Park, S. H., Roy, A., Beaupre, S., Cho, S., Coates, N., Moon, J. S., Moses, D., Leclerc, M., Lee, K., and Heeger, A. J., Bulk heterojunction solar cells with internal quantum efficiency approaching 100%, *Nature Photonics* **3**, 297–302, 2009.

552. Moon, J. S., Jo, J., and Heeger, A. J., Nanomorphology of PCDTBT:PC$_{70}$BM bulk heterojunction solar cells, *Advanced Energy Materials* **2**(3), 304–308, 2012.

553. Liu, J. G., Chen, L., Gao, B. R., Cao, X. X., Han, Y. C., Xie, Z. Y., and Wang, L. X., Constructing the nanointerpenetrating structure of PCDTBT:PC$_{70}$BM bulk heterojunction solar cells induced by aggregation of PC$_{70}$BM *via* mixed-solvent vapor annealing, *Journal of Materials Chemistry A* **1**(20), 6216–6225, 2013.

554. Gutzler, R. and Perepichka, D. F., π-Electron conjugation in two dimensions, *Journal of the American Chemical Society* **135**(44), 16585–16594, 2013.

555. Bässler, H., Charge transport in disordered organic photoconductors a Monte Carlo simulation study, *Physica Status Solidi (B)* **175**(1), 15–56, 1993.

556. Krinichnyi, V. I., Troshin, P. A., and Denisov, N. N., The effect of fullerene derivative on polaronic charge transfer in poly(3-hexylthiophene/fullerene compound, *Journal of Chemical Physics* **128**(16), 164715/01–164715/07, 2008.

557. Krinichnyi, V. I., Troshin, P. A., and Denisov, N. N., Structural effect of fullerene derivative on polaron relaxation and charge transfer in poly(3-hexylthiophene)/fullerene composite, *Acta Materialia* **56**(15), 3982–3989, 2008.

558. Krinichnyi, V. I., Yudanova, E. I., and Spitsina, N. G., Light-induced EPR study of poly(3-alkylthiophene)/fullerene composites, *Journal of Physical Chemistry C* **114**(39), 16756–16766, 2010.

559. Konkin, A., Ritter, U., Scharff, P., Mamin, G., Aganov, A., Orlinskii, S., Krinichnyi, V. I., Egbe, D. A. M., Ecke, G., and Romanus, H., Multifrequency X-, W-band ESR study on photo-induced ion radical formation in solid films of mono- and di-fullerenes embedded in conjugated polymers, *Carbon* **77**, 11–17, 2014.

560. Ginder, J. M., Epstein, A. J., and MacDiarmid, A. G., Ring-torsional polarons in polyaniline and polyparaphenylene sulfide, *Synthetic Metals* **43**(1–2), 3431–3436, 1991.

561. Lee, J. M., Park, J. S., Lee, S. H., Kim, H., Yoo, S., and Kim, S. O., Selective electron- or hole-transport enhancement in bulk-heterojunction organic solar cells with N- or B-doped carbon nanotubes, *Advanced Materials* **23**(5), 629–633, 2011.

562. Park, J. S., Lee, B. R., Lee, J. M., Kim, J.-S., Kim, S. O., and Song, M. H., Efficient hybrid organic-inorganic light emitting diodes with self-assembled dipole molecule deposited metal oxides, *Applied Physics Letters* **96**, 243306–243308, 2010.

563. Padinger, F., Rittberger, R. S., and Sariciftci, N. S., Effects of postproduction treatment on plastic solar cells, *Advanced Functional Materials* **11**(13), 1–4, 2003.

564. Chirvase, D., Parisi, J., Hummelen, J. C., and Dyakonov, V., Influence of nanomorphology on the photovoltaic action of polymer-fullerene composites, *Nanotechnology* **15**(9), 1317–1323, 2004.

565. Chu, C. W., Yang, H. C., Hou, W. J., Huang, J. S., Li, G., and Yang, Y., Control of the nanoscale crystallinity and phase separation in polymer solar cells, *Applied Physics Letters* **92**(10), 103306–103308, 2008.

566. Dante, M., Peet, J., and Nguyen, T. Q., Nanoscale charge transport and internal structure of bulk heterojunction conjugated polymer/fullerene solar cells by scanning probe microscopy, *Journal of Physical Chemistry C* **112**(18), 7241–7249, 2008.

567. Montanari, I., Nogueira, A. F., Nelson, J., Durrant, J. R., Winder, C., Loi, M. A., Sariciftci, N. S., and Brabec, C., Transient optical studies of charge recombination dynamics in a polymer/fullerene composite at room temperature, *Applied Physics Letters* **81**(16), 3001–3003, 2002.

568. Nogueira, A. F., Montanari, I., Nelson, J., Durrant, J. R., Winder, C., and Sariciftci, N. S., Charge recombination in conjugated polymer/fullerene blended films studied by transient absorption spectroscopy, *Journal of Physical Chemistry B* **107**(7), 1567–1573, 2003.

569. Krebs, F. C., *Stability and Degradation of Organic and Polymer Solar Cells*. Chichester, U.K.: John Wiley & Sons, 2012, 360p.

570. Dyakonov, V., Zoriniants, G., Scharber, M., Brabec, C. J., Janssen, R. A. J., Hummelen, J. C., and Sariciftci, N. S., Photoinduced charge carriers in conjugated polymer-fullerene composites studied with light-induced electron-spin resonance, *Physical Review B* **59**(12), 8019–8025, 1999.

571. Brabec, C., Dyakonov, V., Parisi, J., and Sariciftci, N. S., *Organic Photovoltaic: Concepts and Realization*. Berlin, Germany: Springer, 2003, 297p.

572. Marumoto, K., Muramatsu, Y., and Kuroda, S., Quadrimolecular recombination kinetics of photogenerated charge carriers in regioregular poly(3-alkylthiophene)/fullerene composites, *Applied Physics Letters* **84**(8), 1317–1319, 2004.

573. Sensfuss, S., Konkin, A., Roth, H. K., Al-Ibrahim, M., Zhokhavets, U., Gobsch, G., Krinichnyi, V. I., Nazmutdinova, G. A., and Klemm, E., Optical and ESR studies on poly(3-alkylthiophene)/fullerene composites for solar cells, *Synthetic Metals* **137**(1–3), 1433–1434, 2003.

574. Krinichnyi, V. I., Roth, H. K., and Konkin, A. L., Multifrequency EPR study of charge transfer in poly(3-alkylthiophenes), *Physica B* **344**(1–4), 430–435, 2004.

575. Takeda, K., Hikita, H., Kimura, Y., Yokomichi, H., and Morigaki, K., Electron spin resonance study of light-induced annealing of dangling bonds in glow discharge hydrogenated amorphous silicon: Deconvolution of electron spin resonance spectra, *Japanese Journal of Applied Physics Part 1—Regular Papers Short Notes & Review Papers* **37**(12A), 6309–6317, 1998.

576. Yanilkin, V. V., Nastapova, N. V., Morozov, V. I., Gubskaya, V. P., Sibgatullina, F. G., Berezhnaya, L. S., and Nuretdinov, I. A., Competitive conversions of carbonyl-containing methanofullerenes induced by electron transfer, *Russian Journal of Electrochemistry* **43**(2), 184–203, 2007.

577. Poluektov, O. G., Filippone, S., Martín, N., Sperlich, A., Deibel, C., and Dyakonov, V., Spin signatures of photogenerated radical anions in polymer-[70]fullerene bulk heterojunctions: High frequency pulsed EPR spectroscopy, *Journal of Physical Chemistry B* **114**(45), 14426–14429, 2010.

578. Krinichnyi, V. I. and Yudanova, E. I., Structural effect of electron acceptor on charge transfer in poly(3-hexylthiophene)/methanofullerene bulk heterojunctions, *Solar Energy Materials and Solar Cells* **95**(8), 2302–2313, 2011.

579. Krinichnyi, V. I. and Yudanova, E. I., Light-induced EPR study of charge transfer in P3HT/PC$_{71}$BM bulk heterojunctions, *Journal of Physical Chemictry C* **116**(16), 9189–9195, 2012.

580. Aguirre, A., Gast, P., Orlinskii, S., Akimoto, I., Groenen, E. J. J., El Mkami, H., Goovaerts, E., and Van Doorslaer, S., Multifrequency EPR analysis of the positive polaron in I$_2$-doped poly(3-hexylthiophene) and in poly[2-methoxy-5-(3, 7-dimethyloctyloxy)]-1,4-phenylenevinylene, *Physical Chemistry Chemical Physics* **10**(47), 7129–7138, 2008.

581. Niklas, J., Mardis, K. L., Banks, B. P., Grooms, G. M., Sperlich, A., Dyakonov, V., Beaupré, S. et al., Highly-efficient charge separation and polaron delocalization in polymer–fullerene bulk-heterojunctions: A comparative multi-frequency EPR and DFT study, *Physical Chemistry Chemical Physics* **15**(24), 9562–9574, 2013.

582. Krinichnyi, V. I., Yudanova, E. I., and Denisov, N. N., The role of spin exchange in charge transfer in low-bandgap polymer:fullerene bulk heterojunctions, *Journal of Chemical Physics* **141**(4), 044906/01–044906/11, 2014.

583. Eaton, S. S. and Eaton, G. R., EPR spectra of C-60 anions, *Applied Magnetic Resonance* **11**(2), 155–170, 1996.

584. Allemand, P. M., Srdanov, G., Koch, A., Khemani, K., Wudl, F., Rubin, Y., Diederich, F., Alvarez, M. M., Anz, S. J., and Whetten, R. L., The unusual Electron-spin-resonance of fullerene C_{60}, *Journal of the American Chemical Society* **113**(7), 2780–2781, 1991.

585. Bietsch, W., Bao, J., Ludecke, J., and van Smaalen, S., Jahn-Teller distortion and merohedral disorder of C-60(–) as observed by ESR, *Chemical Physics Letters* **324**(1–3), 37–42, 2000.

586. De Ceuster, J., Goovaerts, E., Bouwen, A., Hummelen, J. C., and Dyakonov, V., High-frequency (95 GHz) electron paramagnetic resonance study of the photo-induced charge transfer in conjugated polymer-fullerene composites, *Physical Review B* **64**(19), 195206/01–195206/06, 2001.

587. Konkin, A., Ritter, U., Scharff, P., Roth, H.-K., Aganov, A., Sariciftci, N. S., and Egbe, D. A. M., Photo-induced charge separation process in (PCBM-C_{120}O)/(M3EH-PPV) blend solid film studied by means of X and K-bands ESR at 77 and 120 K, *Synthetic Metals* **160**(5–6), 485–489, 2010.

588. Krinichnyi, V. I. and Yudanova, E. I., Light-induced EPR study of charge transfer in P3HT/*bis*-PCBM bulk heterojunctions, *AIP Advances* **1**(2), 022131/01–022131/15, 2011.

589. Sensfuss, S., Blankenburg, L., Schache, H., Shokhovets, S., Erb, T., Konkin, A., Herasimovich, A. et al., Thienopyrazine-based low-bandgap polymers for flexible polymer solar cells, *European Physical Journal—Applied Physics* **51**(3), 303204/01–303204/05, 2010.

590. Deibel, C., Mack, D., Gorenflot, J., Schöll, A., Krause, S., Reinert, F., Rauh, D., and Dyakonov, V., Energetics of excited states in the conjugated polymer poly(3-hexylthiophene), *Physical Review B* **81**(8), 085202–085206, 2010.

591. Wienk, M. M., Kroon, J. M., Verhees, W. J. H., Knol, J., Hummelen, J. C., van Hal, P. A., and Janssen, R. A. J., Efficient methano[70]fullerene/MDMO-PPV bulk heterojunction photovoltaic cells, *Angewandte Chemie—International Edition* **42**(29), 3371–3375, 2003.

592. Yamanari, T., Taima, T., Sakai, J., and Saito, K., Highly efficient organic thin-film solar cells based on poly(3-hexylthiophene) and soluble C-70 fullerene derivative, *Japanese Journal of Applied Physics* **47**(2), 1230–1233, 2008.

593. Boland, P., Sunkavalli, S. S., Chennuri, S., Foe, K., Abdel-Fattah, T., and Namkoong, G., Investigation of structural, optical, and electrical properties of regioregular poly(3-hexylthiophene)/fullerene blend nanocomposites for organic solar cells, *Thin Solid Films* **518**(6), 1728–1731, 2010.

594. Sperlich, A., Liedtke, M., Kern, J., Kraus, H., Deibel, C., Filippone, S., Delgado, J. L., Martin, N., and Dyakonov, V., Photoinduced C(70) radical anions in polymer: fullerene blends, *Physica Status Solidi—Rapid Research Letters* **5**(3), 128–130, 2011.

595. Pénicaud, A., Peréz-Benitez, A., Escudero, R., and Coulon, C., Single crystal synthesis of $[(C_6H_5)_4P]_2[C_{70}][I]$ by electrocrystallization and experimental determination of the *g*-value anisotropy of C_{70}·– and C_{60}·– at 4.2 K, *Solid State Communications* **96**(3), 147–150, 1995.

596. Reed, C. A. and Bolskar, R. D., Discrete fulleride anions and fullerenium cations, *Chemical Reviews* **100**, 1075–1120, 2000.
597. Tanaka, K., Zakhidov, A. A., Yoshizawa, K., Okahara, K., Yamabe, T., Yakushi, K., Kikuchi, K., Suzuki, S., Ikemoto, I., and Achiba, Y., Magnetic properties of TDAE-C-60 add TDAE-C-70 where TDAE is tetrakis(dimethylamino)ethylene, *Physical Review B* **47**(12), 7554–7559, 1993.
598. Friedrich, J., Schweitzer, P., Dinse, K. P., Rapta, P., and Stasko, A., EPR study of radical-anions of C-60 and C-70, *Applied Magnetic Resonance* **7**(2–3), 415–425, 1994.
599. Hase, H. and Miyatake, Y., Comparative ESR study of C-60 and C-70 radical-anions produced in gamma-irradiated organic-solid solutions at 77 K, *Chemical Physics Letters* **245**(1), 95–101, 1995.
600. Adrian, F. J., Spin-orbit effects in fullerenes, *Chemical Physics* **211**(1–3), 73–80, 1996.
601. Stone, A. J., Gauge invariance of the g tensor, *Proceedings of the Royal Society A: Mathematical, Physical and Engineering Sciences* **A271**(1346), 424–434, 1963.
602. Dubois, D., Kadish, K. M., Flanagan, S., Haufler, R. E., Chibante, L. P. F., and Wilson, L. J., Spectroelectrochemical study of the C60 and C70 fullerenes and their monoanions, dianions, trianions, and tetraanions, *Journal of the American Chemical Society* **113**(11), 4364–4366, 1999.
603. Krinichnyi, V. I. and Yudanova, E. I., Influence of morphology of low-band-gap PCDTBT:PC$_{71}$BM composite on photoinduced charge transfer: LEPR spectroscopy study, *Synthetic Metals* **210**, 148–155, 2015.
604. Krinichnyi, V. I. and Yudanova, E. I., Light-induced EPR spectroscopy of charge transfer in low-band-gap PCDTBT:PC$_{71}$BM bulk heterojunctions, *IEEE Journal of Photovoltaics* **6**(2), 506–515, 2016.
605. Krinichnyi, V. I., Roth, H. K., Sensfuss, S., Schrödner, M., and Al Ibrahim, M., Dynamics of photoinduced radical pairs in poly(3-dodecylthiophene)/fullerene composite, *Physica E* **36**(1), 98–101, 2007.
606. Krinichnyi, V. I., Dynamics of charge carriers photoinduced in poly(3-dodecylthiophene)/fullerene composite, *Acta Materialia* **56**(7), 1427–1434, 2008.
607. Krinichnyi, V. I., Dynamics of charge carriers photoinduced in poly(3-dodecylthiophene)/fullerene bulk heterojunction, *Solar Energy Materials and Solar Cells* **92**(8), 942–948, 2008.
608. Krinichnyi, V. I., An ESR study of photoinduced charge transport in the polymer/fullerene system, *High Energy Chemistry* **42**(7), 572–575, 2008.
609. Krinichnyi, V. I. and Yudanova, E. I., LEPR spectroscopy of charge carriers photoinduced in polymer/fullerene bulk heterojunctions, *Journal of Renewable and Sustainable Energy* **1**(4), 043110/01–043110/18, 2009.
610. Krinichnyi, V. I. and Balakai, A. A., Light-induced spin localization in poly(3-dodecylthiophen)/PCBM composite, *Applied Magnetic Resonance* **39**(3), 319–328, 2010.
611. Krinichnyi, V. I., EPR study of charge transfer in the poly-3-dodecylthiophene-fullerene system, *Polymer Science, Series A* **52**(1), 26–32, 2010.
612. Yudanova, E. I. and Krinichnyi, V. I., Influence of ultrasonic, microwave, and thermal effects on photoinduced charge transfer in poly(3-hexylthiophene)-methanofullerene composites: EPR study, *Polymer Science, Series A* **55**(4), 233–243, 2013.

613. Marumoto, K., Muramatsu, Y., Takeuchi, N., and Kuroda, S., Light-induced ESR studies of polarons in regioregular poly(3-alkylthiophene)-fullerene composites, *Synthetic Metals* **135**(1–3), 433–434, 2003.

614. Janssen, R. A. J., Moses, D., and Sariciftci, N. S., Electron and energy transfer processes of photoexcited oligothiophenes onto tetracyanoethylene and C60, *Journal of Chemical Physics* **101**(11), 9519–9527, 1994.

615. Tong, M. H., Coates, N. E., Moses, D., Heeger, A. J., Beaupre, S., and Leclerc, M., Charge carrier photogeneration and decay dynamics in the poly(2,7-carbazole) copolymer PCDTBT and in bulk heterojunction composites with PC70BM, *Physical Review B* **81**(12), 125210–125215, 2010.

616. Kočka, J., Elliott, S. R., and Davis, E. A., AC conductivity and photo-induced states in amorphous semiconductors, *Journal of Physics C—Solid State Physics* **12**, 2589–2596, 1979.

617. Tanaka, K., Sato, T., Kuga, T., Yamabe, T., Yoschizawa, K., Okahara, K., and Zakhidov, A. A., Weak suppression of ferromagnetism in tetrakis(dimethylamino) ethylene-(C60)1-x(C70)x, *Physical Review B, Solid State* **51**(2), 990–905, 1995.

618. Gotschya, B., Gompper, R., Klos, H., Schilder, A., Schütz, W., and Völkel, G., Ferromagnetic versus molecular ordering in C60 charge transfer complexes, *Synthetic Metals* **77**(1–3), 287–290, 1996.

619. Conwell, E. M., Duke, C. B., Paton, A., and Jeyadev, S., Molecular conformation of polyaniline oligomers: Optical absorption and photoemission of three-phenyl molecules, *Journal of Chemical Physics* **88**(5), 3331–3337, 1988.

620. Harigaya, K., Long-range excitons in conjugated polymers with ring torsions: Poly(para-phenylene) and polyaniline, *Journal of Physics—Condensed Matter* **10**(34), 7679–7690, 1998.

621. Yazawa, K., Inoue, Y., Shimizu, T., Tansho, M., and Asakawa, N., Molecular dynamics of regioregular poly(3-hexylthiophene) investigated by NMR relaxation and an interpretation of temperature dependent optical absorption, *Journal of Physical Chemistry B* **114**(3), 1241–1248, 2010.

622. Breiby, D. W., Sato, S., Samuelsen, E. J., and Mizoguchi, K., Electron spin resonance studies of anisotropy in semiconducting polymeric films, *Journal of Polymer Science Part B—Polymer Physics* **41**(23), 3011–3025, 2003.

623. Marchant, S. and Foot, P. J. S., Poly(3-hexylthiophene)-zinc oxide rectifying junctions, *Journal of Materials Science-Materials in Electronics* **6**(3), 144–148, 1995.

624. Al-Ibrahim, M., Roth, H.-K., Schrödner, M., Konkin, A., Zhokhavets, U., Gobsch, G., Scharff, P., and Sensfuss, S., The influence of the optoelectronic properties of poly(3-alkylthiophenes) on the device parameters in flexible polymer solar cells, *Organic Electronics* **6**(2), 65–77, 2005.

625. So, F., Kido, J., and Burrows, P., Organic light-emitting devices for solid-state lighting, *Materials Research Society Bulletin* **33**(7), 663–669, 2008.

626. Sharma, G. D., Advances in nano-structured organic solar cells, in *Physics of Nanostructured Solar Cells*, Badescu, V. and Paulescu, M., Eds. New York: Nova Science Publishers, 2010, pp. 361–460.

627. Krinichnyi, V. I., Yudanova, E. I., and Denisov, N. N., Light-induced EPR study of charge transfer in poly(3-hexylthiophene)/fullerene bulk heterojunction, *Journal of Chemical Physics* **131**(6), 044515/01–044515/11, 2009.

628. Krinichnyi, V. I., Yudanova, E. I., and Denisov, N. N., EPR Study of charge transfer photoinduced in poly(3-hexylthiophene)/fullerene composite, *Polymer Science, Series A* **52**(7), 715–726, 2010.

629. Crochet, J. J., Hoseinkhani, S., Lüer, L., Hertel, T., Doorn, S. K., and Lanzani, G., Free-Carrier generation in aggregates of single-wall carbon nanotubes by photoexcitation in the ultraviolet regime, *Physical Review Letters* **107**(25), 257402–257406, 2011.
630. Obrzut, J. and Page, K. A., Electrical conductivity and relaxation in poly(3-hexylthiophene), *Physical Review B* **80**(19), 195211/01–195211/07, 2009.
631. Shklovskii, B. I. and Efros, A. L., *Electronic Properties of Doped Semiconductors*. New York: Springer-Verlag, 1984, p. 393.
632. Tycko, R., Dabbagh, G., Fleming, R. M., Haddon, R. C., Makhija, A. V., and Zahurak, S. M., Molecular dynamics and the phase transition in solid C_{60}, *Physical Review Letters* **67**(14), 1886–1889, 1991.
633. Denisov, N. N., Krinichnyi, V. I., and Nadtochenko, V. A., Spin properties of paramagnetic centers photogenerated in crystals of complexes between C_{60} and TPA, in *Fullerenes—Recent Advances in the Chemistry and Physics of Fullerenes and Related Materials*, vol. 4, Kadish, K. and Ruoff, R., Eds. Pennington, NJ: The Electrochemical Society Inc., 1997, pp. 139–147.
634. Pike, G. E., AC conductivity of scandium oxide and a new hopping model for conductivity, *Physical Review B* **6**(4), 1572–1580, 1972.
635. Elliott, S. R., On the super-linear frequency dependent conductivity of amorphous semiconductors, *Solid State Communications* **28**(11), 939–942, 1978.
636. Agostini, G., Corvaja, C., and Pasimeni, L., EPR studies of the excited triplet states of $C_{60}O$ and $C_{60}C_2H_4N(CH_3)$ fullerene derivatives and C_{70} in toluene and polymethylmethacrylate glasses and as films, *Chemical Physics* **202**(2–3), 349–356, 1996.
637. Morosin, B., Hu, Z. B., Jorgensen, J. D., Short, S., Schirber, J. E., and Kwei, G. H., Ne intercalated C-60: Diffusion kinetics, *Physical Review B* **59**(9), 6051–6057, 1999.
638. Geru, I. and Suter, D., *Resonance Effects of Excitons and Electrons: Basics and Applications*, vol. 869, 2013, 283p.
639. Krinichnyi, V. I., Roth, H. K., Schrödner, M., and Wessling, B., EPR study of polyaniline highly doped by *p*-toluenesulfonic acid, *Polymer* **47**(21), 7460–7468, 2006.
640. Krinichnyi, V. I., Tokarev, S. V., Roth, H. K., Schrödner, M., and Wessling, B., EPR study of charge transfer in polyaniline highly doped by *p*-toluenesulfonic acid, *Synthetic Metals* **156**(21–24), 1368–1377, 2006.
641. Kapil, A., Taunk, M., and Chand, S., Preparation and charge transport studies of chemically synthesized polyaniline, *Journal of Materials Science—Materials in Electronics* **21**(4), 399–404, 2010.
642. Wessling, B., New insight into organic metal polyaniline morphology and structure, *Polymers* **2**, 786–798, 2010.
643. Kahol, P. K., Magnetic susceptibility of polyaniline and polyaniline-polymethylmethacrylate blends, *Physical Review B* **62**(21), 13803–13804, 2000.
644. Adrain, F. J. and Monchick, L., Theory of chemically induced magnetic polarization. Effects of S–$T_{\pm 1}$ mixing in strong magnetic fields, *Journal of Chemical Physics* **71**(6), 2600–2610, 1979.
645. Yan, B., Schultz, N. A., Efros, A. L., and Taylor, P. C., Universal distribution of residual carriers in tetrahedrally coordinated amorphous semiconductors, *Physical Review Letters* **84**(18), 4180–4183, 2000.

646. Petta, J. R., Johnson, A. C., Taylor, J. M., Laird, E. A., Yacoby, A., Lukin, M. D., Marcus, C. M., Hanson, M. P., and Gossard, A. C., Coherent manipulation of coupled electron spins in semiconductor quantum dots, *Science* **309**, 2180–2186, 2005.
647. Kersting, R., Lemmer, U., Deussen, M., Bakker, H. J., Mahrt, R. F., Kurz, H., Arkhipov, V. I., Bassler, H., and Gobel, E. O., Ultrafast field-induced dissociation of excitons in conjugated polymers, *Physical Review Letters*, **73**(10), 1440–1443, 1994.
648. Simmons, S., Brown, R. M., Riemann, H., Abrosimov, N. V., Becker, P., Pohl, H.-J., Thewalt, M. L. W., Itoh, M. K., and Morton, J. J. L., Entanglement in a solid-state spin ensemble, *Nature* **470**, 69–72, 2011.

Index